TICS

pe of the Series

rch approaches from biology, philosophy and linguistics, the emerging field of biose- that animals, plants and single cells all engage insemiosis—the conversion of physical entional signs. This has important implications and applications for issues ranging from to animal behaviour and human psychology, leaving biosemiotics at the cutting edge of he fundamentals of life.

r book series Biosemiotics draws together contributions from leading players in inter- iotics, producing an unparalleled series that will appeal to all those interested in the ori- ion of life, including molecular and evolutionary biologists, ecologists, anthropologists, philosophers and historians of science, linguists, semioticians and researchers in artificial n theory and communication technology.

volumes:
v.springer.com/series/7710

Peirce and Biosemio

BIOSEMIO

VOLUME 11

Series Editors

Aims and Sco

Combining resea
miotics proposes
signals into conv
natural selection
the research on
The Springe
national biosem
gins and evolut
psychologists,
life, informatio

For furthe
http://ww

Vinicius Romanini • Eliseo Fernández
Editors

Peirce and Biosemiotics

A Guess at the Riddle of Life

Springer

Editors
Vinicius Romanini
Escola de Comunicações e Artes
University of São Paulo
São Paulo
São Paulo
Brazil

Eliseo Fernández
Linda Hall Library of Science
and Technology
Kansas City
Missouri
USA

ISSN 1875-4651
ISSN 1875-466X (electronic)
ISBN 978-94-007-7731-6
ISBN 978-94-007-7732-3 (eBook)
DOI 10.1007/978-94-007-7732-3
Springer Dordrecht Heidelberg New York London

Library of Congress Control Number: 2014932414

Printed on acid-free paper

Springer is part of Springer Science+Business Media (www.springer.com)

Contents

1 **Introduction** ... 1
Vinicius Romanini and Eliseo Fernández

2 **The Intelligible Universe** ... 9
Nathan Houser

3 **The Continuity of Life: On Peirce's Objective Idealism** ... 33
Ivo A. Ibri

4 **Peircean Semiotic Indeterminacy and Its Relevance for Biosemiotics** ... 51
Robert Lane

5 **Peircean Habits, Broken Symmetries, and Biosemiotics** ... 79
Eliseo Fernández

6 **Semeiotic Causation and the Breath of Life** ... 95
Menno Hulswit and Vinicius Romanini

7 **The Ineffable, the Individual, and the Intelligible: Peircean Reflections on the Innate Ingenuity of the Human Animal** ... 127
Vincent Colapietro

8 **Instinct and Abduction in the Peircean Informational Perspective: Contributions to Biosemiotics** ... 151
Lauro Frederico Barbosa da Silveira and Maria Eunice Quilici Gonzalez

9 **The Life of Symbols and Other Legisigns: More than a Mere Metaphor?** ... 171
Winfried Nöth

10 Signs Without Minds 183
John Collier

11 Dicent Symbols and Proto-propositions in Biological Mimicry 199
João Queiroz, Frederik Stjernfelt and Charbel Niño El-Hani

12 Semeiosis as a Living Process 215
Vinicius Romanini

Bibliography 241

Index 243

Contributors

Vincent Colapietro Pennsylvania State University, University Park, USA

John Collier University of Kwazulu-Natal, Durban, South Africa

Eliseo Fernández Linda Hall Library of Science and Technology, Kansas City, USA

Maria Eunice Quilici Gonzalez São Paulo State University, Marilia, Brazil

Charbel Niño El-Hani Federal University of Bahia, Salvador, Brazil

Nathan Houser Indiana University, Indianapolis, USA

Menno Hulswit Radboud University, Nijmegen, The Netherlands

Ivo Assad Ibri Pontifical Catholic University of São Paulo, São Paulo, Brazil

Robert Lane University of West Georgia, Carollton, Georgia, USA

Winfried Nöth Catholic University of São Paulo (PUC), São Paulo, Brazil

João Queiroz Federal University of Juiz de Fora, Juiz de Fora, Brazil

Vinicius Romanini University of São Paulo, São Paulo, Brazil

Frederik Stjernfelt University of Aarhus, Aarhus, Denmark

Lauro Frederico Barbosa da Silveira São Paulo State University, Marilia, Brazil

About the Authors

Charbel N. El-Hani is associate professor at the Institute of Biology, Federal University of Bahia, Brazil, where he coordinates the History, Philosophy, and Biology Teaching Lab. He received his Bs C in Biology from Federal University of Bahia, and his PhD in Education from University of São Paulo. He is currently the Book Review Editor of *Science & Education*. His areas of research are: Science education research, Philosophy of Biology, and Evolutionary Biology.

Eliseo Fernández was born and educated in Argentina. He works at the Linda Hall Library of Science and Technology as a scientific consultant and taught at the University of Missouri at Kansas City for 18 years at the Physics Department. He has lectured and published papers and reviews on topics in the philosophy and history of science and on the thought of C. S. Peirce. His current research focuses on the application of some ideas of Peirce to current philosophical problems in physics and biosemiotics.

Frederik Stjernfelt is Professor, Ph.D. at the Center for Semiotics, Aarhus University, Denmark. Member of the Royal Danish Academy of Sciences and Letters. PI of the research projects "Joint Diagrammatical Reasoning in Language" and "Humanomics". Recent books: *Diagrammatology. An Investigation on the Borderlines of Phenomenology, Ontology, and Semiotics* (2007), *Semiotics. Critical Concepts* I-IV (ed. with P. Bundgaard, 2010), *The Democratic Contradictions of Multiculturalism* (with J.M.Eriksen 2012).

Ivo Assad Ibri is full professor of philosophy at the Pontifical Catholic University of São Paulo (PUCSP), Brazil. He is the founder and director of the Center for Pragmatism Studies of PUCSP and editor of *Cognitio – Journal of Philosophy.* His main research interest is American pragmatism, especially the work of Charles Peirce, as well its connections with German idealism. He published several essays on pragmatism and a book on Peirce's metaphysics titled Kósmos Noétos. He is researcher of the Brazilian agency CNPq (National Counsel of Technological and Scientific Development) and member of the board of consultants of the Peirce Edition Project at Indiana University.

Joao Queiroz (www.semiotics.pro.br) is a professor at the Institute of Arts and Design, Federal University of Juiz de Fora, Brazil. Queiroz earned a Ph.D. in Communication and Semiotics from the Catholic University of São Paulo (PUC-SP), and received a post-doctoral fellowship in Intelligent Systems at the School of Electrical and Computer Engineering (FEEC-DCA), State University of Campinas (UNICAMP). His research interests include: Peircean Semiotics, Biosemiotics and Cognitive Science, as well as South-American and Brazilian literature.

John Collier was educated at MIT, UCLA and Western Ontario in Philosophy and Philosophy of Science. He started studying Peirce on his own as an undergraduate. His dissertation was on theory change, but it was strongly influenced by Peircean thought, especially the Pragmatic Maxim and pragmatics in general. He has worked on five continents and is glad there are no universities in Antarctica. For the past ten years he has been at the University of KwaZulu-Natal in Durban, South Africa. His current work is on complex organization in Biology, most recently ecology at the Federal University of Bahia in Salvador, Brazil, where he is visiting.

Lauro Frederico Barbosa da Silveira, doctor in Philosophy by the Pontifical Catholic University of São Paulo, Brazil. Professor at the Graduate Program of Philosophy of the State University of São Paulo (UNESP) Campus of Marilia, SP. Honorary member of the Psychoanalytic Nucleus of Marilia and Region. Associate to the Charles S. Peirce Society, USA. Member of the Research Group on Cognitive Studies, Marilia and of the Self-organization Group at the Center of Logic and Epistemology (CLE),Unicamp, Campinas , SP. Author of Curso de Semiótica Geral and of several articles on Philosophy and Semiotics.

Maria Eunice Q. Gonzalez has a BSc in Physics (1977), and an MSc in Logic and Philosophy of Science (1984). Her PhD thesis "A cognitive approach to visual perception" was completed in 1989 at the University of Essex, UK. She is a founder member of the Brazilian Society for Cognitive Science, and the Research Group on Cognitive Studies at UNESP, and member of the CLE research group on self-organization at UNICAMP, and the Ecological Psychology Society. She is a Latin American representative of the International Complex Systems Society.

Menno Hulswit teaches philosophy at Radboud University (Nijmegen, The Netherlands) and biology and science at Arentheem College; he completed his PhD in philosophy at Radboud University in 1998. His research interests include process philosophy and pragmatism, especially the philosophy of C.S. Peirce, and most specifically, the concept of causation. He is the author of From Cause to Causation. A Peircean Perspective (Philosophical Studies Series 90, Dordrecht, Kluwer Academic Publishers, 2002), and co-editor of Living Doubt: Essays Concerning the Epistemology of Charles Sanders Peirce (Dordrecht, Kluwer Academic Publishers, 1994). He has written a number of articles on the concept of causation.

Nathan Houser is Professor Emeritus of Philosophy at Indiana University in Indianapolis (IUPUI) and President of the Peirce Foundation. He has served as Director of the Peirce Edition Project and the Institute for American Thought and as

President of the Charles S. Peirce Society and the Semiotic Society of America. From 1993 to 2009 he was General Editor for the Indianapolis critical edition of Peirce's writings and he co-edited the two-volume *Essential Peirce* and *Studies in the Logic of Charles Sanders Peirce.* He is the author of many articles on Peirce including "The Scent of Truth" (2005); "Peirce in the 21st Century" (2005), "Peirce's Contrite Fallibilism" (2006), "Peirce as a Sign of Himself" (2009), "Peirce, Phenomenology, and Semiotics" (2009), and "Signs and Survival" (2013).

Robert Lane is Professor of Philosophy at the University of West Georgia. He is editor (for Peirce submissions) of the *Transactions of the Charles S. Peirce Society* and associate editor of the textbook *Pragmatism, Old and New* (ed. Susan Haack). His work has appeared in the *Journal of the History of Philosophy, History of Philosophy Quarterly, Contemporary Pragmatism, Cognitio, Bioethics* and other journals.

Vincent Colapietro is Liberal Arts Research Professor of Philosophy at the Pennsylvania State University (University Park Campus). One of his main areas of historical specialization is American pragmatism, with special attention to Charles Peirce, while one of his principal foci of theoretical research is semiotics. Recently he has become increasingly interested in sketching a convincing portrait of the human animal, a portrait at once informed by evolutionary theory and attentive to the definitive abilities of human agents. In conjunction with this interest, he is drawn to a number of perspectives or traditions (phenomenology, hermeneutics, and psychoanalysis in addition to pragmatism).

Vinicius Romanini (www.minutesemeiotic.org), Ph.D., is Professor of Communication Studies at the School of Communications and Arts, University of Sao Paulo (USP), Brazil. His research interest is focused on the development of Peirce´s semeiotic into a transdisciplinary theory of Communication. He is the scientific editor of *Semeiosis – Transdisciplinary Journal of Semiotics,* member of the CLE research group on self-organization at UNICAMP, and of the International Advisory Committee (IAC) for the Peirce Society.

Winfried Nöth is Professor of Cognitive Semiotics at the Catholic University of São Paulo. Until 2009, he was Professor of Linguistics and Semiotics and Director of the Interdisciplinary Center for Cultural Studies at the University of Kassel. Nöth is an honorary Member of the Int. Ass. for Visual Semiotics and the former President of the German Ass. for Semiotic Studies. Nöth's more than 300 articles and 31 authored or edited books (such as: Handbook of Semiotics, Self-Reference in the Media) are on cognitive semiotics, linguistics, literature, media, images, maps, systems theory, culture, evolution, and Charles S. Peirce.

Abbreviations of Peirce's Works

CN Peirce's Contributions to The Nation, 4 parts, eds. Kenneth L. Ketner and James E. Cook. Lubbock: Texas Tech Press, 1975–1988. References use *CN* followed by volume and page numbers.

CP *Collected Papers of Charles Sanders Peirce*. Edited by C. Hartshorne, P. Weiss (volumes 1–6) and A. Burks (volumes 7–8). Cambridge: Harvard University Press, 1931–1958. References appear as *CP* followed by volume and paragraph numbers.

EP *Essential Peirce. Selected Philosophical Writings*. Vols. 1–2, various editors at the Peirce Edition Project. Indiana University Press, 1992 an 1998. References appear as EP followed by volume and page numbers.

HP Historical Perspectives on Peirce's Philosophy of Science. Carolyn Eisele, ed. New York: Mouton, 1985.

MS Charles Sanders Peirce manuscripts in Houghton Library at Harvard University—usually followed by catalogue number and sheet number.

NEM *The New Elements of Mathematics by Charles S. Peirce*. Edited by Carolyn Eisele, 4 volumes in 5 books. The Hague: Mouton, 1976. References appear as *NEM* followed by volume and page numbers.

RLT *Reasoning and the Logic of Things. The Cambridge Conferences Lectures of 1898 by Charles Sanders Peirce*. Edity by Kenneth Laine Ketner, with an introduction by Kenneth Laine Ketner and Hilary Putnam. Cambridge: Harvard University Press, 1992. References appear as RLT followed by page numbers.

R Peirce's manuscripts in the Houghton Library of Harvard University, as cataloged in Richard Robin, *Annotated Catalogue of the Papers of Charles S. Peirce*, Amherst, MA: University of Massachusetts Press, 1967, and in Richard Robin, "The Peirce Papers: A Supplementary Catalogue," *Transactions of the Charles S. Peirce Society* 7 (1), 1971, 37–57. These manuscripts are available in a microfilm edition, *The Charles S. Peirce Papers*, produced by Harvard University Library. Except where otherwise noted, references are by Robin's manuscript number and, when available, page number.

SS Semiotic and Significs: The Correspondence between Charles S. Peirce and Victoria Lady Welby. Edited by Charles Hardwick. Bloomington: Indiana University Press, 1977.

W Writings of Chalres S. Peirce: A Chronological Editon. Edited by Max H. Fisch and others. Bloomington University Press, 1982. References use volume and page numbers after *W*.

Introduction

Vinicius Romanini and Eliseo Fernández

The aim of this volume is to offer the reader a collection of representative contemporaneous views about the relevance of Peirce's philosophy in general, and his semeiotic[1] in particular, to the development of the field of biosemiotics. Not all of the authors are practicing biosemioticians, but all of them are Peirce scholars with something to say about how Peirce's ideas can contribute to the discussion of the origin of life and the diversification, evolution and functioning of living systems. Each chapter opens with an excerpt from one of Peirce's works to offer the reader a taste of his method and style. These excerpts introduce some crucial quotes that serve to justify the importance of Peirce's ideas to biosemiotics.

We therefore felt it appropriate to write this introduction in the same vein, showing how Peirce himself evaluated his work regarding the mysteries of life and living organisms after 40 years of continuous dedication to science and philosophy, mostly from the point of view of a broader conception of logic as semeiotic. What strikes us most in the passage below is how the phenomenon of life turns into a paradigmatic example of genuine triadic relations. So we think the quote below puts the reader of this volume in the correct mindset:

> For forty years, that is, since the beginning of the year 1867, I have been constantly on the alert to find a ***genuine*** triadic relation—that is, one that does not consist in a mere collocation of dyadic relations, or the negative of such, etc. (I prefer not to attempt a perfectly definite definition)—which is not either an intellectual relation or a relation concerned with

[1] Semeiotic, semiotic and semiotics are the three spellings usually employed by semioticians. Peirce himself used several diffent versions (including semeiotics and semeotic, but apparently never semiotics). Max Fisch (1986, pp. 321–322) stressed that semeiotic was Peirce's most usual and prefered but this is still a matter of scholar discussions. We then decided to let the authors of each chapter use their own prefered version. The same goes for semeiosis and semiosis.

V. Romanini (✉)
University of São Paulo, São Paulo, Brazil
e-mail: vinicius.romanini@usp.br

E. Fernández
Linda Hall Library of Science and Technology, Kansas City, USA
e-mail: fernande@lindahall.org

V. Romanini, E. Fernández (eds.), *Peirce and Biosemiotics,* Biosemiotics 11,
DOI 10.1007/978-94-007-7732-3_1, © Springer Science+Business Media Dordrecht 2014

> the less comprehensible phenomena of life. I have not met with one which could not reasonably be supposed to belong to one or other of these two classes. As a case as nearly brute and inorganic as any, I may mention the form of relationship involved in any screw-form which is definitely of the right-hand, or occidental, mode, or is definitely of the Japanese, or left-handed, mode. Such a relation exists in every carbon-atom whose four valencies are saturated by combination with four atoms of as many different kinds. But where the action of chance determines whether the screw be a right-handed or a left-handed one, the two forms will, in the long run, be produced in equal proportions, and the general result will not be definitely, or decisively, of either kind. We know no case of a definitely right-handed or left-handed screw-phenomenon, where the decision is not certainly due to the intervention of a definitely one-sided screw in the conditions of that decision, except in cases where the choice of a living being determines it; as when Pasteur picked out under the microscope the two kinds of crystals of a tartrate, and shoved those of one kind to the right and those of the other kind to the left. We do not know the mechanism of such choice, and cannot say whether it be determined by an antecedent separation of left-handed screws from right-handed screws or not. No doubt, all that chance is competent to destroy, it may, once in a long, long time, produce; but it is a question whether absolute chance—pure tychism—ought not to be regarded as a product of freedom, and therefore of life, not necessarily physiological. It could not be caused, apparently, by the inorganic action of dynamical law. For the only way in which the laws of dynamics involve triadic relations is by their reference to second differentials of positions. But though a second differential generally involves a triadic relation, yet owing to the law of the conservation of energy, which has been sufficiently proved for purely inorganic phenomena, the dynamic laws for such phenomena are expressible in terms of first differentials. It is, therefore, a non-genuine, or, as I phrase it, a "degenerate" form of triadic relationship which is involved in such case. **In short, the problem of how genuine triadic relationships first arose in the world is a better, because more definite, formulation of the problem of how life first came about; and no explanation has ever been offered except that of pure chance, which we must suspect to be no explanation, owing to the suspicion that pure chance may itself be a vital phenomenon. In that case, life in the physiological sense would be due to life in the metaphysical sense.** [bold added by us] Of course, the fact that a given individual has been persuaded of the truth of a proposition is the very slenderest possible argument for its truth; nevertheless, the fact that I, a person of the strongest possible physicistic prejudices, should, as the result of forty years of questionings, have been brought to the deep conviction that there is some essentially and irreducibly other element in the universe than pure dynamism may have sufficient interest to excuse my devoting a single sentence to its expression. For you may be sure that I had reasons that withstood severe, not to say hostile criticism; and if I live to do it, I shall embody them in a volume. (Some Amazing Mazes, Fourth Curiosity, CP 6.322, c. 1909)

There are a few ideas that we would like to link to this passage since they will show up in discussions in most of the chapters. The first is the hypothesis that all genuine triadic relations are of a semeiotic nature (see also CP 2.241-242). But what does Peirce mean by a genuine triadic relation? Following his own definitions, they are necessarily composed of three correlates which participate in a causational relation in which the first of them (the sign) immediately determines the third (the interpretant) in some particular aspect, and in such a way that this immediate determination of the sign upon the interpretant is also a mediate determination of the second correlate (the object) upon this very same interpretant. In a more direct wording: the sign is a medium of information that professes to represent its object in some aspect so as to produce an effect, which is its interpretant.

Another important idea is that the laws of nature, which are all expressible by differential equations, might be semeiotic at the bottom, although of a degenerated kind. By a degeneration of a genuine triadic relation Peirce means the possibility of its translation into dyads, that is, reactive instantiations of the kind we usually describe as reversible mechanical causality. A semeiotic grounding of every natural process would imply the whole universe as it appears to our senses is intelligible *per se*, independently of any intelligent organic mind, human or not. This is the basis of Peirce's extreme realism. Since the type of causation that rules every natural process is irreversible due to a chance aspect, it is precisely this irreversibility that produces the flow of time, the growth of complexity and the syntax that must be embodied in any natural language (human or not) able to produce a narrative about reality.

The chance aspect of nature is explained by Peirce's doctrine of tychism, while the continuous flow of causation that embraces the whole of reality is explained by his doctrine of synechism. Together, they depict a reality depending on some sort of evolutionary idealism demanding an also ideal community of possible interpreters to reveal its objective nature in an ever distant conditional future.

The chapters of this volume follow a plan that begins with the more general discussion of Peirce's cosmology and evolutionary idealistic philosophy and gradually brings these concepts to his general logic, or semeiotic, and its application to phenomena related to organic life and mind. No one could give a proper start better than Nathan Houser, former editor of the Peirce Edition Project. His opening chapter, "*The Intelligible Universe*" tackles Peirce's attempt at explaining our world in semeiotic terms. Nathan comments on an extract from Peirce's 1904 paper, "New Elements" (EP 2, Sel. 22), which the dean of Peirce scholarship Max H. Fisch described as the best statement of Peirce's general theory of signs up to that time. The chosen passage concerns Peirce's much debated cosmological account of the beginning of the universe and, as Peirce himself remarked, it seems very mystical and mysterious. However, after recent accounts of the origin of the universe by some contemporaneous cosmologists interested in the role of information in the development of our physical world, Peirce's account looks more mainstream. Nathan focuses on Peirce's claim that to explain the universe we must assume that it is intelligible and that the origin of the universe must be conceived of semeiotically. What role did signs play in the evolution of the universe out of the primordial nothingness that preceded all determination? To answer this question, or to explore it at least, he considers Peirce's characterization of "nothing" as a "chaos of reactions utterly without any approach to law" and his claim that "creation consisted in mediating between the lawless reactions and the general possibilities by the influx of a symbol." Nathan's conclusion is that Peirce's cosmology forces us to consider the fundamental role of semeiosis, or sign action, at the very beginnings of the physical universe and prior to biological and cultural developments.

The torch is then passed to Ivo Ibri, head of the Center for Pragmatic Studies in São Paulo, Brazil. In his chapter, "*The Continuity of Life: on Peirce's Objective Idealism*," Ibri explains that Peirce's metaphysics is a complex system of concepts constitutive of a *theory of the world*. His *synechism* is deeply connected to his *Objective Idealism*, as the former claims that all reality is somehow continuous, notwithstanding an imperfect or incomplete continuity, and the latter refuses any ontological splitting between matter and mind, affirming that both are manifestations of the one and only substance, namely, *mind.* As Ibri states, Peirce, conceives nature as a living organism that, from the stone to the human being, scales life in degrees of vivacity measured by its capacity of spontaneity and habit changing. Ibri's chapter shows how different degrees of life interact dialogically by interchanging signs. This will require that the new and important science of biosemiotics takes into account the grounds and meaning of that dialogue.

In the fourth chapter, "*Peircean Semiotic Indeterminacy and its Relevance for Biosemiotics*," Robert Lane presents a detailed explanation of Peirce's early and late views on semeiotic indeterminacy and then considers how those views might be applied within biosemiotics. Peirce distinguished two different forms of semeiotic indeterminacy: generality and vagueness. He defined each in terms of the "right" that indeterminate signs extend, either to their interpreters in the case of generality or to their utterers in the case of vagueness, to further determine their meaning. In Peirce's view, no sign is absolutely determinate, i.e., every sign is indeterminate to at least some degree and so exhibits some degree of generality or vagueness. If Peirce was right about this, then no instance of biosemiosis is completely determinate—every biosign must be general or vague to some degree. Lane shows that in Peirce's view, whether a sign is general or vague depends on its immediate object, "the idea which the sign is built upon," and explains how Peirce would go about identifying the immediate object of a sign lacking both a minded utterer and a minded interpreter—an identification that must be possible if any biosign is indeterminate.

The fifth chapter, "*Peircean Habits, Broken Symmetries, and Biosemiotics*," written by one of the editors of this volume (Eliseo Fernández), is dedicated to bringing to light the crucial role of habits in Peirce's philosophy and semeiotic and its connection to discoveries unknown to Peirce, such as the continuous expansion of the universe; the roles of the genetic and other organic codes in the maintenance, propagation and evolution of organisms; the existence of objective randomness in the physical world and in mathematics; and the relations between physical symmetries, their breaking, and the emergence of new laws in nature. Fernandez outlines how these physical, biological and mathematical discoveries may help us to articulate, unify and generalize Peirce's insights into the evolutionary character of the laws of nature, the semeiotic underpinnings of living systems, and the relations between semeiosis and physical causation. On the way to bringing these ideas together, Fernandez shows the paramount role of Peirce's conception of the habit of taking habits in binding the otherwise separate notions of law, generalization, and growth, and draws some tentative conclusions on the possible import of a generalized notion of habit upon the future of biosemiotics, including its relations to other scientific disciplines.

Another editor of this volume (Vinicius Romanini) is the co-author of its sixth chapter alongside Menno Hulswit, one of the most authoritative scholars on Peirce's original concept of causation. In "*Semeiotic Causation and the Breath of Life*" they argue that the distinctive character of life is its being a semeiotic process grounded by the flow of causation: living organisms are characteristic examples of processes directed toward general end states, that is, states that represent habitual dispositions to behave in the future according to successful past experiences. Familiarity and memory provide the necessary information to denote correctly the objects of attention in daily experience, without which no living being would survive. To be able to have such dispositions to act coherently, the teleology of living beings must involve a combined action of final causation, efficient causation, and chance. All of this is an expression of semeiosis, which provides the formal aspect of causation by which transmission of forms from causes into effects becomes possible. Thus, symbols are living signs capable of gathering information during perception and of conveying it to its interpretants so as to produce general habits of conduct. The conclusion is that the flow of causation that we perceive in reality is the predicate of every true proposition; and any living species is analogous to a true proposition as much as it is attuned to the flow of causation that grounds the real and allows its permanence.

The seventh chapter is authored by Vincent Colapietro, famous for his studies on the semeiotic self and the communicative nature of knowledge. In the "*Instinctual Bases of Experimental Inquiry*," Colapietro faces the subtleties of human inquiry and concludes that its most sophisticated character (e.g., theoretical physics, analytic chemistry, or evolutionary biology) traces its roots to the instinctual dispositions of the human animal. Such, at least, are Peirce's most emphatic claims regarding scientific investigation. In most respects, his mature position is that of a thoroughgoing naturalism, hence a thoroughgoing evolutionism, but his vehement rejection of reductionism needs to be considered in connection with his characteristic espousal of naturalism. A consideration of how experimental inquiry is rooted in the instinctual dispositions of human beings, but evolves far beyond its biological bases, enables us to appreciate Peirce's naturalistic, but non-reductive account of human inquiry and, indeed, also of human agency. In this chapter Colapietro highlights features of Peirce's account not yet explored in depth or detail, trying to push beyond what has been accomplished therein, consolidating their insights but also proffering novel hypotheses of his own. Human intelligence is an embodied, social, and symbolific capacity and all of its characteristic achievements (including the greatest discoveries by scientific inquirers) bear the marks of this intelligence. However far human intelligence culturally evolves beyond its instinctual bases, these are always present and operative in even its most refined activities and stunning accomplishments. However deeply rooted is human intelligence in human instincts, its transcendence of its origins cannot be gainsaid. The self-transformative and self-transcending character of human intelligence, even when understood as an instinctive capacity, calls for an explanation. Peirce's writings provide us with the resources for explaining this along with much else.

Chapter number eight is co-authored by Lauro Silveira and Maria Eunice Gonzalez, both from the State University of São Paulo (Unesp), Brazil. Their chapter "*Instinct and Abduction in the Peircean Informational Perspective: Contributions to Biosemiotics*" analyzes the Peircean notion of instinct and its relationship with abductive reasoning in the context of Peirce's reflections about the nature of information. They present the Peircean concept of abductive reasoning, emphasizing those passages where Peirce mentions instinct as an anchor for this type of reasoning. They then relate this to one of the last works of Peirce, *An Essay Toward Improving Our Reasoning in Security and in Uberty* (1913), in which he discusses how the instinctual habits of animal species, including humans, serve as inductive proof of semeiotic. They propose the hypothesis that instinct, grounded on information, may express a manifestation of abduction in the unfolding of the semiotic process that characterizes life. In its myriad of interconnected levels of communication, life is depicted as a complex self-organizing informational process that causes dynamic webs of self-referential signs. An explanation of the role played by the Peircean concept of information in the understanding of how abduction and instinct may come together in the life process constitutes the main theme of their chapter.

The Peircean scholar Winfried Nöth is the author of the ninth chapter, "*The Sign as a Living Being: More than a Mere Metaphor?*" where he examines the biosemeiotic implications of the thesis put forward by Peirce in 1901 that "every symbol is a living thing, in a very strict sense that is no mere figure of speech" (CP 2.222). Nöth demonstrates that this thesis is typical of Peirce's synechistic method of arguing. What it means is that on a continuum between life and the lifeless, symbols are much closer to the former than to the latter. By the "very strict sense" in which symbols are living beings Peirce cannot have meant that symbols are of flesh and blood. On the other hand, Nöth argues, Peirce found it necessary to emphasize that symbols do not only live metaphorically, which would mean that they are only life-like although in fact lifeless. According to Nöth, Peirce did not believe that symbols and ideas depend solely on the agency of human symbol makers and that "ideas are [...] all mere creations of this or that mind." Instead, he argued that symbols are endowed with the capacity "of finding or creating their [own] vehicles" of propagation (CP 2.217, 1901) and procreation. Among these and other characteristics of living organisms which Peirce ascribes to symbols and to "ideas" in a more or less strict sense (sometimes only "in a certain sense") are: an autonomous (but in a sense vicarious) agency, self-reference, creativity, self-replication, growth, "the power of bringing things to pass" (CP 2.217, 1901), survival, and death. The chapter also examines why Peirce mostly ascribes these characteristics to symbols and not to signs in general.

The biosemiotician John Collier wrote chapter number ten, "*Signs without Minds*." Collier is explicitly non-sympathetic with Peirce's objective idealism, but acknowledges the necessity of enlarging the realm of semeiosis to encompass phenomena of signification that occur beyond the traditional conception of mind as a property connected to brains. The purpose of his chapter is to explore and explain the cases in which non-mental signs are signs for their bearers, which Collier claims to be especially important to endo-biosemiotics. The author relies specially on Peirce's famous case of the sunflower, and discusses whether it is a sign of the direction of the sun. Looking at this and other similar examples, Collier tries to determine what properties something non-mental must have in order to serve as a sign for its bearer. The answer, he claims, lies in teleology, specifically in functionality. He then argues that candidate non-mental signs must be functional for their bearer, and that the grounds of functionality ensure that there is a Peircean interpretant, which completes the triad, which becomes a sign with a further interpretant, and so on.

João Queiroz, Frederik Stjernfelt and Charbel Niño El-Hani have worked together to produce a good example of interdisciplinary approach while applying Peirce's semeiotic to the biological domain. In the eleventh chapter of this volume, "*Dicent Symbols and Proto-propositions in Biological Mimicry*", they show that Peirce's mature theory of signs brings an important contribution to the building of a general semeiotic theory of mimicry, since it is quite helpful in addressing semantic and pragmatic aspects of biological information. As they explain, the semeiotic processes involved in biological mimicry most often do not result from learning processes taking place in the individual semeiotic system. They arise instead from the fine-tuning by natural selection of inherited capacities among variants, over thousands or millions of generations. Still, the concrete sign exchange takes place within the lifetime of a single individual, and those signals, which indicate and describe at the same time, should be conceived of as dicisigns. This calls for an investigation of the Peircean notion of the dicisign, which is a generalization of the notion of proposition. On the one hand, it liberates assertion from the confines of language and points to its appearance also in pictures, gestures, etc. That is, it generalizes propositions from being a human privilege so as to embrace also simpler dicisigns found in biology. As a corollary to their previous analysis of firefly signaling (El-Hani et al. 2010), they analyze the capacity of producing propositions (i.e., dicisigns) as a general requisite for semeiotic systems to act as mimics.

But how far can we go down the ladder of complexity and still find genuine triadic relations? Is the white cell chasing a microorganism moving by a purpose, as much as the dog is so moving when told by his master to fetch a book? A generalization of semeiosis to encompass all sorts of living and mental phenomena is sought by Vinicius Romanini in the twelfth and last chapter. Based on Peirce's last attempts to build a taxonomy of all possible classes of signs from what he was then calling the "respects," or fundamental relational components of every sign, the chapter "*Semeiosis as a living process*" presents a general logical structure for semeiosis, called solenoid, made of recursive loops from which periods and phases can be identified. The author claims that semeiosis does not need agency of any kind other than the action of the sign itself. It is the general habit embodied by a symbol that governs the action in both examples. If this is the case, Romanini supposes, the white cell doesn't need a cortex to act accordingly to the rules of the immune system. All it needs is to participate in a meaningful process that allows the recognition of a threat by previous familiarity gained during past experiences. Learning and evolution are semeioticaly identical process based on habit-taking. They are logical before being biological. Mind-like phenomena are needed in any biosemeiosis, he concludes.

The quote from Peirce that we have cited at length in this introduction suggest that chance, dynamics, life and mind are so profoundly connected as to evoke a sort of co-naturality between mind and matter through semeiosis. In this connection some metaphysical explanation must be given. It should not be a cheap metaphysics based on dogmatic assumptions but one based on sound phenomenological and logical premises. Peirce's theory of categories, which are real and universal, should be the guide for an answer about the interconnection of many phenomena that are usually studied by science only after they have been chopped to pieces as if with an ax, "leaving as the ultimate elements, unrelated chunks of being" (CP 7.570). The fundamental categories should provide a means to unify the whole of reality into an intelligible continuous process. An explanation for the phenomenon of life should be also an explanation for the phenomenon of inteligence and meaning, for every genuine triadic relation pressuposes meaning (CP 1.345). Since science is a form of inquiry depending on our production and sharing of meaning, as Peirce advocates, we must assume the possibility that the biosemiotic program of research must not be separated from the metaphysical possibity that our minds partake of some generalizing process, genuinely triadic, going on in the universe:

> As to protoplasm, what the three ***cenopythagorean categories,*** as I call them, do, and what they are limited to doing, is to call attention to three very different characters of this chemical body. The first is a ***posse*** which it has in itself; for the ***priman*** stops at ***can-bes*** and never reaches to existence, which depends on interaction, or ***secundanity.*** This internal power which the category merely suggests, we recognize as that of feeling. Though it is priman, it is without any doubt dependent upon the extreme complexity of the protoplasmic molecule, if the word molecule can be applied to so intricate, unstable, and ununified a system. But it is the law of high numbers that extreme complication with a great multitude of independent similar results in a new simplicity. Next there is reactive force, a twoness, which is emphasized in the nerve cells together. It is the property by which any state of high cohesiveness tends to spread through the albuminoid matter. We usually call the property contractility. Thirdly, the categories suggest our looking for a synthetizing law; and this we find in the power of assimilation, incident to which is the habit-taking faculty. This is all the categories pretend to do. They suggest a way of thinking; and the possibility of science depends upon the fact that human thought necessarily partakes of whatever character is diffused through the whole universe, and that its natural modes have some tendency to be the modes of action of the universe. (CP 1.351)

If this hypothesis is worth entertaining, instead of "the science of signs in living systems" (Kull 1999, p. 386), a Peircean approach to biosemiotics should regard it as the science of signs *as* living systems.

References

El-Hani, C., et al. (2010). Firefly femmes fatales: A case study in the semiotics of deception. *Biosemiotics, 3,* 33–55.

Fisch, M. (1986). *Peirce, semeiotic and pragmatism: Essays by Max H. Fisch.* K. Ketner and C. Kloesel (Eds.). Bloomington: Indiana University Press.

Kull, K. (1999). Biosemiotics from the twentieth century: A view from biology. *Semiotica, 127*(1/4), 385–414.

The Intelligible Universe

Nathan Houser

A symbol is something which has the power of reproducing itself, and that essentially, since it is constituted a symbol only by the interpretation. This interpretation involves a power of the symbol to cause a real fact; and although I desire to avoid metaphysics, yet when a false metaphysics invades the province of logic, I am forced to say that nothing can be more futile than to attempt to form a conception of the universe which shall overlook the power of representations to cause real facts. What is the purpose of trying to form a conception of the universe if it is not to render things intelligible? But if this is to be done, we necessarily defeat ourselves if we insist upon reducing everything to a norm which renders everything that happens, essentially and *ipso facto* unintelligible. That, however, is what we do, if we do not admit the power of representations to cause real facts. If we are to explain the universe, we must assume that there was in the beginning a state of things in which there was nothing, no reaction and no quality, no matter, no consciousness, no space and no time, but just nothing at all. Not determinately nothing. For that which is determinately not *A* supposes the being of *A* in some mode. Utter indetermination. But a symbol alone is indeterminate. Therefore, Nothing, the indeterminate of the absolute beginning, is a symbol. That is the way in which the beginning of things can alone be understood. What logically follows? We are not to content ourselves with our instinctive sense of logicality. That is logical which comes from the essential nature of a symbol. Now it is of the essential nature of a symbol that it determines an interpretant, which is itself a symbol. A symbol, therefore, produces an endless series of interpretants. Does anybody suspect all this of being sheer nonsense? *Distinguo.* There can, it is true, be no positive information about what antedated the entire Universe of being; because, to begin with, there was nothing to have information about. But the universe is intelligible; and therefore it is possible to give a general account of it and its origin. This general account is a symbol; and from the nature of a symbol, it must begin with the formal assertion that there was an indeterminate nothing of the nature of a symbol. This would be false if it conveyed any information. But it is the correct and logical manner of beginning an account of the universe. As a symbol it produced its infinite series of interpretants, which in the beginning were absolutely vague like itself. But the direct interpretant of any symbol must in the first stage of it be merely the *tabula rasa* for an interpretant. Hence the immediate interpretant of this vague Nothing was not even determinately vague, but only vaguely hovering between determinacy and vagueness; and its immediate interpretant was vaguely hovering between vaguely hovering between vagueness and determinacy and determinate vagueness or determinacy, and so on, *ad infinitum.* But every endless series must logically have a limit.

N. Houser (✉)
Indiana University, Indianapolis, USA
e-mail: nhouser@iupui.edu

V. Romanini, E. Fernández (eds.), *Peirce and Biosemiotics,* Biosemiotics 11,
DOI 10.1007/978-94-007-7732-3_2, © Springer Science+Business Media Dordrecht 2014

> Leaving that line of thought unfinished for the present owing to the feeling of insecurity it provokes, let us note, first, that it is of the nature of a symbol to create a *tabula rasa* and therefore an endless series of *tabulae rasae*, since such creation is merely representation, the *tabulae rasae* being entirely indeterminate except to be representative. Herein is a real effect; but a symbol could not be without that power of producing a real effect. The symbol represents itself to be represented; and that representedness is real owing to its utter vagueness. For all that is represented must be thoroughly borne out.
>
> For reality is compulsive. But the compulsiveness is absolutely *hic et nunc*. It is for an instant and it is gone. Let it be no more and it is absolutely nothing. The reality only exists as an element of the regularity. And the regularity is the symbol. Reality, therefore, can only be regarded as the limit of the endless series of symbols.
>
> A symbol is essentially a purpose, that is to say, is a representation that seeks to make itself definite, or seeks to produce an interpretant more definite than itself. For its whole signification consists in its determining an interpretant; so that it is from its interpretant that it derives the actuality of its signification.
>
> A *tabula rasa* having been determined as representative of the symbol that determines it, that *tabula rasa* tends to become determinate. The vague always tends to become determinate, simply because its vagueness does not determine it to be vague (as the limit of an endless series). In so far as the interpretant is the symbol, as it is in some measure, the determination agrees with that of the symbol. But in so far as it fails to be its better self, it is liable to depart from the meaning of the symbol. Its purpose, however, is to represent the symbol in its representation of its object; and therefore, the determination is followed by a further development, in which it becomes corrected. It is of the nature of a sign to be an individual replica and to be in that replica a living general. By virtue of this, the interpretant is animated by the original replica, or by the sign it contains, with the power of representing the true character of the object. That the object has at all a character can only consist in a representation that it has so,—a representation having power to live down all opposition. In these two steps, of determination and of correction, the interpretant aims at the object more than at the original replica and may be truer and fuller than the latter. The very entelechy of being lies in being representable. A sign cannot even be false without being a sign and so far as it is a sign it must be true. A symbol is an embryonic reality endowed with power of growth into the very truth, the very entelechy of reality. This appears mystical and mysterious simply because we insist on remaining blind to what is plain, that there can be no reality which has not the life of a symbol.
>
> How could such an idea as that of *red* arise? It can only have been by gradual determination from pure indeterminacy. A vagueness not determined to be vague, by its nature begins at once to determine itself. Apparently we can come no nearer than that to understanding the universe.
>
> That is not necessarily logical which strikes me today as logical; still less, as mathematics amply exemplifies, is nothing logical except what appears to me so. That is logical which it is necessary to admit in order to render the universe intelligible. And the first of all logical principles is that the indeterminate should determine itself as best it may.
>
> A chaos of reactions utterly without any approach to law is absolutely nothing; and therefore pure nothing was such a chaos. Then pure indeterminacy having developed determinate possibilities, creation consisted in mediating between the lawless reactions and the general possibilities by the influx of a symbol. This symbol was the purpose of creation. Its object was the entelechy of being which is the ultimate representation.
>
> EP2: 322-24

The text quoted above is from one of Peirce's most problematic and obscure unfinished writings. It is unclear when Peirce began composing "Καινὰ στοιχεῖα" or for what purpose but there is reason to place it around 1904 and to think that he was returning to a shelved project on the foundations of mathematics. Surprisingly, though, Peirce moved into semiotics and even turned to metaphysics and the cosmological speculations of the quoted text.[1] There are passages of "Καινὰ στοιχεῖα" that don't seem to cohere smoothly with Peirce's best known writings of the time and this led the distinguished Peirce scholar, Joseph Ransdell, to question the 1904 composition date, but he came to accept it as a good approximation.[2]

The editors of *The Essential Peirce,* vol. 2, the source for the above quotation, have translated Peirce's Greek title as "New Elements," which conforms with Peirce's apparent intent to associate his aim with that of Euclid in the first book of his *Elements:* "[t]he dominating idea of Euclid in writing his first book was plainly that the first elements of geometry can only be comprehended by understanding the logical structure of the doctrine" (EP 2: 301, 1904).[3] About a decade earlier, Peirce had written a textbook entitled "New Elements of Mathematics" in which he rejected the standard order of mathematical instruction which he believed to be out of accord with the more natural conceptual development and organization essential for effective mathematics learning. Peirce believed, for example, that ordinal numbers should be taught before cardinals and he contrived new ways to teach counting and set theory. He believed, contrary to standard practice, that the logical way to teach geometry was to begin with topical geometry, followed by perspective, before taking up metrical geometry. According to Carolyn Eisele, Peirce wanted to give mathematics education "a new look," to take account of the pioneering work of such consequential figures as Cantor, Klein, and Lobatchewsky (NEM II: x ff.). But Peirce could not convince textbook publishers to take a chance on his new approach; according to Arthur Burks, Peirce was simply "too far ahead of his time."[4] So after failing to place his book with three publishers, Peirce gave up writing textbooks and his manuscript, the object of so much creative effort, was set aside and was eventually lost.[5] Nevertheless, a decade later, Peirce was ready to resume his

[1] Peirce's cosmological speculations in "Καινὰ στοιχεῖα" might better be said to be cosmogonical but it has become the general practice to refer to Peirce's "cosmology" even when "cosmogony" would be the more explicit term.

[2] An example of the not so smooth fit of "Καινὰ στοιχεῖα" with other writings of the time is Peirce's treatment of his sign classification in the two sections of his 1903 Syllabus for his Lowell Lectures that are published immediately before "Καινὰ στοιχεῖα" in EP 2. In those sections of his Syllabus (EP 2: sels. 20 & 21), Peirce added two trichotomies of signs that enabled him to make distinctions in his discussion of symbols that, surprisingly, he did not carry into his discussion in "Καινὰ στοιχεῖα." Joseph Ransdell's thoughts on "Καινὰ στοιχεῖα" were expressed on the online Peirce forum, Peirce-L, in the early months of 2006. The Peirce-L archives can be accessed through the Arisbe website at www.cspeirce.com. A new and lively Peirce-L discussion of "Καινὰ στοιχεῖα" started in late 2012 and continued into the Spring of 2013.

[3] Presumably what Peirce means by "logical structure" with reference to Euclid's work is the conceptual organization and the deductive architecture and integrity of the *Elements.*

[4] Burks 1978, p. 917.

[5] Carolyn Eisele spent years working with Peirce's mathematical manuscripts and concluded that the "lost manuscript" had in fact been disassembled but mainly preserved and was mostly reassembled as published in her 1976 edition of *The New Elements of Mathematics* (NEM II). See Eisele's introduction to NEM II for her account of Peirce's travails with his editors. It is hoped that the Indianapolis critical edition of *New Elements* will shed more light on the "lost manuscript" and the compositional history of the pages that have survived.

project or, at any rate, to return to the issue of the logical structure of mathematics.[6] But Peirce's return to his "New Elements" project appears to have been short-lived, perhaps because his thought was still charged with the intellectual focus and energy of his recent stay in Cambridge and Boston where he had composed and delivered his 1903 Harvard and Lowell Lectures, or maybe because it was around this time that the focus of his concern turned to explicating and defending his version of pragmatism in the pages of *The Monist.*[7]

Peirce began "Καινὰ στοιχεῖα" with a comment on the Euclidean style in mathematical writing, a style he had employed in his *New Elements* textbook, but that he had given up: "with advancing years I have lost the power of writing about logic in mathematical style… [and] in losing the power of writing this style, I have equally lost my admiration of it" (EP 2: 301, 1904).[8] In particular, it seems that Peirce no longer wanted to adhere to the Euclidean constraint requiring that mathematics be presented logically but without any consideration, let alone critique, of its underlying logic. So it appeared that Peirce was embarking on a consideration of the logic of mathematics, beginning with a review of the key elements of Euclid's method: definitions, postulates, axioms, corollaries, diagrams, letters, theorems, and scholiums (EP 2: 302-03, 1904). After briefly treating the Euclidean method, Peirce distinguished theoretical from practical propositions and parallel distinctions between *definite* and *vague propositions* and *individual* and *general* propositions. He continued with a discussion of the three ways a sign connects with the truth: (1) by signifying characters or qualities (the Aristotelian *Form*), (2) by *denoting* its object (the

[6] An indication that Peirce's interest in this matter had been rekindled can be found in several of his book reviews for *The Nation* from around this time, for example: his 29 January 1903 "Review of Thomas Smith's *Euclid*" (N 3: 111-13); his 21 May 1903 "Review of J. W. Mellor's *Higher Mathematics for Students of Chemistry and Physics*" (N 3: 121-22); and his 23 July 1903 "Review of J. I. D. Hinds' *Inorganic Chemistry*" (N 3: 132-33). Also of note is Peirce's report on the 23 November 1904 meeting of the National Academy of Sciences which appeared in *The Nation* in December 1904 (N 3: 192-96). Peirce reported on his presentation of his own memoir on topical geometry and made the following remark: "Mr. Peirce remarked that this condition of preserving the connection of parts belongs to vacuous space itself, while it is demonstrable that these properties of space which are investigated by metrics and by graphics have nothing corresponding to them in vacuous space itself. Accordingly, Topics, or topical geometry, is alone the science of space itself, and all graphics, and *à fortiori* all metrics, can be regarded as a special problem of topical geometry." Perhaps it is not such a stretch to suppose that Peirce's focus in 1903 and 1904 on topical geometry, "alone the science of space itself," would have led him to cosmological speculations. It is also of interest that in this same report Peirce remarked that "there are no points upon a line until they are in some way marked; and indeed there is no multitude of points that could be marked without leaving room for a greater multitude to be marked." This recapitulates views on continuity that Peirce first worked out in detail for his Cambridge Conferences Lectures of 1898 (RLT).

[7] For an account of Peirce's intellectual history during this period see the introduction to EP 2, p. xxv ff.

[8] Peirce's mature views about highly technical mathematical logic (or about writing in that style) seem to be echoed by another famous logician, Bertrand Russell, who wrote of his disenchantment with mathematics and logic in *My Philosophical Development:* "I think that the timelessness of mathematics has none of the sublimity that it once seemed to me to have …. I cannot any longer find any mystical satisfaction in the contemplation of mathematical truth" (Russell 1959, pp. 211–212).

Aristotelian *Matter*), and (3) by determining an interpretant, a new sign of the same object with the same signification or *meaning* (ultimately the expression of the very fact in question—Aristotle's *entelechy*). Using his conception of sign, Peirce identified the two great tasks of humanity as theory and practice, stipulating that theory "sets out from a sign of a real object with which it is *acquainted*, passing from this, as its *matter*, to successive interpretants embodying more and more fully its *form*, wishing ultimately to reach a direct *perception* of the entelechy," and that practice sets "out from a sign signifying a character of which it *has an idea*, passes from this, as its *form*, to successive interpretants realizing more and more precisely its *matter*, hoping ultimately to be able to make a direct *effort*, producing the entelechy." Logic, he says takes the "movement" of theory as primary (EP 2: 304-05, 1904).[9]

Peirce went on to discuss the distinction between meaning and reference and to explain the ideas of logical depth (connotation), logical breadth (denotation), and information: "the total of fact (true or false) that in a given state of knowledge a sign embodies." As early as 1867 Peirce had developed these distinctions and had then equated information, or area, to breadth × depth (W2: 78-86).[10] He distinguished between relations as reactions (real relations) and sign relations and he identified three kinds of sign relations, icons, indexes, and symbols corresponding to the three ways signs "connect with" the truth: icons by possessing the very quality signified, indexes by reacting directly with the object denoted, and symbols by actually determining the interpretant sign. Peirce noted that icons, because they bring their interpreters "face to face with the very character signified," are "the mathematical sign par excellence" (EP 2: 307, 1904). Of symbols, Peirce said that they alone express laws and that they are essential for language and all abstract thought and are the only kind of sign that can express an argument. He noted that a "sign has its being in its adaption to fulfill a function" and that a symbol is adapted to fulfill its function simply by fulfilling it, by determining the interpretant it was constituted to determine: "An interpretant of a symbol is an outgrowth of the symbol" (EP 2: 322, 1904). Peirce identified three main kinds of symbols, terms, propositions, and arguments, and explained how an argument can be transformed into a compound proposition by eliminating "every monstration of its special purpose" (EP 2: 308, 1904), and how a proposition can be transformed into a term (or rhema) by eliminating the indexes (the "parts that separately denote its objects").[11]

---

[9] Peirce's 1902 entry on "Matter and Form" in Baldwin's *Dictionary of Philosophy and Psychology* (CP 6.353-63) provides background for what he says about matter and form in this paragraph and is also suggestive of some of his cosmological ideas.

[10] See Fernández's contribution to this volume, p. 99, n. 11, for an encapsulation of how "information" is understood by semioticians and for some secondary references relating to Peirce's understanding of information.

[11] In "Sundry Logical Conceptions," Sect. 3 of his Syllabus for his 1903 Lowell Lectures (EP 2: sel. 20), Peirce added the trichotomy *sumisigns, dicisigns, and arguments* to his previous one-trichotomy classification of signs (*icons, indexes, and symbols*). Soon afterwards, in Sect 5 of his Syllabus (EP 2: sel. 21), he added a third trichotomy, *qualisigns, sinsigns, and legisigns,* and renamed *sumisigns,* now calling them *rhemes,* clearly related to the *rhema* of "Καινὰ στοιχεῖα." With his trichotomy of *rhemes, dicisigns, and arguments* in place, Peirce could now make a precise clas-

Peirce had much more to say about symbols, especially about propositions, which he pointed out should not be confused with judgments. He noted that the key distinctions of vague and distinct, and general and individual, so important for logic, are propositional distinctions, and he maintained that, although every proposition must involve an icon to give it content, the vital spark is its indexical reference, its reaching out to the external object it signifies. In themselves, symbols have no real existence and thus cannot exert real force. They exist only in replica, embodied in words or gestures or other instantiations. But Peirce insisted that symbols, though without force, are by no means powerless: "I maintain that every sufficiently complete symbol governs things, and that symbols alone do this. I mean that though it is not a force, it is a law" (EP 2: 313, 1904). This led Peirce to a consideration of the nature of law and its causal efficacy.

Consonant with the semiotic framework of "Καινὰ στοιχεῖα" Peirce described a law as a symbol, or formula, "to which real events truly conform." More specifically, he said that a law is "an asserted symbolical proposition" (EP 2:314, 1904) though he declined to consider whether this implied that laws are always utterances.[12] What Peirce emphasized was that laws are not mere uniformities but are rightly understood to be reasons for predicting the character of relevant types of events.[13] To explain the causal efficacy of laws, Peirce gave an elaborated account of Aristotle's four causes—material, formal, efficient, and final:

> The individuating internal cause is called the *material cause*. Thus the integrant parts of a subject or fact form its *matter*, or material cause. The individuating external cause is called the *efficient,* or *efficient cause*; and the *causatum* is called the *effect*. The defining internal cause is called the *formal cause*, or *form*. All these facts which constitute the definition of a subject or fact make up its form. The defining external cause is called the *final cause,* or end. (EP2: 315-16, 1904)

Applying these distinctions to the conception of law, Peirce noted that the truth of a law (which consists in its being a symbol) is the defining external cause of its agreement with the relevant facts (symbolic expressions of the effects). Accordingly, "a symbol may be the cause of real individual events and things" (EP 2: 316, 1904). This is evident, according to Peirce, if we note that a cause is the premiss of an argument and that only symbols can be arguments. "Every sufficiently complete

---

sification of symbols into rhematic symbols, dicent symbols, and argument symbols, obviously related to his older and less technical distinction of terms, propositions, and arguments. Notice that all symbols of whatever kind are legisigns (laws).

[12] About 3 years later, however, in his famous Manuscript 318, Peirce argued forcefully that utterers are not essential (EP 2: 404, 1907). In his contribution to this volume, pp. 51–78, Robert Lane explicitly treats cases of signs without "minded utterers" or "minded interpreters."

[13] In a 1904 draft of a review of Herbert Nichols' *A Treatise on Cosmology* (CP 8.191-193), Peirce contrasted the conception of uniformity with that of law: "For while uniformity is a character which might be realized, in all its fullness, in a short series of past events, law, on the other hand, is essentially a character of an indefinite future; and while uniformity involves a regularity exact and exeptionless, law only requires an approach to uniformity in a decided majority of cases." (See Sect. VI of Eliseo Fernández's contribution to this volume, pp. 79–94, for discussion of different types of laws referred to by Peirce).

symbol is a final cause of, and 'influences,' real events, in precisely the same sense in which my desire to have the window open, that is, the symbol in my mind of the agreeability of it, influences the physical facts of my rising from my chair, going to the window, and opening it" (EP 2: 317, 1904). Peirce had expressed this point more poetically in June 1903 in an unpublished review of Baldwin's *Dictionary of Philosophy and Psychology:* "Minds cannot reconcile themselves to the notion that consciousness stands [an] idle spectator of human conduct" (CP 8.168).

As one reads through the text of "Καινὰ στοιχεῖα" it is easy to lose track that Peirce's subject is the logic of mathematics. However, bearing in mind Peirce's conception of logic as "the study of the essential nature of signs" (EP 2: 311, 1904), one can imagine how Peirce might have brought his discussion back to an explicit treatment of mathematical foundations had he not abandoned this work.[14] It is difficult to imagine, though, how Peirce could have supposed that the "cosmological story" in the opening quotation to this paper would contribute to the logic of mathematics. It is easier to imagine that Peirce's treatment of his theory of signs and, in particular, his concentration on symbols and their law-like nature sidetracked him by suggesting a new approach to the beginning of things. I say "cosmological story" to emphasize the rather puzzling nature of this account which, taken literally, is problematic. For one thing, Peirce's 'story' reads more like poetry or allegory than scientific metaphysics. It reminds one of the biblical creation tale from the Gospel of John which proclaims that "In the beginning was the Word," though in Peirce's rendering the *word* (logos) became the *symbol*. Notice that in the opening quotation Peirce emphasized that the universe is intelligible and therefore explicable, and so "it is possible to give a general account of it and its origin." This "general account," Peirce said, is a symbol, which of course an account of anything must be. But Peirce did not seem to mean only that cosmology, or any account of the origin of the universe, qua account, must be symbolic but more fundamentally that the universe, per se, is an outgrowth of the influx of a symbol into the primeval chaos from which the world emerged. But how could that be? How could a symbol, or even a sign of lesser complexity, sprout from the germinal chaos? That is a mystery.

More problematic than the enigmatic tone of Peirce's account are some of its apparent inconsistencies. He wrote that in order to explain the universe "we must assume that there was in the beginning a state of things in which there was nothing, no reaction and no quality, no matter, no consciousness, no space and no time, but just nothing at all." But this primordial *nothing* must have been a state of utter indetermination rather than a determinate nothing. Thus, according to Peirce, since "a symbol alone is indeterminate," the indeterminate nothing that was the absolute beginning of the universe had to be a symbol. But later, toward the end of the opening quotation, he wrote that the original state of nothingness was a "chaos of reactions utterly without any approach to law" and, as just mentioned in the previous paragraph, that "creation consisted in mediating between the lawless reactions and

[14] For some idea of how Peirce might have developed these themes in the context of the foundations of mathematics, see his "The Logic of Mathematics; an Attempt to Develop My Categories from Within" (CP 1.417-520, c. 1896).

the general possibilities by the influx of a symbol." So the universe at its beginning was said to have been an indeterminate nothingness void of qualities and reactions but was then also described as "a chaos of reactions." This adds to the mystery of the influx of the premier symbol and suggests that Peirce, who was careful with his science, might have been composing a metaphorical or figurative cosmological story. Notice that Peirce, himself, remarked that his readers might suspect what he was writing to be "sheer nonsense" and he pointed out that any account of the origin of the universe must necessarily be general and vague. Yet we must be wary of discounting too quickly Peirce's speculations by writing them off as too vague and figurative.[15]

Many commentators are not so charitable in their assessment of Peirce's cosmological speculations, which began at least as early as 1878 when he wrote: "What sort of a conception we ought to have of the universe, how to think of the ensemble of things, is a fundamental problem in the theory of reasoning" (W3: 307, 1878). By 1884, Peirce would maintain that "the postulate that things shall be explicable extends itself to laws as well as to states of things" and that "all known laws are due to chance and repose upon others far less rigid themselves due to chance and so on in an infinite regress…" (W4: 548, 551, 1883–1884). A growing consensus among Peirce scholars is that Peirce's aim was to explain how natural laws could have evolved from a lawless primordial chaos but there is disagreement about his motivation. Some believe that he was principally driven by his desire to defend his scholastic realism and its claim that laws are real against the nominalist precepts of the science of his day.[16] Others suppose he was primarily interested in filling out the system of philosophy that he referred to as his guess at the riddle, which he hoped would succeed the systematic philosophy of Aristotle.[17] No doubt Peirce's

[15] See Robert Lane's contribution to this volume, pp. 51–78, for an excellent treatment of Peirce's views on semiotic generality and vagueness.

[16] Paul Forster argues in his recent book (Forster 2011) that Peirce's battle against nominalism, in defense of his realism, was central to his philosophical program and that Peirce's "alternative to nominalism was predicated on a positive answer" to the crucial question: "are laws real and how can they be accounted for?" Forster's Chap. 9 and 10 treat Peirce's cosmology and his answer to this crucial question. Not everyone agrees that Peirce developed his cosmology in defense of his realism (see, for example, Short 2012, p. 386), but it is beyond doubt that Forster is right about Peirce being in a life-long battle against nominalism. In 1904, Peirce wrote William James that he and Schiller "carried pragmatism too far" but that its "most important consequence… on which I have always insisted" is that under the pragmatic conception of reality "we must abandon nominalism. That in my opinion is the great need of philosophy" (CP 8. 258). Two years later Peirce wrote: "In the fourteenth century Nominalism was rendered a respectable opinion by the halting realism of Scotus and by the extravagant unpragmatism of his followers. But after physical science has discovered so many general principles in Nature, nominalism becomes a disgraceful habitude of thought" (CP 6.175, 1906).

[17] Murray G. Murphey, for one, seems to hold this view. See his chapter on cosmology in Murphey 1993 (Chap. XVI) and note, especially, p. 355 where he writes that Peirce seems to have regarded the development of an architectonic theory as "the true purpose of philosophy and that it "was just this hope of creating an all embracing system which would serve as the framework for all future discovery and knowledge which seems to have motivated Peirce's work." Andrew Reynolds (Reynolds 2002, p. 113) offers a similar but somewhat divergent motivation by suggesting that Peirce was simply doing the work of philosophy as he understood it. Peirce was contributing to

motivations were complex, perhaps including both of these aims, but it seems certain that he was driven in part by the more straightforward age-old curiosity that had propelled the Pre-Socratic philosophers to ask what the world was made of and thus to conceive of primal matter, their *arche* (W5: 295, 1886; W6: 181, 1887–1888). So although Peirce may well have hoped to improve and strengthen his system of philosophy, it might be more accurate to ascribe a more direct scientific motivation for his cosmological inquiries. By the summer of 1886, he was ready to declare that "If the universe is thus progressing from a state of all but pure chance to a state of all but complete determination by law, we must suppose that there is an original, elemental, tendency of things to acquire determinate properties, to take habits.... Here then is a rational physical hypothesis, which is calculated to account, or all but account for everything in the universe except pure originality itself"—and he went on to say that the next step "would be to attempt to verify this hypothesis by seeing how far it would account for and explain the observed characteristics of the laws of nature" (W5: 293, 1886). Peirce never abandoned his hypothesis that natural laws are emergent, the result of evolution, though he periodically revised his cosmological speculations to accord more smoothly with his classification of the sciences and his developing metaphysics and, presumably, to enable easier verification. He never reached a final theory nor established any scientific results from his cosmological hypothesis and no part of his speculative thought has been more decidedly rejected as a failed program.[18] Peirce's former student, Christine Ladd Franklin, thought that his cosmological speculations of the early 1890's showed that he was losing his mind,[19] and many otherwise laudatory commentators on his work have concluded with W. B. Gallie, that Peirce's cosmology is the "white elephant" of his philosophy.[20] Not many dismiss Peirce's cosmology as crazy, as Ladd Franklin did, but many do reject it as a serious, or at any rate a viable, scientific hypothesis.

---

what he believed philosophy ultimately hopes to accomplish: "to find that there is some intelligible truth, some absolutely valid reasonableness, to ascertain how far this reasonableness governs the universe, and to learn how we may best do its service" (CN2: 208, 1899).

[18] It was not without trying, however, that Peirce failed to establish scientific results. Peirce believed that to demonstrate that space was hyperbolic would lend support to his cosmological hypothesis. In early 1891 Peirce began to correspond with Allen Risteen about how to conduct experiments to measure the curvature of space and he acquired instruments from the U. S. Coast & Geodetic Survey for curvature observations. Toward the end of 1891, Peirce sought a grant from the Office of the Nautical Almanac to conduct experiments to investigate the curvature of space. Peirce wrote to Simon Newcomb, Superintendent of the Office of the Nautical Almanac: "The discovery that space has a curvature would be more than a striking one: it would be epoch-making. It would do more than anything to break up the belief in the immutable character of mechanical law, and would thus lead to a conception of the universe in which mechanical law should not be the head and centre...." But Newcomb turned Peirce down. (See "Methods of Investigating the Constant of Space," W8, sel. 36, and the editorial note to that selection, W8: 424-425; also see the introduction to W8: xlvi & lxiii–lxiv).

[19] See the introduction to W8, pp. xcvi–xcvii, for a brief account of reactions to Peirce's 1890–1892 *Monist* cosmological series. Also see *Houser* 2009.

[20] Gallie 1952, p. 215. Some others who regard Peirce's cosmology as unsuccessful, if not confused or mistaken, include Rulon Wells (1964), Bernard Suits (1979), and Andrew Reynolds (2002). Reynolds concludes his excellent treatment of Peirce's cosmology with the overall assess-

In an important recent study, T. L. Short concluded that not only did Peirce never succeed in formulating a proper cosmology but he failed even to achieve his primary goal of explaining how the laws of nature could have evolved out of a primordial chaos.[21] Short traced Peirce's cosmological program through a number of stages beginning with his early attempt to develop a statistical account of natural law, through his middle period, roughly from 1891 to 1902, dominated by his self-proclaimed and Shelling-inspired objective idealism when he gave mind, as feeling, priority and viewed matter as "mind hidebound in habit," and on into his late period after he had given up his law of mind (that feelings spread and coalesce into ideas) for a teleological account of mind as driven by purpose.[22] Short briefly considered the cosmological story of "Καινὰ στοιχεῖα" that we are considering here and distinguished it as transitional between Peirce's middle period and his teleological period—it contains echoes of the middle cosmological period "but with crucial differences."[23] The "initial indeterminacy" that Peirce postulated to have obtained at the beginning of the universe had shifted from a chaos of feelings (a state of firstness in Peirce's scheme of categories) to a chaos of symbols (a state of thirdness in Peirce's scheme) which seems incompatible with the idea of law evolving out of a lawless chaos (since symbols are themselves laws). So, according to Short, the cosmological story of "Καινὰ στοιχεῖα" has the universe beginning not with firstness but with thirdness and proceeding "not by a law of spreading, or generalization, but by a teleological process of fulfilling a purpose, of becoming more concrete."[24] Short tells us that Peirce was unable to resolve the inconsistencies between the successive versions of his cosmology and his developing theory of categories and he let his cosmological program fade away. But Short acknowledges that Peirce denied that his cosmology suffered from irresolvable internal contradictions even though it is true that he never carried his program forward to resolution.[25] It is well to remember that cosmology abounds with apparent contradictions which are tolerated because of the uniqueness and inscrutability of the circumstances at issue and one must be prepared to give more weight to nuanced distinctions than probably one normally would. In particular, in Peirce's case, one must be careful not to equate the beginning of the universe with a beginning of the timeless chaos, the indeterminate nothingness, out of which the world emerged. Bearing this distinction in mind, it would not be contradictory for Peirce to say that the universe began with the influx of a symbol (perhaps the genesis of thirdness) into a primordial chaos of "general possibilities and lawless reactions."

ment that it falls short of the mark but says that "if it is not quite good science, it remains at least very interesting metaphysics, which in its effects can act something like good poetry, providing a guiding vision for future research of the kind such as the cosmological theories of the early Ionian philosophers, which proved to be the seeds of an extremely fruitful scientific research tradition" (Reynolds 2002, p. 183).

[21] The results of Short's study appear mainly in two papers, Short 2010a and 2010b. But see also Short 2011, as well as n. 12 from Short 2007, pp. 138–139.

[22] Short discusses this development principally in Short (2010b), see especially pp. 533, 539.

[23] Short 2010b, p. 541.

[24] Short 2010b, p. 542.

[25] Ibid.

Still, it would be disingenuous to deny Short's characterization of Peirce's cosmological account as vague and figurative[26] with little to offer as a basis for exact science. Yet Short acknowledges that we "may grant the vague and figurative an important place in thought, even in scientific thought,"[27] and it is instructive that Peirce is by no means alone among scientists who begin their cosmologies with a vague and figurative attempt to account for the beginning of the universe. In fact it is generally agreed that anything other than a vague account is not possible. Referring to the original chaos, Tufts University astrophysicist, Eric Chaisson, writes: "One cannot inquire about what happened at the exact moment of the Bang (precisely zero time).... The composition of the Universe at this time was indescribable, and its dominant action unimaginable."[28] In a similar vein, Stephen Hawking claims that, from the combined theoretical perspectives of general relativity and quantum theory, "the question of what happened before the beginning of the universe is rendered meaningless."[29] According to Paul Davies, theoretical physicist from Arizona State University, the simple answer to the question 'What caused the big bang?' is "Nothing." Why? "[N]ot because there was a mysterious state of nothing before the big bang, but because time itself began then."[30] But even though it may be convenient to dismiss talk of "a mysterious state of nothing before the big bang" as somehow out of bounds for science proper, that vague enigma looms large in the concerns of many cosmologists. New York University philosopher of physics, Tim Maudlin, maintains that the "question of accounting for what we call the 'big bang state'—the search for a physical explanation of it—is probably the most important question within the philosophy of cosmology," but he says that there is no consensus on how to answer it.[31] Some cosmologists argue that the beginning of our universe must have arisen in a law-governed way out of some previous condition and others speculate that "there might be special sorts of laws, or special sorts of explanatory principles, that would apply uniquely to the initial state of the universe."[32]

Just as they did for Peirce, concerns about the origin and pervasiveness of natural law engage contemporary cosmologists. According to astrophysicist Michael Murphy from Swinburne University, whether the laws of physics are "the same everywhere in the universe and throughout its entire history" is "one of the biggest questions of modern science."[33] Of course one must bear in mind that the very idea of natural law is vague and connotes rather different meanings for different

[26] Peirce said, himself, that "Our conceptions of the first stages of the development, before time yet existed, must be as vague and figurative as the expressions of the first chapter of Genesis" (W6: 209, 1887–1888).

[27] Short 2010b, p. 531.

[28] Chaisson 1987.

[29] Hawking and Mlodinow 2010, p. 135.

[30] See Davies 2010.

[31] Interview with Maudlin (Anderson 2012).

[32] Ibid.

[33] The quotation from Michael Murphy is from a news release, "Laws of Physics Vary Throughout the Universe, New Study Suggests," published in *Science Daily* on 9 Sept 2010 (http://www.sciencedaily.com/releases/2010/09/100909004112.htm).

cosmologists. Perhaps the greatest divergence (ignoring explicitly religion-based conceptions) is between the conception of natural laws as necessary and timeless in the way laws of pure mathematics are necessary and timeless and the conception of natural laws as contingent and emergent.[34] Cosmologists who are committed to scientific determinism, the doctrine that "[g]iven the state of the universe at one time, a complete set of laws fully determines both the future and the past,"[35] usually conceive of laws of nature rather like the laws of pure mathematics—as absolute, holding without exception. Hawking, a champion of scientific determinism, claims that it is "the basis of all modern science" and he emphasizes that it excludes "the possibility of miracles or an active role for God."[36] But it also excludes the possibility of the emergence of laws by cosmic evolution.

Cosmologists like Hawking who conceive of physical laws as timeless and absolute are faced with the problem of accounting for the surprising fact that of the billions of possible sets of physical laws, most of which are incompatible with the emergence of life, our universe was so incredibly lucky to have just the laws it has. This has led some to embrace a strong form of the so-called anthropic principle, the idea that the existence of life in our universe "imposes constraints not just on our environment but on the possible form and content of the laws of nature themselves,"[37] and even to suppose that our set of physical laws was fine-tuned intentionally for the emergence of life. But since the idea that our universe was intentionally designed to support life is not tolerable to scientific determinists like Hawking, or to modern scientists generally, another explanation is needed. Hawking's preferred solution, inspired by the work of such famous predecessors as Richard Feynman and Arthur Eddington, is that our universe is merely one of the innumerable possible universes, *all* of which appeared spontaneously with the big bang, and each in a perfectly law governed way (though with different assortments of natural laws).[38] Within the context of this multiverse,[39] this complete set of parallel universes, it is not a surprise to find our universe with its "fine-tuned" set of life-supporting natural laws—as a possible universe it obviously had to be included in the multiverse. But the very idea of a multiverse of all possible universes, coexisting in parallel, seems quite bizarre and strikes one as sounding rather like the "sheer nonsense" Peirce feared his readers would ascribe to him. Nevertheless, most contemporary cosmologists accept the idea of the multiverse in one form or another.

One theoretical physicist who is skeptical about the multiverse idea is Perimeter Institute cosmologist, Lee Smolin. He questions the now almost standard idea that universes flash into existence with a random but fixed set of absolute physical

[34] Relevant to this distinction, Einstein stated that "as far as the laws of mathematics refer to reality, they are not certain; and as far as they are certain, they do not refer to reality" (Einstein 1923).

[35] As stated in Hawking and Mlodinow 2010, p. 30.

[36] Ibid.

[37] Hawking and Mlodinow 2010, p. 155.

[38] Hawking and Mlodinow 2010, p. 136.

[39] Apparently the term "multiverse" was coined in 1895 by William James; see James 1897, p. 43; see also James 1909, p. 43, 197).

laws and promotes the view that the laws of nature are emergent. He is attracted to Peirce's approach to cosmology and believes that Peirce was right in holding (1) that the supposition that there are universal laws of nature that can be known but not explained is not a justifiable position, (2) that uniformities, especially laws, "are precisely the sort of facts that need to be accounted for," and (3) that "the only possible way of accounting for the laws of nature, and for uniformity in general, is to suppose them results of evolution."[40] So Smolin's way of explaining how our universe ended up with a fine-tuned set of physical laws that support Earth's life forms is not to maintain that ours just happens to be a universe, among the vast multitude of universes that make up the multiverse, with laws of nature that support living organisms, but that our good fortune to have such a fine-tuned set of natural laws is not so surprising if we abandon the notion that the laws we have are absolute and eternal and accept that they are the result of evolution by cosmic natural selection. Although this is the briefest non-technical sketch of Smolin's cosmology, ignoring many crucial components,[41] it highlights key elements of his approach, which were also key elements in Peirce's cosmology: the rejection of absolutes and the embrace of universal evolutionary growth.

But, as Smolin stresses, how to account for emergent natural laws is a great puzzle in its own right. Since the initial conditions giving rise to the emergence of law in the universe cannot be known, it is impossible to give operational, or experimental, meaning to the basic terms necessary for a standard scientific explanation of emergent law.[42] To attempt to do so one would commit the cosmological fallacy, "the mistake of extending a method that is designed to study small subsystems of the universe that come in many copies to the universe as a whole."[43] Compounding the puzzle is the logical problem of how to explain the emergence of law without resorting to some more general law. If the evolution of law is, itself, supposed to be governed by law, then we have just shifted the problem from one law to another, a metalaw that also requires explanation. "So," as Smolin puts it, "we still lack sufficient reason for laws."[44] If, on the other hand, we suppose that the evolution of laws is not itself law-governed, then again we "lack sufficient reason for laws." Smolin calls this "the metalaws dilemma."[45]

It appears to be the view of T. L. Short that Peirce got caught in the horns of this dilemma (although he doesn't mention it). As Short points out, a chief motivation for Peirce's cosmological program was to account for lawfulness in the universe and that required explaining how law emerged, or could have emerged, from a law-

[40] Smolin 2011; the reference to Peirce is from Peirce's 1891 "The Architecture of Theories" (EP1: 288).

[41] Smolin explicates his theory of cosmological natural selection in Smolin 1997. His most recent general treatment of cosmology is Smolin 2013.

[42] Smolin 2011.

[43] Ibid.

[44] Ibid.

[45] Ibid. See also Smolin 2013, pp. 242–247, for discussion of the metalaws dilemma and for some approaches to its solution. Smolin's work on the metalaws dilemma was done in collaboration with Roberto Mangabeira Unger.

less primordial chaos. But scientific explanation, at least typically, requires that the explanandum (what is to be explained) be shown to follow logically from the explanans (the designated initial conditions together with a covering law) and, as Smolin pointed out, an appeal to law in the explanans would cancel out the explanatory power of a purported explanation of lawfulness. Peirce was well aware of this so in most of his cosmological narratives he avoided appeals to metalaws[46] and conjectured that natural law could be explained as the outgrowth of a primal tendency for habit formation. Now at one point Short seems to accept that Peirce may have found a way around the metalaw dilemma because, as Short points out, a tendency is caused by a type of outcome (and thus is a form of final causation) and not by a law (which is characteristic of efficient causation): "[a] final cause, being general, takes the place of law but it is not a law."[47] But then he quotes Peirce's claim that "final causality cannot be imagined without efficient causality" and argues that "[i]f there is no final causation without efficient causation, then final causation cannot come first and explain the evolution of laws of efficient causality."[48] So Short concludes that Peirce was forced to resort to a law to explain law and thus was impaled on a horn of the metalaw dilemma.[49] It might be argued, however, that Peirce did escape this fate because on his account the germ of a tendency for habit formation could only have arisen by pure chance—though once introduced, or at least once it took hold, it would have grown by its own action (CP 8.317, 1891). If in its first instance the germ of a tendency for habit formation occurred by pure chance then whatever supposed underlying efficient causality was instrumental in bringing it about must also have involved only chance happenings—the lawless reactions of the primal chaos that Peirce refers to at the end of the opening quotation. In fact, the lawless reactions of the primal chaos would not be causes at all, strictly speaking, precisely because they were lawless events, brute instances of secondness with no subordination to law whatsoever. However, as mentioned earlier, Short gave another reason for believing that Peirce ran afoul of the metalaw dilemma, one that pertains specifically to the cosmological story given in the opening quotation. In that account Peirce substituted a tendency for representation for his more usual appeal to a tendency for habit formation and speculates that it was the influx of a symbol, rather than of an incipient habit, into the lawless primal chaos that provided the necessary germ of regularity that set the universe going.[50] But as Short correctly

---

[46] An exception seems to be Peirce's 1884 statement quoted above where he wrote that "all known laws are due to chance and repose upon others far less rigid themselves due to chance and so on in an infinite regress…" But even in this case Peirce did not suppose that any metalaw was absolute and eternal.

[47] Short 2010b, p. 538.

[48] Short 2010b, pp. 539–540. The Peirce text quoted is from EP 2: 121.

[49] Short 2010b, p. 535. Paul Forster also understood that, in attempting to explain lawfulness in the universe, Peirce faced "an intractable dilemma," essentially the same as Smolin's metalaw dilemma. Unlike Short, Forster believes that Peirce's evolutionary account does resolve this dilemma by explaining the emergence of law teleologically (Forster 2011, pp. 181–184).

[50] See Fernández's contribution to this volume for an excellent discussion of habits and symbols in Peirce's semiotics (see, especially, pp. 79–94).

points out, symbols are themselves laws requiring explanation and so cannot be used innocently to explain the emergence of lawfulness. Probably Peirce would have countered by explaining that the germ of the tendency for representation (the germ of semiosis) must have occurred purely by chance producing the broadest and least informative symbol possible, a mere tabula rasa "entirely indeterminate except to be representative." Once a symbol, however vague, is produced, it will reproduce itself in new and more determinate interpretants, themselves symbols. So by treating the tendency to represent analogously to the tendency for habit formation, a stronger case can be made for the cogency of Peirce's cosmological story than Short's analysis indicates. I will try in the remainder of this paper to strengthen the case for Peirce's cosmology although to fully make the case would be quite complicated and would require an in-depth study of Peirce's many relevant texts as well as a far more thorough examination of Short's critique than can be undertaken here.[51] But hopefully at least it has been shown that Peirce's cosmological speculations, if not cognate with the speculations of contemporary cosmologists, are anyway resonant with them in key respects and are worthy of serious scientific consideration. The ongoing relevance of Peirce's cosmological thought is demonstrated by the fact that a respected physicist such as Smolin refers to Peirce's ideas in support of his own.[52]

The cosmological story told in the opening quotation is of special relevance for biosemiotics precisely because Peirce chose to frame it in semiotic terms and to emphasize the crucial role of semiosis in the formation of the universe. This puts biosemiotics in the context of semiosis on a cosmic scale and raises the question whether it is viable to regard the emergence of semiosis as virtually synonymous with the emergence of life as many biosemioticians, perhaps beginning with Thomas Sebeok, have been inclined to suppose. This semiosis-as-life view was recently encapsulated by Terrence Deacon in his claim that "the origin of life is the origin of semiosis" and that the "defining attribute" of a biological organism is that it is "a representation-creating subject."[53] Of course there is no denying that Peirce repeatedly attributed life and growth to signs, to symbols in particular, even going so far as to claim, as he did in the opening quotation to this paper, that "there can be no reality which has not the life of a symbol." But wasn't Peirce speaking metaphorically when he attributed life to a symbol? Eliseo Fernández insists that Peirce "is not simply using a metaphor; rather, he is contemplating the possibility of generalizing the

[51] I mention only T. L. Short here because he is such a respected Peirce scholar and his recent studies concluding that Peirce's cosmology is a failed program must be considered and answered by anyone hoping to make the case that Peirce's cosmology was a success as a scientific theory. Another Peirce scholar whose conclusions need close attention is Andrew Reynolds: see Reynolds 2002, p. 175, for a statement of "the greatest difficulties afflicting Peirce's cosmology."

[52] Another physicist of note, Ilya Prigogine, has recognized Peirce's introduction of absolute chance into his cosmology as "a pioneering step" (Prigogine and Stengers 1984, pp. 302–303), and philosopher of science, Ian Hacking, has noted Peirce's success in putting "emergentism together with ideas of statistical mechanics, to form a new and vigorous indeterminism" (Hacking 1983, pp. 455–475). See also Fernández contribution to this volume, p. 82, n. 2.

[53] Deacon 2012, p.163.

conception of a living being so as to include signs as well as oaks or capybaras."[54] Lucia Santaella championed this position as early as 1999 in her pioneering paper "A new causality for the understanding of the living," where she claimed that semiosis is "a synonym of life, not only life in the philosophical, but actually in the biological, sense."[55] And in his masterful contribution to this volume (pp. 171–181), Winfried Nöth defends Peirce's claim that "every symbol is a living thing, in a very strict sense that is no mere figment of speech" (EP 2: 264, 1903). Following Peirce, Nöth delineates the distinctive biological characteristics of symbols:[56] symbols are complex signs that combine qualitative feelings with an intrinsic energy to react with external symbols in a way that tends to influence them or, as Peirce wrote, "bring along other ideas [symbols] with it" (W8: 148, 1892). Symbols exhibit purposiveness and act principally by final causation with the aim of self-reproduction and self-replication. The characteristic purpose of a symbol, according to Nöth, is to represent its object and "to convey some further information concerning it" (CP 2.231, 1910) and in fulfilling this purpose, over time, a symbol undergoes a process of replication and growth (taking on different and more developed meanings). In these ways, Nöth insists, symbols really are living beings although they "live in symbiosis" having "an agency of their own" but, like parasites, they can only live in host organisms and need symbol users to enact their purposes and to replicate—to provide the efficient causation necessary for their final causation. Peirce made this clear with his famous example: "law, without force to carry it out, would be a court without a sheriff; and all its dicta would be vaporings" (CP 1.212, 1902).[57] So again we must confront the paradoxical big bang instant of Peirce's cosmology, the influx of the premier symbol into primordial chaos from which the world emerged. How is the premier symbol itself to be explained and are we compelled to admit that its influx signaled not only the birth of the universe but also the genesis of life?

Consistent with Peirce's cosmological story, how could there have been a "host organism" to give substance—real forcefulness—to the premier symbol whose influx into the primeval chaos signaled the beginning of the historical universe? It could only have been the primeval chaos, itself, that state of pure indeterminacy that Peirce ascribed to the universe in its pre-history, which somehow provided the means necessary for the premier symbol to make its grand entry. Of course, as noted above, we are greatly limited in what we can know about the earliest stages of the universe and even more so about what there was before the big bang (or before the influx). But we do not have to surrender to our ignorance even though, as Peirce reminded us, "[o]ur conceptions of the first stages of the development, before time yet existed, must be as vague and figurative as the expressions of the first chapter of Genesis" (W8: 209, 1887–1888). But those vague and figurative conceptions can give us a foothold, as it were. Even Hawking, who advised that our usual ideas "do

[54] Fernández 2010, p. 103.

[55] Santaella 1999, p. 499.

[56] Nöth notes that Peirce's claim that symbols have life applies also to other legisigns, a subclass of signs that Peirce introduced in 1903.

[57] T. L. Short has expressed serious reservations about the usual interpretation of Peirce's court and sheriff metaphor: see note 65 below.

not apply to the very early universe," acknowledged that though it "is beyond our experience" it is "not beyond our imagination." So Peirce asked what must have obtained prior to the chaos, prior in a logical but not strictly temporal sense, and he imagined a state of firstness, characterized as an ancient continuum of qualities with the vaguest of beings (RLT: 259, 1898)—"the ovum of the universe" (RLT: 262, 1898). Subject only to lawless chance or spontaneity, somehow this cradle of continuous undifferentiated feeling randomly took on "a thousand definite qualities" (RLT: 260, 1898). Peirce's example, in the opening quotation, of the gradual emergence of definite redness from the indeterminate color continuum illustrates this point. The next step in the evolution of the pre-historical universe, still speaking logically, for time had not yet got going, was a stage of arbitrary reactions between the feeling-qualities. At some point among the chance occurrences, the brute events,[58] incipient laws sprang up, tendencies for habit formation, and those that didn't die out grew increasingly stronger, as they naturally would by virtue of their inherent natures (CP 8.317, 1891). What developed was a habit of acquiring habits which was, as Fernández has pointed out, the earliest of the laws of nature.[59]

In the opening quotation, Peirce wrote that the way to conceive of this tendency for habit formation is as a symbol and of its emergence into the primeval chaos as the influx of a symbol.[60] At the moment of influx the premier symbol would have been extremely vague, as indeterminate as the chaos from which it arose, but as a general type or law[61] it would have tended to bring instances under its governance and would have had its being in those instances, which would have become more and more determinate (EP 2: 292, 1903). According to Peirce, the object of the original symbol was representation itself and the upshot of that symbol was the creation of a tabula rasa "entirely indeterminate except to be representative" (EP 2: 323, 1903). That was the beginning of semiosis and governance by final causation.[62] But

[58] These are 'brute' reactions, not because of their forcefulness but because they are utterly lawless. According to Peirce, we must think of them as events, for "they had all the here-and-nowness of events," even though they are not truly in time (RLT: 260, 1898).

[59] Fernández 2012.

[60] This brings to mind Peirce's 1906 discussion of the idea of growth and how it requires the congress of the female and male functions. The female is the essential sex and functions as the seed while the male adds "the principle of unrest" and "executes a hunch" (EP 2: 374).

[61] Symbols are always legisigns in Peirce's more detailed classifications of signs.

[62] As was just explained above about the origin of the tendency for habit formation, the tendency to form representations or symbols may have sprung up and died out repeatedly in the lawless happenings of the pre-historical universe but eventually symbols must have grown increasingly stronger until full-fledged semiosis, action by final causation, got going. We can think of what in this paper is referred to as the premier symbol as the first symbol to generate a lasting semiosis—the one, or the first of the ones, that started the universe on a law-governed course and thereby signaled the beginning of time and the historical universe. According to Santaella, semiosis "is the technical version of final causality". She discusses this in some detail in Santaella 1999 (the quotation is from p. 514). T. L. Short also stresses the "end-directedness" of semiosis and attests that it "accounts for the significance of signs" (Short 2007, p. 172). But Short argues that "all significance is relative to potential interpretation" and that there can be no interpretation without intentionality, which is a form of purposeful action subject to possible error. Intentionality, Short maintains,

this account of the beginning of things where a tendency toward representation is supposed to have developed into government by law through semiosis is pragmatically equivalent, or nearly so, to the account where a tendency toward habit formation is supposed to have evolved into government by developed habits. The key to this equivalence is that symbols administer their governance by establishing supporting habits, their final logical interpretants (EP 2: 418, 1907). But where in these beginnings do we find the substance, the real forcefulness, that Nöth reminded us is required to provide the efficient causation necessary to actually implement symbolic actions? We might ask more simply, how was the premier symbol instantiated, for without a substantial body any sign would have been powerless.[63]

We have seen that Peirce attributes at least two continua to the primeval chaos, a continuum of qualities, "themselves mere eternal possibilities" (RLT: 260, 1898), and a continuum of unpersonalized feeling. Ordinarily we would suppose that essential to feeling per se is an embodiment capable of feeling but Peirce insisted that neither a feeling mind nor a physical substratum is necessary (CP 1.305, c. 1907).[64] So Peirce's pre-historical universe is pure firstness to this point. Then, somehow, feelings of definite qualities (sense-qualities) begin to emerge spontaneously from these continua creating discontinuities and moments of secondness—they "spring up in reaction upon one another, and thus into a kind of existence" (RLT: 259, 1898). It is in arbitrary reactions (brute events) between the sense-qualities that symbols begin to appear and the question we confront is whether these reacting feelings have acquired sufficient forcefulness to act as sheriff to implement the dictates of the legislating symbols?[65] Though feelings themselves may not logically depend on embodiment in a physical substratum, surely semiosis does logically require the material embodiment of the operative signs.[66] As Peirce remarked in the opening

"appears to be limited to animals" and this marks an important limitation of the sphere of semiosis. Short warns against the extension of Peirce's semiotic "to all of life or even cosmologically" for "[b]y being extended so far, it explains much less" (Short 2007, p. 177).

[63] Of course the body of a symbol is not part of its essence as a sign but it brings the symbol into actuality where efficient action may be taken in compliance with its aim. See CP 6.455, 1908 & CP 1.213, 1902.

[64] Nevertheless, Peirce did usually associate feeling with excited nerve matter (for example, see W8: 105, 1890) but clearly his conception of "unpersonalized feeling" was distinct from what he conceived of as feeling proper and may have been what he elsewhere called "psychic life."

[65] T. L. Short believes that this interpretation of Peirce's sheriff metaphor is mistaken because he thinks it interposes an unnecessary intermediary efficient agent who must be caused to cause the law to be enforced. Short believes that if law is to really govern actual events then its influence on them must be direct (Short 2007, p. 102). He says that Peirce's metaphor, properly understood, means that "The sheriff is the fact of imposition, not an agent that imposes" (Short 2007, p. 139 n. 12). Although respectful of Short's arguments, which must always be seriously considered, perhaps Peirce's point may be precisely that intermediate between a law and any given instance of its implementation there must be an efficient agent who/which is directly engaged semiotically to act mechanically. This would not preclude the need for a material basis for semiosis itself but stresses that the implementation of law always involves both final and material causation.

[66] For a more complete discussion of the details to be considered before passing judgment on the adequacy of Peirce's cosmological story, it would be necessary to go more deeply into his classification of signs and, in particular, the internal composition of symbols. Symbols are complex

quotation, "[i]t is of the nature of a sign to be an individual replica and to be in that replica a living general." An individual replica is an existing embodied sign; it is the embodiment that gives the sign, in its generality, its life—at least that is what Peirce seems to be saying. But is even embodiment enough to give a symbol the forcefulness necessary to act materially? Surely semiosis that depends on the material causation marshaled by the operative symbols also requires an expenditure of energy. Unlike the continua of quality and feeling which Peirce believed to be eternal possibilities accessible in any possible universe from its beginning, he did not believe that matter and energy are eternal but that they, along with space, are "products of an evolution from a primeval (and infinitely long past) chaos of unpersonalized feeling."[67] To fully answer how symbols could have gained the forcefulness necessary to begin to govern and thus set the universe on a course of lawful growth it would be necessary to work out how matter and energy arose from the arbitrary reactions of sense-qualities, but that is what Peirce tells us happened. Presumably, then, matter and energy emerged from the pre-historical chaos along with symbols.

Finally we return to the question whether the influx of the premier symbol signaled not only the birth of the universe but also the genesis of life? As just noted in the preceding paragraph, Peirce claimed that signs, though general, live in their replicas, their embodiments. The idea that it was the marshaling of evolving matter and energy under the governance of symbols (or signs in general) that signaled not only the beginning of the historical universe but also the origin of life may be an attractive idea but it is appears to be wildly off the mark as far as modern science is concerned. According to current estimates, the big bang that brought our universe into existence occurred about 13.82 billion years ago[68] but life, at least life on Earth, is thought to have originated around 4 billion years ago. By the origin of life most contemporary biologists mean the onset of biopoiesis, the process by which pre-DNA strands of nucleic acids began to catalyze chemical reactions somehow following "a blueprint for its own reproduction."[69] Peirce was well acquainted with the biological science of his time and probably would have concurred with this encapsulation of biopoiesis (which in his day was called abiogenesis).[70] He acknowledged that the "life-slime" is a chemical compound and even claimed that there "is no inherent impossibility in its being formed synthetically in the laboratory, out of its chemical elements" (W8: 179, 1892). But, as Fernández points out, Peirce generalized the conception of life and did not believe that its essential elements had to be limited to protoplasm or strands of nucleic acids. The more general idea is

---

signs always involving icons and indexes. It is reasonable to assume that in the pre-historical universe, before the influx of symbols, semiosis got started in a preliminary way with the emergence of iconic and indexical signs, but this is a question that cannot be examined here. Kelly Parker's chapter on scientific metaphysics (Chap. 8) in his book, *The Continuity of Peirce's Thought* (Parker 1998), is a good place to start.

[67] CP 8 Bibliography [G-c. 1893-5], p. 283.

[68] European Space Agency report on Plank space telescope study, 21 March 2013, www.esa.int.

[69] From an interview with Andy Knoll, Tyson ed. 2004.

[70] See Peirce's 1892 paper, "Man's Glassy Essence" (W8: sel. 29; EP 1: sel. 24), for the best account of his general understanding of biological science.

that of any process by which a being (of whatever sort) somehow prompts external energetic reactions to implement something like a blueprint for its own replication. The general conditions for life are the ones detailed by Nöth that apply to symbols as appropriately as they apply to protoplasm. But Peirce's conception of life seems to have been even more general than that. He speaks of a vague background psychic life of the cosmos prior to all evolutionary development: "I will not trouble you with any disquisition on the extreme form of realism which I myself entertain that every true universal, every continuum, is a living and conscious being" (RLT: 162, 1898). According to Peirce, wherever there is feeling there is conscious life—"whatever is First is *ipso facto* sentient" (RLT: 260, 1898)—and, as we have seen, he held that even before the beginning, in the eternal realm of pre-historical possibility, there was a continuum of unpersonalized feeling to which he attributed at least a slumbering consciousness (CP 6.221, 1898). This is a panpsychism. But could Peirce have been exaggerating, hoping to make a strong impression on the Harvard University students who were attending his 1898 Cambridge Conferences Lectures, from with the above quotations were taken?[71]

While it is quite likely that Peirce did hope to shock Harvard's philosophy students into new paths of speculative thought, it is unlikely that he let his rhetorical flourishes carry him beyond what he was willing to fully defend. Even so, we must be careful to not draw too hasty or too broad a conclusion about what follows from Peirce's attribution of sentience, living conscious being, to the unpersonalized feeling of the pre-historical cosmos. For one thing, Peirce did not straightforwardly equate mind with feeling even though, as Short pointed out, in his middle cosmological period, especially during the early 1890s, Peirce sometimes seemed to do so (Short 2010b, p. 532). But already in 1890, at the beginning of his middle period, Peirce recognized that feeling was at most only one of three "elementary phenomena of mind" (W8, p. 103), the other two being sensations of reaction (disturbances of feeling) and general conceptions (awareness of being governed by a habit). It seems that even then Peirce understood that mind, certainly in its definitive form, consists most fundamentally of interrelated general conceptions and associated interpretative habits, not the concomitant feelings that make us conscious of our mental operations (our thought). By 1902 Peirce stated clearly his conviction that unconsciousness mind exists from which it follows that consciousness, or feeling, cannot be an essential attribute of mind: "feeling is nothing but the inward aspect of things, while mind on the contrary is essentially an external phenomenon" (CP 7.364, 1902). The essential attribute of mind, according to Peirce, is final causation which, as we have seen, made its first appearance with the influx of the original symbol. So the development of mind began with the inception of semiosis and habit formation.

Supposing then that the genesis of semiosis was the genesis of mind, we might characterize fully developed mind as a complex dynamic system of signs

[71] The quotation from CP 6. 221 is not, strictly speaking, from the 1898 Cambridge Conferences Lecture series but it was originally intended to be used for those lectures—see Ketner's and Putnam's introduction to RLT, p. 1.

fundamentally involving symbols and generally accompanied by a consciousness of learning and mediation. Could there have been some level of mentality before the influx in the pre-symbolic universe, in that primordial nothingness that Peirce called the ovum of the universe? We know that Peirce attributed psychic life, and even a slumbering consciousness, to the ancient continuum of qualities even in its eternal state of absolute indeterminacy. But there could have been no developed mind then. Perhaps as definite qualities began to emerge and after secondness sprang forth in arbitrary reactions between feeling-qualities, incipient iconic and indexical representations might also have begun to form with consciousness gradually awaking from its primordial slumber and working its way out a completely isolated feeling (a consciousness of firstness) toward a sense of resistance (a consciousness of secondness). But, as Peirce made clear, what was original was the background psychic life, not consciousness per se: "Most of us are in the habit of thinking that consciousness and psychic life are the same thing and otherwise greatly to overrate the functions of consciousness" (EP 2: 447, 1908). So Peirce did not hold that the universe was suffused with mind from the beginning, nor with a living consciousness, but rather with a generalized psychic life. Peirce may have conceived of this background psychic life as an attractive force or as a sort of sympathetic attractiveness, which would fit with his theory of agapastic evolution, but this is among the most obscure parts of Peirce's metaphysics and is not well understood. Clearly, though, Peirce's panpsychism is not at all ordinary but it is a panpsychism all the same and is not congenial to today's science. Arthur Burks, the editor of volumes 7 and 8 of the Harvard edition of Peirce's writings and a principal designer of the ENIAC, studied Peirce's panpsychic cosmic theory of evolution and professed it to be mistaken but, nevertheless, declared it to be ingenious, "the best of the nineteenth century philosophical theories of evolution.... [and] the best non-reductive philosophical theory of evolution to this date."[72] Regardless of whether Peirce's cosmology is accepted whole, it has proved itself useful to modern cosmologists as we have seen, especially for its pioneering embrace of chance and evolution of physical law. But Peirce's synechism and agapasm have yet to be discovered by modern science.[73] It remains to be seen how future science will regard Peirce's

[72] Burks 1997, pp. 531–532. Also see Burks 1996.

[73] Hopefully the efforts of some pioneering biosemioticians are beginning to remedy this. I have not mentioned synechism and agapasm in this paper, which has grown beyond its ideal size already. Nevertheless, I should mention that Peirce believed in three principal kinds of evolution which he called agapasm, tychasm, and anancasm (but Peirce held that there is no sharp line of demarcation between them). Agapasm was evolutionary growth driven by attraction or sympathy—Peirce said it was evolution by creative love. Tychasm was evolution driven by fortuitous variation and anancasm was evolution driven by mechanical necessity. These kinds of evolution are related to the philosophical doctrines, agapism, tychism, and anancism, according to which love, chance, and mechanical necessity are each operative in the cosmos (W8: 194, 1892). Peirce's fullest treatment of his agapasm and the related law of love was in his fifth paper for his first *Monist* series, "Evolutionary Love" (W8: sel. 30, 1892), where he claimed that agapastic evolution depends on the continuity of mind and is closely linked to his synechism. See the discussion of the *Monist* series papers in the introduction to W8 for a fuller treatment of this topic; also see Parker

generalized view of life and mind and how it will fit into more inclusive future cosmologies.

In conclusion it may be useful to review key ideas from Peirce's cosmology that are of special relevance for biosemiotics. First and foremost, his cosmology is a thoroughgoing naturalism that emphasizes the ubiquity and prime importance of growth and recognizes even the laws of nature to be outcomes of evolutionary development. To account for increasing variety in the universe, Peirce believed it necessary to postulate spontaneity as an agency for he understood that physical law, or what he called mechanical law, was unsupportive of proliferation of kinds. Peirce's acceptance of absolute chance as a factor in the growth of the universe was a pioneering step for cosmology and one not yet applied to physical law by many contemporary scientists who are committed to determinism. Another pioneering step taken by Peirce was his recognition of the power of symbols to affect the course of actual events and his bold inclusion of semiosis as an operative form of final causation. Of course the importance of signs in biological processes has long been recognized and in recent years Peirce's semiotic ideas have permeated the flourishing science of biosemiotics.[74] But as shown above, Peirce conceived of sign action as fundamental even in bringing the universe out of its pre-historic chaos of possibility and indetermination and into actuality. Accordingly, the laws that first emerged concerned the operation of signs and were laws of mind (or laws of habit). As adherence to these laws became more and more regular (as habits became more firmly entrenched) events became more determinate and mechanical. As Peirce put it, mind hidebound in habit became physical matter subject to physical (mechanical) law. This is the basis for Peirce's objective idealism.[75] It took Peirce's genius to conceive a cosmology that assumed and accounted for underlying continuities between mind and matter and between laws of mind and physical laws. Finally, considering that in Peirce's cosmology it was an act of semiosis that launched the universe, we see that his conception of life is generalized far more broadly than views current in the biological sciences. Presumably this would extend the field of relevance for biosemiotic research across the entire history of the universe. Peirce's scientific philosophy, far from being a relic of nineteenth century thought, continues to be a rich resource for cosmologists and for biosemioticians.

1998, pp. 215–219. For a helpful discussion of the connection between Peirce's synechism and his objective idealism, see Ivo Ibri's contribution to this volume, pp. 33–49.

[74] See Favareau et al. (2012), for a helpful review of how Peirce's ideas have influenced biosemiotics, especially through the influence of Hoffmeyer.

[75] See Short 2010a and Ibri's contribution to this volume for a discussion of Peirce's objective idealism. Also see the exchange between David A. Dilworth and Short concerning Peirce's objective idealism in the pages of *Cognitio:* Dilworth 2011; Short 2011; Dilworth 2012.

References

Anderson, R. (2012). What happened before the big bang? The New Philosophy of Cosmology (interview with Tim Maudlin). *The Atlantic* (on line edition), 19 January 2012.

Burks, A. (September 1978). Review: Charles S. Peirce, *The new elements of mathematics. Bulletin of the American Mathematical Society, 84*(5), 913–918.

Burks, A. (1996). Peirce's evolutionary pragmatic idealism. *Synthese, 106,* 323–372.

Burks, A. (1997). Logic, learning, and creativity in evolution. In N. Houser, D. D. Roberts, & J. Van Evra (Eds.), *Studies in the logic of Charles S Peirce* (pp. 497–535). Bloomington, Ind.

Chaisson, E. (1987). The three eras of cosmic evolution. *World Futures: The Journal of Global Education, 23*(1–2), 11–29.

Davies, P. (3 September 2010). Interview with Paul Davies. *The Guardian* (Friday).

Deacon, T. W. (2012). Information. In Favareau, Cobley, & Kull (Eds.), (pp. 161–164).

Dilworth, D. A. (2012). Peirce's objective idealism: A reply to T. L. Short's "What was Peirce's objective idealism?" *Cognitio 12*(1), pp. 53–74.

Einstein, A. (1923). *Sidelights on relativity*. New York: E.P. Dutton & Co (trans. G.B. Jeffery and W. Perrett).

Favareau, D., et al. (Eds.). (2012). *A more developed sign: Interpreting the work of Jesper Hoffmeyer*. Tartu: Tartu University Press.

Fernández, E. (2010). Peircean habits and the life of symbols. *Semiotics 2010* (pp. 98–109). Ottawa: Legas Publishing.

Fernández, E. (2012). Habit and generalization, *V Jornadas Peirce en Argentina*. Academia Nacional de Ciencias de Buenos Aires. http://www.lindahall.org/services/reference/papers/fernandez/habit_and_generalization.pdf. Accessed 23–24 Aug 2012.

Fernández, E. (2014). Peircean habits, broken symmetries, and biosemiotics. Vol. (pp. 79–94).

Forster, P. (2011). *Peirce and the threat of nominalism*. New York: Cambridge University Press.

Gallie, W. B. (1952). *Peirce and pragmatism*. Harmondsworth: Penguin Books.

Hacking, I. (1983). Nineteenth century cracks in the concept of determinism. *Journal of the History of Ideas, 44,* 455–475.

Hawking, S. and Mlodinow, L. (2010). *The grand design*. New York: Bantam Books.

Houser, N. (2009). Peirce as a sign to himself. In J. Deely & L. Sbrocchi (Eds.), *Semiotics 2008* (pp. 387–395). Ottawa: Legas Publishing.

Ibri, I. A. (2014). The continuity of life: on Peirce's objective idealism. Vol. (pp. 33–49).

James, W. (1897). *The will to believe and other essays in popular philosophy*. New York: Longmans Green and Co.

James, W. (1909). A pluralistic universe. New York: Longmans Green and Co.

Lane, R. (2014). Peircean semiotic indeterminacy and its relevance for biosemiotics, Vol. (pp. 51–78).

Murphey, M. G. (1961). The development of Peirce's philosophy. Cambridge: Harvard University Press. Reprinted 1993, Indianapolis: Hackett.

Nöth, W. (2014). The life of symbols and other legisigns: More than a mere metaphor? Vol. (pp. 171–181).

Parker, K. A. (1998). *The continuity of Peirce's thought*. Nashville: Vanderbilt University Press.

Prigogine, I., & Stengers, I. (1984). *Order out of Chaos: Man's new dialogue with nature*. New York: Bantam.

Reynolds, A. (2002). *Peirce's scientific metaphysics*. Nashville: Vanderbilt University Press.

Russell, B. (1959). *My philosophical development*. New York: Simon and Schuster.

Santaella, L. (1999). A new causality for the understanding of the living. *Semiotica, 127*(1/4), 497–519.

Short, T. L. (2007). *Peirce's theory of signs*. Cambridge: Cambridge University Press.

Short, T. L. (2010a). What was Peirce's objective idealism? *Cognitio, 11*(2), 333–346.

Short, T. L. (2010b). Did Peirce have a cosmology? *Transactions of the Charles S. Peirce Society, 46*(4), 521–543.

Short, T. L. (2011). Reading Peirce differently: A response to David Dilworth. *Cognitio, 12*(2), 257–271.
Short, T. L. (2012). Review of Paul Forster's Peirce and the threat of nominalism. *Transactions of the Charles S. Peirce Society*, *48*(3), 385–387.
Smolin, L. (1997). *The life of the cosmos*. Oxford: Oxford University Press.
Smolin, L. (2011). Unification of the state with the dynamical law. Address delivered to the Perimeter Institute for Theoretical Physics, Waterloo, Ontario. http://pirsa.org/11100113. Accessed 28 Oct 2011.
Smolin, L. (2013). *Time reborn*. New York: Houghton Mifflin Harcourt Pub. Co.
Suits, B. (1979). Doubts about Peirce's cosmology. *Transactions of the Charles S. Peirce Society, 15*(4), 311–321.
Tyson, P. (Ed.). (2004). Interview with Andy Knoll conducted 3 May 2004 by J. McMaster for "Origins: How Life Began," NOVA online.
Wells, R. (1964). The true nature of Peirce's evolutionism. In E. C. Moore & R. S. Robin (Eds.), *Studies in the philosophy of Charles Sanders Peirce* (2nd Series, pp. 304–322). Amherst: University of Massachusetts Press.

The Continuity of Life: On Peirce's Objective Idealism

Ivo A. Ibri

> "Now, in obedience to the principle, or maxim, of continuity, that we ought to assume things to be continuous as far as we can, it has been urged that we ought to suppose a continuity between the characters of mind and matter, so that matter would be nothing but mind that had such indurated habits as to cause it to act with a peculiarly high degree of mechanical regularity, or routine." (CP-6.277).

A Short Foreword

Peirce's metaphysics is a complex system of concepts constitutive of a *theory of the world*. Among them, there is one of continuity, or *Synechism* as Peirce called it, and another of *Objective Idealism*. Both are interconnected doctrines, as the former claims that all reality is somehow continuous, notwithstanding an imperfect or incomplete continuity, and the latter refutes all ontological splitting between matter and mind, affirming that both are manifestations of one and the same substance, namely, *ideality*.

Synechism is, in fact, a synthesis of Peirce's idealism and realism, in the way that it is possible to conceive a reality constituted by general relations and possibilities under only one substance, viz., *eidos* or ideality.

Reader of German Romanticism authors, mainly Schiller and Schelling, Peirce was also committed to the conception of life, so dear to that philosophical movement. With his own metaphysics he achieved the Romantic targets, namely, highlighting Nature as a living organism, in which, from the stone to the human being, life was scaled in degrees of vivacity measured by its capacity of spontaneity. In this regard, this essay will show how Peirce's conception of different degrees of life interact dialogically by interchanging signs that require the new and important science of Biosemiotics for taking into account the grounds and meaning of that dialogue.

I. A. Ibri (✉)
Pontifical Catholic University of São Paulo, São Paulo, Brazil
e-mail: ibri@uol.com.br

V. Romanini, E. Fernández (eds.), *Peirce and Biosemiotics*, Biosemiotics 11,
DOI 10.1007/978-94-007-7732-3_3,

On Philosophical Terminology—Setting the Stage for a Better Understanding of Some Peircean Concepts

A long acquaintance with a great author of the history of philosophy whom we admire, allows us to develop, along with an absorption of his concepts and vocabulary, a kind of instinct for guessing[1], that not only tries to decipher what there is of *spirit* behind the *letter,* hence something that hovers beyond the text as unsaid, but also that which the author said, but did not make sufficiently clear.

Charles Peirce is one of these authors, and certain peculiarities of his work spur the emergence of that guessing instinct, given the fact that it is comprised mainly of manuscripts that he never saw published in his lifetime, nor was able to review, suppress and, I suppose, add passages that the imminence of a publication stimulates for the sake of quality and precision. Clear terminology, however, was always a cause of worry for him[2], nearly obsessed with giving new names, often unprecedented, to the new concepts he created during the development of his work, many rooted in tradition but receiving from him an original approach, coherent with the theoretical edifice of his philosophy.

This concern with a biunivocal terminology of concepts probably derives—and here is an example of guessing—from his education and practice in natural sciences.[3] Indeed, the so-called sciences of Nature, and even their derivatives such as the applied sciences, enabled the practice of a biunivocal terminology between terms and their objects, which supposedly contributed very much to the advancement of knowledge in these fields. Unfortunately, the same cannot be said, for various reasons, of the so-called human sciences, especially philosophy.

Among the causes of the *illness of language* of which many contemporary authors accuse philosophy, there has rarely been, prioritarily, the *virus of bad terminology* that infected doctrines with the vagueness and polysemy that preclude the clarity of concepts, but rather, the illness would lie in the metaphysical intentions inherent in the use of philosophical language. It must be acknowledged that the insistence in an indispensable analysis of the logical soundness of the arguments, to which philosophical theories ought to be subjected, should be preserved from this criticism.

It is not uncommon to say that the objects of human sciences in general, and philosophy in particular, are of greater complexity than those of the sciences of Nature, requiring, thus, the renouncing of terminological biuniqueness in favor of a language that expresses such complexity, thereby justifying a certain (and allowed) semantic vagueness. However, if this line of argument, on the one hand, errs in ignoring the complexity of the objects of the countless fields of knowledge of natural sciences, on the other hand, it seems to want to preserve the right to use philosophi-

---

[1] Incidentally, a Peirce expression, as a necessary faculty for the success of a heuristic logic, defined by him, as we all know, as abduction.

[2] See his *Ethics of Terminology* (CP- 2.219-226).

[3] On Peirce's education in Natural Sciences and how it forms part, in a special manner, in his Philosophy, interesting approaches may be found in (Fisch 1986, pp. 376–383) and (Lenzen 1964, pp. 33–50).

cal language in such a way as to maintain, to an unsuspecting reader, the dangerous ambiguity between conceptual difficulty and linguistic obscurity, in which a terminologically confusing discourse often takes on a sham guise of profundity.

The history of philosophy shows that each author, and one could rarely generalize for a whole period, keeps his own vocabulary, which, if it cannot be agreed in favor of a biunivocal terminology, should be at least respected in its own context. This is a way of being fair to philosophical concepts, namely, to seek in each theoretical system the clearest sense that an author wished to give to a concept, since a similar term may appear under distinct meanings in distinct authors.

Without a doubt, this respect for a terminological context entails a non-trivial study of the history of philosophy that will be beneficial in many ways. A first would allow one to become terminologically acquainted with the respective author concerned. A second would be to understand the concepts in their historical context. It is not uncommon to fall into a kind of *historical parallax*—if I may be allowed this metaphor—when judging concepts in light of contemporary knowledge. Finally, a third benefit of this study would be to avoid the proposition of pseudo-new concepts that had been already formulated, often with admirable depth, by an author or school of a much earlier period.

Justifying a claim of originality in philosophy involves, undoubtedly, an honest examination of the history of ideas. In *exact* sciences, knowledge is cumulative, and past knowledge always appears updated, somehow, in current theories. Human sciences have areas in which knowledge with difficulty is increasingly acquired, on a step-by-step basis, in a historical process. Philosophy is one of them and a terminological non-agreement would require constant remission to the history of ideas, and only then become aware that something *new* is being proposed.

Let us then pass on to the specific issue of terminology that gave rise to these initial considerations. It relates to the terms *realism* and *idealism,* as they appear in Peirce's vocabulary. Contemporaneously, as we know, these terms are considered as meaning opposing philosophical positions: realists would be those who admit the existence of an external world, independent of the mind that represents it; idealists, in turn, state that the world is constituted by mind and its perceptive capacity, in such a way that there would not be anything like a reality independent of its representation. This question, thus posed, one can say, does not belong to the theoretical context of Peircean works. Nevertheless, many uninformed scholars ask themselves how Peirce could be considered, simultaneously, a realist and a nominalist.

I previously touched on the necessary study of the history of philosophy because it would be the only resource capable of resolving this question, by becoming aware that realism, such as proclaimed by Peirce, is of a Scholastic extraction[4] and, so, it is not solely confined to the predicate of otherness in relation to its representation, but rather that this reality is also organized in general relations, which, in the Peircean vocabulary, are called *laws* or natural *habits*.

[4] Interested readers may consult (Ibri 1992, Chapter 2). Other perspectives can be found in (Boler 1963), (Dileo 1991), (Almeder 1975) and (Michael 1988).

In the case of *idealism,* it should be distinguished historically in its two nuances, viz., subjective and objective. Generally, the former is associated with the conception of a reality confined to subjectivity, while the latter refers to a substantial predication of reality, affirming it as of an ideal nature. In accordance with this conceptualization, Berkeley and Fichte can exemplarily be distinguished as subjective idealists, while both Plato's and Schelling's positions, as well as Peirce's, can be considered under an objective idealism.[5]

This is a clear case where unfamiliarity with the historical context of philosophical terms seriously jeopardizes the understanding of the conceptual position of some authors. Surprisingly, there are scholars focused on philosophical language analysis who get entangled in issues that, ultimately, are of a linguistic nature, notwithstanding being dependent on knowledge of the history of philosophy, often disdained by schools that preach that such a history is nothing more than a delirious parade of metaphysical doctrines.

These considerations certainly do not resolve satisfactorily the possible conceptual interlacing between realism and objective idealism, which we shall address in the course of this essay.

Why Realism and Idealism in Peirce? Genesis Conjectures

The study of the sciences of Nature and, chiefly, its practice in the investigation of phenomena in light of theories so as to confront them with experimental results, can teach a lot under a philosophical viewpoint. It is not by chance that Peirce mentions in his famous *What Pragmatism Is*[6] how his mind was trained in a laboratory, and how this distinguished him from other philosophical thinkers. The fact that he admits this distinction stimulates, somehow, our efforts at *guessing* what that text does not explicitly say. In this guessing activity, there is nothing that does not meet the principles that Peirce formulated in his logic of abduction, namely, that we must be stimulated to conjecture, if we are to advance in any research intended to explain facts.[7]

Indeed, scientific research places theoretical representations in contact with their objects through phenomena derived from experiments, which are nothing more than a direct contact with the otherness of such phenomena. Objects can, genuinely, *object* to theoretical predictions through their experimental replicas, or *indexes,* in the Semiotic vocabulary—that is to say, as symbols, represented by the general character of the theories, are confronted by them, namely, the indicative signs. To a mind trained in such practice, it is trivial to say that processes of theory validation occur through some level of *adherence* with experimental results, and when they

[5] Details of this may be found in (Ibri 1992, Chapter 4). For other approaches, we suggest consulting: (Guardiano 2011), (McCarthy 1984) and (Tiercelin 1998).

[6] CP 5.411-437, EP 2.331-359.

[7] Readers may wish to consult, as examples, the texts *The Nature of Meaning*, EP 2.208-225 and *Pragmatism as the Logic of Abduction*, EP 2.226-241.

do not, they should be parametrically readjusted or, even, be radically changed. For this reason, it must also sound trivial to say to a man of science that the last *word* that justifies a theory belongs to the object.[8] So, in the final analysis, this practice is, one may say, a *lesson in otherness*, in its philosophical meaning.

From another angle, the obligation to intervene in the structure of theories, adjusting forms and parameters when predictions derived from them do not adhere to experimental data, leads to this question: what strange power does the contingency of the experiment have, in its particularity and finitude, to *disallow* a general theory? Would not these data *necessarily* be replicas of something equally general that would manifest itself through its indexing or, in other words, would not such manifestation reveal itself as a sequence of indexes that organize themselves according to some general rule? Without delving here into what would be a good induction, suffice it to say that some redundancy of the sequence of indexes—its continuance in similar sequential experiments—reinforces the idea that it is not some merely contingent order, present in phenomena.

This would be a second lesson, namely of the awakening of the belief that investigation is, indeed, of the nature of a *dialogue* with general objects endowed with logical rules, with syntactic structures that belong to them. This presupposition is, admittedly, that there is some corresponding form between an investigated object and its theoretical representation, and that this object is definitely not a particular, but something of the nature of a symbol[9], as much as the theories that seem to represent it. *Science is science of the universal*, as Plato and Aristotle affirmed—one seeks the symbols and not merely the indexes that existentialize them.

From these considerations, it seems inappropriate to support a theory of mirrors.[10]According to this theory, every investigation supposedly seeks a genuine picture of the phenomenon, a perfect image. Now, we know very well that this expectation belongs to a period in history when, justifiably, in face of the success of the scientific revolution in the Renaissance, the thinkers of that time formulated the idea of a universe similar to a clock, whose mechanism would be possible to discover and represent to its perfection. But here, one must revert to history, to again do justice to the ideas of each era, and not simply condemn them as metaphysical nonsense. Reasons of a theological nature, among others evident in the culture of that time, stimulated this deterministic world view that was continuously propagated through the first half of the nineteenth century[11]. But this determinism was abandoned in philosophy during the second half of the nineteenth century, and in science in the second decade of the twentieth century. Darwin set a milestone when he resorted to the principle of Chance for his theory of evolution. However, he was

[8] As Peirce himself said: "[…] If the facts won't agree with the Theory, so much the worse for them. They are bad facts. This sounds to me childish, I confess." (CP 5.116).

[9] To further explore the ontological conception that Peirce developed regarding the symbol, see his *New Elements*, EP 2.300-324, particularly the last pages. Different approaches and perspectives can be found in (Parker 1994) and (Nöth 2010).

[10] As Rorty (1979) does.

[11] Incidentally, this determinism remained as belief to some contemporary scientists, like Einstein.

not exactly a philosopher. At the end of the nineteenth century, this principle, pioneered by Peirce, was formulated in philosophy[12].

Renouncing the mirror-theory was never a problem to Peirce. World images in the plane of knowledge, if required to represent it, should, in his philosophy, always be considered somewhat faded, often out of focus, showing its object roughly, albeit under a predictive viewpoint—an essential purpose of knowledge—sufficiently clear to show what future conduct could reasonably be expected of such object. The mythological search for an absolutely certain knowledge was, fortunately, abandoned, and in this task Peirce was definitely a pioneer, as Popper explicitly acknowledged[13].

His theory of Chance, introducing indetermination in terms of the ontology of the object, was epistemologically accompanied by the doctrine of fallibilism[14], namely, that our knowledge is fallible, approximated, subject to a permanent correction process. The mirror and the mirrored object are indeterminate in various degrees. Uncertainty definitely shapes human rationality, and the entire task of knowledge is to reduce, to an ontologically feasible point, the fadedness of the *images*. They are the ones that substantiate thought in the form of signs and, through them, we try to adjust the *focus* of immediate objects in relation to dynamic ones[15]. Here we continue to use the Rortyan metaphor of the mirror only to show how, contrary to the use that Rorty makes of it, it would be feasible in relation to Peirce's philosophy.

Returning to the possibility of conjecturing on the *spirit* of a philosophical work, a hypothesis seems to suggest itself out of the long familiarity with Peirce's works: in what way did his education in science and its experimental practice possibly inspire Peirce to advocate a realism, not only of the existence of an external world of particulars, but also constituted of relations of order between them? Would he be so vividly interested in the work of Scotus as he was in his youth, if the seeds of realism had not been already instilled in his mind by the precocious practice of an experimental science?[16] It seems plausible to think that scientific practice, the

[12] As Popper (1972) acknowledges: "Thus Peirce conjectured that the world was not only ruled by the strict Newtonian laws, but that is was also at the same time ruled by laws of chance, or of randomness, or of disorder: by laws of statistical probability. This made the world an interlocking system of clouds and clocks, so that even the best clock would, in its molecular structure, show some degree of cloudiness. So far as I know, Peirce was the first post-Newtonian physicist and philosopher who thus dared to adopt the view that to some degree all clocks are clouds, or in other words, that only clouds exist, though clouds of very different degree of cloudiness."

[13] In the essay mentioned above, Popper (1972) describes deterministic hope as *nightmare*.

[14] See for example: CP 1.13-14; and particularly CP 1.141-175.

[15] In the conceptualization proposed by Peirce, the immediate object is that which appears in the representation, while the dynamic one is the real object, possessing otherness in relation to its representations. One of the best Peircean descriptions of the different types in interpretants can be found in a letter from Peirce dated March 14, 1909 (SS. 108–119). In its entirety, this letter also touches on points that we have addressed throughout these pages, such as Peirce's education in natural sciences and his determined realistic position, and we strongly recommend its reading. See also, CP 4.536 and CP 8.315.

[16] Provocative analyses on the relation between realism and experimental practice in science in Peirce can be found in (Forster 2011) and (Delaney 1993, particularly Chapter 1 and 2).

search for an adjustment of theories in terms of the otherness of phenomena and the relation of order they display, suggests, to those who make this practice an object of philosophical reflection, as it did to Peirce, a realism of a Scotist nature, that is, the acknowledgment that some universals are real.

Now as we continue this reflection, asking ourselves about the *nature* of these universals, which Peirce called *laws*[17]and, later, real *continuities*[18], we enter the domain of objective idealism.[19] I conjecture that Peirce's idea of idealism would not have originated from his practice in science—for from it, it would seem plausible to say, originated a realism constituted by the predicates of *otherness* and *generality*. The history of philosophy at the end of the nineteenth century had already paraded many idealist schools, whether of a subjective or objective nature. Peirce was a serious reader of the history of ideas, and his objective idealism, as he explicitly confessed, was drawn from Schelling, whence derives the complexity of speculating not only on the nature of the universals, but also on the substantial relations between matter and ideality, a topic which we will address further along. The nature of sense qualities, such as colors, sounds, odors etc., should be also added to his reflection on objective idealism, and not only on the fabric of the general relations recognized as part of the object.

The practice of a science that deals with phenomena and has a specific field of experimentation, by instilling in the mind of the investigator the notion of objective reality constituted by a generalized otherness, does not, I suppose, induce subjective attitudes toward philosophy. The act of constantly dealing with the object, with that which can always, potentially, object to its representation, must lead to the admission of an objective reality, indifferent to any constituting appropriation of a subjective nature. Under this hypothesis, realism, as a philosophical posture, appears to harmonize with objective predications. In this regard, although in the current state of our reflection in which the development of the possible relations between realism and idealism in Peirce's philosophy is still incipient, it should be noted that this idealism, if simultaneously viable with a realism, must be of an objective nature.[20] It would be strange indeed to accept a realism of the universals, *necessarily* of an objective nature, simultaneously with an idealism centered on a constituting subject. The acceptance of a universe of real symbols, that is, of natural laws that are formed by themselves, independently of the way in which they can be represented, means accepting concurrently that such laws have the same eidetic nature[21]—such universe of real symbols is potentially representable in a theoretical universe of symbols, which immediately suggests a connaturality between object

[17] EP 2.67-74.

[18] NEM 4.343-344.

[19] See (Ibri 1992, pp. 55–69) for details.

[20] On other occasion, I addressed this theme also on its interfaces with semiotics. See (Ibri 2006).

[21] We use the term *eidetic* in the Greek meaning of *eidos*, designating something of the nature of ideality.

and sign.[22] The acknowledgment of this connaturality, as suggested here, derives only from a philosophical conjecture on the consequences of scientific practice. In fact, the concept of connaturality is harmoniously present in what I call the *symmetry of categories*[23], notwithstanding its formulation from the conception of this symmetry.

I consider *connaturality* an important concept. It opens many doors for ontological reflection, which we can no longer avoid if we truly wish to understand what Peirce intended with his philosophy. But, as a scholar of Kant, Peirce would not propose a *bad* metaphysics, but rather one that, in the Kantian style, is strictly grounded on a Phenomenology, with the precaution of bearing in mind the broad concept of the phenomenon proposed by Peirce. Ultimately, both these authors agree on what would be a sound metaphysics, but would certainly differ on what, after all, can be considered *phenomenon.* To Peirce, dreams are phenomena, no less than the irregularity of a tree canopy.[24] Kant would hardly admit these examples, especially because it would not be of theoretical interest to his epistemology, in which the condition of semantic possibility of a phenomenon should be in its submission to a strict rule of understanding, i.e., that of *causality.* It is known, however, that phenomena that possess a high degree of spontaneity, and that repudiate being subjected to causal rules, are fundamental both to Peircean epistemology and to its always correlate ontology. These are, ultimately, consequences of two distinct philosophical presuppositions, namely, *determinism* and *indeterminism.*[25]

Questions gradually emerge in this heuristic process of possible reconstruction of Peirce's realism-idealism, that have no other purpose than to show that these doctrines do not necessarily emerge from metaphysical nonsense, but from a reflection on the ontological nature of the objects of scientific investigation. It will be seen that such an ontology will not ask an innocuous question about what is the *being* of things, but rather it will conjecture on what constitutes it, in view of its *practical consequences.*[26] Here, pragmatism is called upon not only as a rule for the meaning of theories, but also as a logical rule capable of entering the universe of meaning of an ontological nature, namely, the plane of real objects.

However, this aspect will be addressed further along, and I now return to a concept that is precious to Semiotics, that of the *dialogue,* to emphasize the importance of the notion of connaturality, deeply related to objective idealism.

---

[22] See, for example, CP 5.549-573, where Peirce explains the relationship between sign and object, within the context of the theory of truth provided by pragmatism.

[23] See (Ibri 1992, Chapter 1), (Ibri, 2009, 2010, 2011).

[24] See CP 1.284. For other approaches and perspectives, see, for example, (Anderson 1995), (Blachowicz 1972) and (De Tienne 1993).

[25] For further details, see (Ibri 1992, Chapter 3). For other approaches, see (Cosculluela 1992).

[26] I addressed the issue of practical consequences in pragmatism in (Ibri 1992, Chapter 6).

Connaturality, Semiotic Dialogy and Pragmatism

Earlier I said that scientific investigation may be understood as a dialogue derived from the adaptation of theories to phenomena, in which the objects reveal what they are by the way they *appear.* Here, *appear* and *being* are logical correlates, and not aspects that possibly contradict each other, as in ancient metaphysics.

A proposal to reflect on the notion of dialogue implies considering the possibility of modifying conduct through experience. This would mean some sort of interpretation of the phenomenal world in which one is inserted. In turn, *interpreting* requires an extraction of pragmatic meaning, namely, one that possibly determines future conduct. To be sure, the planning of future conduct is nothing more than to nourish the hope that a new form of action, guided by the reformulation of concepts, may perform the role of an efficient mediation in relation to the otherness of phenomena, i.e., to the *reality* of the objects. In a universe of existents, this circulation of meanings capable of modifying conduct, or even, of strengthening pre-established conducts in the form of *habits*[27], involves a signic, or semiotic, dialogue in which the notion of language[28] extends to a realistic signic context of philosophy. Meaning, here, transcends its mere consideration as a property of languages practiced by men, rather extending toward the entirety of the universe of phenomena, so as to acquire its *pragmatic legitimacy.* I refer to this latter expression for the purpose of establishing that *meaning,* to pragmatism, implies a necessary stage of existential interaction, an insertion into a *theater of reactions,* according to Peirce's vocabulary, of transformation of the indeterminate to a determination that involves a possible contradiction in a spatial-temporal context. I mention the word *stage* to stress that the meaning will be always consummated in a general instance, i.e., involving a reflexive return of experience to the plane of the concept. Under the perspective of the categories, secondness of experience is a necessary phase for general meaning on the plane of thirdness.

When admitting a realism of the universals, we must accept its objective ideality as a consequence. On this point, idealism is nothing more than an acknowledgment of the very nature of what constitutes real thirdness. Connaturality, on the plane of ideality, also derives from the acknowledgment that the object is of the same nature as its representations. Logical structures equally permeate sign and object.

An idealism of this kind is only a *backgrounding doctrine,* namely, a mere acknowledgment of a substantial connaturality between the cognoscible objects in their generality and the theories that represent them. In the language of Semiotics, sign and object are connatural, and this philosophical acknowledgment simply justifies how a dialogue of theories with experience can be possible. *Backgrounding doctrine,* an expression I propose here, thus represents a condition of epistemological possibility, and not an efficient instrument (either direct or explicit) for research, albeit it must be admitted that it establishes a world view whose consequences will transcend the mere effectiveness of the consolidation of knowledge. I will endeavor to enunciate some of these consequences at the end of this essay.

---

[27] On this notion of habit, important to the line of argument of this essay, I will return later.

[28] See (Ibri 2011).

Mind and Matter—Idealism in Light of Pragmatism

Peirce's reflection on the relations between mind and matter[29] already appears in some of his essays on this theme[30]. In them, he speculates on the possible relations between both, ruling out Cartesianism, first of all, as a dualism unsuitable for philosophies that seek to generally overcome them, by unifying concepts in light of the simplification recommended by Ockham's razor. On the other hand, a relationship between matter and mind in which the former is taken as genetically primordial and the latter its special case, would incur a *materialism.* A pragmatic analysis of the concept of materialism would, however, lead to mistakenly characterize it as ontological determinism[31], which would mean reducing phenomena of a psychic nature to a physicalism ascertainable by laws of matter.

Peirce emphatically rejects this alternative relationship between matter and mind, imagining, exemplarily, how absurd it would be to conceive a mechanism capable of *feeling.*[32] It would be, in other words, reducing feelings to the mechanical laws that created them, which, to Peirce, would be an inexplicable ultimate regularity. As a consequence, this would mean reducing firstness to thirdness, breaking the independence of the former in relation to the latter.[33]

There remains, therefore, as a possible relation between matter and mind, the subsumption of the former in relation to the latter, characterizing what he called *idealism.* But what kind of idealism would this be, derived from a logical analysis of the Cartesian substantial dualism? Without a doubt, given the ontological tenor of the issue, Peirce refers to an objective idealism that would affirm that physical laws are of an eidetic nature. However, we must remember that this consequence had already been extracted from his adoption of realism, in his conclusion that natural law must be understood as a system of logical relations of the nature of semiotic symbols, and, thus, eidetic.[34]

However, we must reflect more deeply on the ever disturbing, not to say unprecedented to many scholars, sentence enunciated by Peirce on *what is matter*. Summarizing his arguments in favor of an objective idealism, the passage quoted below concludes with such a sentence:

> The materialistic doctrine seems to me quite as repugnant to scientific logic as to common sense; since it requires us to suppose that a certain kind of mechanism will feel which would be a hypothesis absolutely irreducible to reason—an ultimate, inexplicable regularity; while the only possible justification of any theory is that it should make things clear and reasonable. Neutralism is sufficiently condemned by the logical maxim known as Ockham's razor, i.e., that not more independent elements are to be supposed than necessary.

[29] CP 6.24-25 (EP 1.292-293/W 8.105-106).

[30] See CP 2.272-277; W 8.135-157 (EP 1.312-333/CP 6.102-163) and W 8.165-183 (EP 1.334-351/CP 6.238-271).

[31] See (Ibri 1992, pp. 59–62).

[32] See (Ibri 1992, pp. 47–49, 104–105).

[33] This independence is phenomenologically and cosmologically justified. I have addressed this theme in Ibri (2010) and Ibri (1992).

[34] See also (Ibri 1992, chapter 4 and 5).

> By placing the inward and outward aspects of substance on a par, it seems to render both primordial. *The one intelligible theory of the universe is that of objective idealism, that matter is effete mind, inveterate habits becoming physical laws.*[35]

It is a well-known fact that Peirce's idealism derives from Schelling[36], as he himself confesses, including the sentence that states that matter is *mind exhausted by inveterate habits*. The most significant passage of the Schellingean work that bears similarity to the section of Peirce's work quoted above is:

> Matter is, indeed, nothing else than mind viewed in an equilibrium of its activities. There is no need to demonstrate at length how, by means of this elimination of all dualism, or all real opposition between mind and matter, whereby the latter is regarded merely as mind under a condition of dullness, or the former, conversely, as matter merely in becoming [...][37]

Under the light of pragmatism, to say what matter *is,* is to say how it *behaves* phenomenologically, or, in other words, what are the *practical consequences* of the concept of matter, considering such consequences as the experimentally verifiable predicates of matter, capable not only of affecting human conduct, but also the interactive conduct of the very universe of material events. I am here evidently proposing to use the expression *practical consequences* also in a radically realistic and ontological sense, extending the notion of conduct to all beings that cohabit a given real universe, albeit respecting the logical genesis of the doctrine that, in the most general way possible, is a necessary passage from indetermination to determination, from the general to the particular, implying the existence of some theater of reactions that can become the condition of possibility of such determination.[38] Keeping to this amplified line of understanding of the idea of pragmatism, the passage of the indeterminate general to the determinate particular and, as such, reactive to other particulars, is the necessary correlative passage of an internal[39] to an external world, in which the former comprises *continuity* and the latter *discontinuities.* A vocabulary of the theory of continuity[40] is shown here almost naturally, to provide support to a better understanding of the necessary relations between the general and the particular. I avail myself of it not only as an alternative for the general/ particular terms, but also in a context explained further on, relative to the ontological meaning of the Peircean categories.

---

[35] CP 6.24-5; my italics.

[36] I described the influence of Schelling on Peirce in (Ibri 2010). See also, (Dilworth 2011).

[37] Schelling (1978, p. 92).

[38] This condition arises from a logic of possibilities where something generically possible should cease to be so, in order for the term possibility assumes some significance. See, CP 6.219.

[39] I here use the word 'internal' in a general sense, in which interiority would be distinguished from subjectivity, thus acquiring a realistic logical meaning. Subjectivity would be a special case of interiority. In this sense, interior objects are those that can only be known or represented indirectly, by logical inference. Outside world, in turn, is the only one that can be directly observed—under the categorical point of view, it is under the secondness.

[40] See Chapter 4 of Ibri (1992). Other approaches regarding the concept of Continuity in Peirce can be consulted in (Havenel 2008), (Lane 2011 and 2011a), (Moore 2007), (Potter et al. 1977) and (Rosa 2003).

Although these considerations apparently extrapolate the boundaries of pragmatism[41] beyond its purpose of constituting a semantic rule associated with human conduct, in fact what is sought here is to interface the vocabularies of logic and ontology, endeavoring to create, in light of the background of realism, more powerful heuristic tools for an understanding of what we have defined as *symmetry of the categories*.[42] Indeed, these relations between the particular and the general, internal and external worlds, continuities and discontinuities, are consummated in the inter-relations between Peirce's three categories. While the first and third categories are of a continuous nature, or, in other words, harbor the mode of being of what is continuous, general and internal, secondness is the mode of being of the theater of reactions that harbor the discontinuous, the particular, the external side of the other two categories. These considerations on the three Peircean categories should apply—resorting again to the background of the author's realism-idealism—both phenomenologically and ontologically, and, for this reason, justify the use of the expression *symmetry of the categories,* meaning three modes of being that indifferentiate or symmetrize logical rights between subject and object, Man and Nature, radically breaking dualisms of genesis and proposing the abandonment of anthropocentric postures of philosophy.

Idealism Evolutionarily Grounded

One of the features of Peirce's objective idealism, according to my previous comments, is to break the mind-matter dualism in order to consider the material universe as a special form of mind, whose conduct would be driven by ingrained habits. An awareness of this proposition implies understanding natural laws as *habits* of conduct of matter, and the use of this terminology gains significance by enabling a reflection on *how* these habits were acquired, opening a conceptual door for an overlapping of idealism with some form of evolutionism. Peirce's hypothesis on the formation of habits of Nature, that is, on the origin of the natural laws, is, incidentally, one of his most brilliant ideas, according to William James.[43] On the one hand, Peirce proposed an evolutionism that would permeate his entire philosophy.[44] On the other, in the same line of argumentation, he conjectures on the formation of the laws of Nature. Here is a crucial passage of his work on this topic:

> What kind of explanation can there be then? I answer, we may still hope for an evolutionary explanation. We may suppose that the laws of nature are the result of an evolutionary process.[45] [Further:] But if the laws of nature are the result of evolution, this evolutionary process must be still in progress. For it cannot be complete as long as the constants of the

[41] See (Houser 2003), for a very interesting approach of Pragmatism. See also, (Ibri 2013 and 2013a).

[42] See note 23.

[43] NEM-III/2, pp. 872–874.

[44] See (Ibri 1992, Chapter 3).

[45] CP 7.512.

> laws have reached no ultimate possible limit. But if the laws of nature are still in process of evolution from a state of things in the indefinitely distant past in which there were no laws, it must be that events are not even now absolutely regulated by law.[46]

Continuing, Peirce speculates on an objective tendency for the acquisition of habits:

> But if the laws of nature are the result of evolution, this evolution must proceed according to some principle; and this principle will itself be of the nature of a law. But it must be such a law that it can evolve or develop itself.
> Evidently it must be a tendency toward generalization—a generalizing tendency [...] Now the generalizing tendency is the great law of mind, the law of association, the law of habit-taking [...] Hence I was led to the hypothesis that the laws of universe have been formed under a universal tendency of all things toward generalization and habit-taking.[47]

This always seemed to me the strongest argument for his objective idealism. Indeed, questions of genesis, such as the one that Peirce formulated on natural laws, can only be made if the theoretical system of philosophy bears a sound metaphysics, i.e., one grounded on phenomenology and logic, as it occurs in the Peircean philosophical system.

Natural laws, then, are habits acquired by a typical tendency of the mental universe. This argument evidently proposes an objective idealism, as it conjectures on Nature's capacity to generalize, viz., an inductive capacity. Indeed, the more radical evolution of Peirce's realism-idealism shows that two logical forms meet objectively in natural processes, namely, abduction and deduction. Although there is no space here for a deeper presentation of this point, briefly it can be said—just to satisfy the curiosity of those who had no contact with this part of Peirce's works—that natural abductions are evidenced by the insertion of diversity and growing complexity of the universe[48], while the laws of nature act deductively on its pertinent events, that is, enabling them to be conducted out, by necessity, from the rules that constitute them.

Thus, there is a substantial eidetic monism associated with realism, which is, indeed, its condition of possibility in light of the evolutionary argument on the origin of the natural laws. Laws emerge as acquired habits, which can only occur through a tendency of an eidetic nature. Idealism remains as a *backgrounding doctrine,* enabling the reality of the continuities which constitute the first and third categories. The universe of discontinuities of the second category should not, in turn, be substantially foreign to ideality, since, in the grounding of the categories, it is the *locus* for appearing as an exteriority of firstness and thirdness. Upon acknowledging this condition of substantial harmony between the categories as modes of being of reality, matter should be thus considered as possessing an eidetic nature, as it is that which constitutes the existence of the objects in their particularity.

It is interesting to recall, under this line of evolution in Peirce's ontology, how he states that "Pragmatism is a correct doctrine only insofar as it is recognized that material action is the mere husk of ideas [...] *But the end of thought is action*

[46] CP 7.514.

[47] CP 7.515.

[48] See (Ibri 2006)

only insofar as the end of action is another thought."[49], evidencing how thought and action are connatural and interactive, as relations that establish themselves between the general and the particular, or between interiority and exteriority. Thus, deliberate *action* would be nothing more than thought endowed with purpose that enters a universe of existential reactivity.

Idealism and the Peircean Notion of Life

Matter, as explained, is then a form of mind ruled by inveterate habits, meaning that its predicates have a high degree of stability, and its conduct a high degree of redundancy. However, these habits are not fully crystallized and determinant of an exact behavior that would constitute a universe of strict laws, conducive to a deterministic concept of world. Peirce affirmed that the presence of firstness in the conduct of matter, acting in it through Chance, promotes a certain level of erraticity, making those laws or habits of a probabilistic nature. The regularities of the universe are, thus, merely approximate, and natural events reveal, on the one hand, the presence of the firstness that it always inserts through the principle of *Chance,* dispersion, irregularities and asymmetries, while, thirdness, in turn, acts on existence, making it redundant, habitual, albeit not in a strict way.

It is the presence of real thirdness, incidentally, that provides all and any cognitive operation, which, to be possible, always depends on the condition that its objects appear phenomenologically organized in a spatial-temporal form. A chaotic world would be absolutely incognoscible—such a need for a real order as a condition of epistemological possibility is, perhaps, the most vigorous proof of realism.[50] Resorting to the vocabulary of Synechism, cognition requires real *continuities* that can be represented in the *continuum* of thought. In Peirce's philosophy, we may say, epistemology and ontology interlace in such a way that one would hardly be able to consider one without the other.

In light of these considerations, we may define, between what is called mind and matter, a gradation of vivacity, or *life,* associating a greater degree to mind and a lesser one to matter, which would be the most important criterion for distinguishing both, after undoing the substantial dualism responsible for a split concerning their nature. But what would define that vivacity? What would be its most important predicates?

It seems reasonable to say that the ability to break inefficient habits that should act as mediations regarding the otherness, with which it cohabits, is an adequate way of measuring these degrees of vivacity. A certain capacity thus considered involves purposes and, therefore, if the achievement of such purposes are obstructed by an inadequate existential interaction or conduct, then will be justified the rupture of the correspondent habits that should work as guides of action.

---

[49] CP 8.272. my italics.

[50] See (Ibri 1992) for details.

Based on this, it would seem right to call upon the concept of *learning,* in the sense of deliberate change of habits, forsaking those that no longer serve as efficient mediations for the purpose of conduct. Inefficient habits for the intended purposes, if remaining as such, wind up interrupting the semiotic dialogue of the mind with its existentially interactive environment, a dialogue that is the very mainstay of what can be called learning. To learn, therefore, is to be potentially capable of changing conduct, and the capacity for doing so defines the main determining parameter of the degree of mental vivacity. A *living* mind must keep that dialogue and, in so doing, in light of fallibilism, may frequently incur in errors that will require it to be even more agile in marshaling resources to correct conduct until new habits can be acquired, always with the attribute of efficient mediations in the face of otherness.

The use of the word *mind* in a realistic sense, as we do here, is not exclusively confined to being human, but extends to all beings that cohabit some semiotic universe, i.e., in which there is circulation, commerce of signs and pragmatic meanings, capable of affecting conduct. Under such a broad meaning, then, not only the human universe is considered, but also the animal and vegetable, each one distinctly inserted into the gradation of vivacity defined by its respective competence of learning.

Continuity of Life—Final Considerations

The Peircean concept of the universe emerges as something essentially dynamic, an unfinished reality in which an eidetic nature is all-pervading, and as one which, under a background of substantial ideality, the actors who cohabit in it share the modes of being of the three categories.

Yet again, it must be repeated that idealism is not an operating doctrine in the sense of being essential to the successful course of scientific investigation: that is, once grounded and acknowledged, it no longer needs to be called upon to solve problems that imply choices of mediations and conduct. I suppose that for this reason, after formulating it, Peirce did not need to refer to it as a support to other doctrines.[51] It remained in the background, as the scenario of a vital theater of a semiotic and pragmatic nature. This nature emerges in Peirce's universe as the continuous production of signs, of interpretants that precede the adoption of a conduct, and in it are consummated as their external side, in the form of deliberate and purposeful action.

The special sciences, in Peirce's classification of sciences, depend on philosophy. In a more directly evident way, this dependency occurs with regard to phenomenology and logic or semiotic, but is even less evident as regards the other normative sciences, namely, aesthetics and ethics, and perhaps even less, concerning metaphysics as ontology. However, awareness and recognition of this *non-evident* dependency, eventually lead the investigations of a special science to a higher heuristic level, not only by the introduction of an enriching vocabulary which will enable the perception of new aspects of phenomena, but also by affording the philosophical acknowledgment of the equality of logical rights introduced by the symmetry of categories.

[51] For an opposite perspective, see (Short 2010a, b).

A science such as Biosemiotics will have its horizons expanded not only for the reasons mentioned, but also for the possibility of considering its primordial object, *life,* in light of an objective idealism determiner of a realism. Reality seen under the lens of these doctrines enables the justification of natural procedures that have a purpose, and are, therefore, necessarily interpretative, showing intelligence in the creation of mediations concerning the surrounding environment, and involving cognition processes that seek to represent the conduct of otherness in order to better suit its own conduct.

If we regard as *intelligent* every procedure that, in some way, involves deliberate purpose and conduct, we would see that Nature is full of them[52], and the beings that cohabit it share a general and common purpose, namely, the *continuity of life.* Knowing this enables us, together with the acknowledgment of the substantial ideality of Nature, to regard it with admiration, the ultimate finality of an esthetic nature that will determine a desirable ethics of investigation.

References

Almeder, R. (1975). The epistemological realism of Charles Peirce. *Transactions of the Charles Sanders Peirce Society, 11*(1), 3–17.

Anderson, D. R. (1995). *Strands of system: The philosophy of Charles Peirce*. Indiana: Purdue University Press.

Blachowicz, J. (1972). Realism and idealism in Peirce's Categories. *Transactions of the Charles Sanders Peirce Society, 8*(4), 199–213.

Boler, J. F. (1963). *Charles Peirce and scholastic realism: A study of Peirce's relation to John Duns Scotus*. Seattle: University of Washington Press.

Cosculluela, V. (1992). Peirce on Tychism and determinism. *Transactions of the Charles Sanders Peirce Society, 28*(4), 741–755.

De Tienne, A. (1993). Peirce's definition of the phaneron. In E. C. Moore (Ed.), *Charles Sanders Peirce and the philosophy of science: Papers from the Harvard sesquicentennial Congress*. Tuscaloosa: The University of Alabama Press.

Delaney, C. (1993). *Science knowledge, and mind: A study in the philosophy of C. S. Peirce.* Notre Dame, Indiana: University of Notre Dame Press.

Dileo, J. (1991). Peirce's haecceitism. *Transactions of the Charles Sanders Peirce Society, 27*(1), 79–109.

Dilworth, D. (2011). Peirce's objective idealism: A Reply to T. L. Short 'What was Peirce's Objective Idealism.' *Cognitio, São Paulo, 12*(1), 53–74. (jan./jun. 2011).

Fisch, M. (1986). *Peirce, semeiotic and pragmatism: Essays by Max H. Fisch*. Bloomington: Indiana University Press (In K. Ketner & C. Kloesel (Eds.)).

Forster, P. (2011). *Peirce and the threat of nominalism*. New York: Cambridge University Press.

Friedrich, W. J. (1978). *System of transcendental idealism.* Charlottesville: University Press of Virginia (translation by P. Heath).

Guardiano, N. (2011). The intelligibility of Peirce's metaphysics of objective idealism. *Cognitio, São Paulo, 12*(2), 187–204 (jul./dez).

Havenel, J. (2008). Peirce's clarifications of continuity. *Transactions of the Charles Sanders Peirce Society, 44*(1), 86–133.

Houser, N. (2003). Pragmatism and the loss of innocence. *Cognitio, São Paulo, 4*(2), 197–210.

[52] See CP 4.551; CP 6.17.

Ibri, I. A. (1992). *Kósmos Noetós: a arquitetura metafísica de Charles S. Peirce*. São Paulo: Perspectiva.
Ibri, I. A. (2006). The heuristic exclusivity of abduction in Peirce's philosophy. In R. F. Leo & S. Marietti (Eds.), *Semiotics and philosophy in C. S. Peirce*. Cambridge: Cambridge Scholars Press.
Ibri, I. A. (2009). Reflections on a poetic ground in Peirce's philosophy. *Transactions of the Charles Sanders Peirce Society, 45*, 3.
Ibri, I. A. (2010). Peircean seeds for a philosophy of art. In K. Haworth, J. Hogue, & L. G. Sbrocchi (Eds.), *Semiotics 2010 "The Semiotics of Space"* (pp. 1–16). New York: Legas Publishers.
Ibri, I. A. (2011). Semiotics and epistemology: The pragmatic ground of communication. In R. M. Calcaterra (Ed.), *New perspectives on pragmatism and analytic philosophy* (pp. 71–82). Oed Amsterdam: Rodopi.
Ibri, I. A. (2013a). Neopragmatism viewed by pragmaticism. *European Journal of Pragmatism and American Philosophy, 1*, 181–192.
Ibri, I. A. (2013b). Choices, dogmatisms and bets: Justifying Peirce's realism. *Veritas, 57*(2), 51–61.
Lane, R. (2011). The final incapacity: Peirce on intuition and the continuity of mind and matter. (Part 1).*Cognitio, São Paulo, 12*(1), 105–119.
Lane, R. (2011a). The final incapacity: Peirce on intuition and the continuity of mind and matter. (Part2). *Cognitio, São Paulo, 12*(2), 237–256 (jul./dez).
Lenzen, V. F. (1964). Charles S. Peirce as astronomer. In E. Moore & R. Richard. *Studies in the philosophy of Charles Sanders Peirce* (2nd Series). Cambridge: Harvard University Press.
McCarthy, J. (1984). Semiotic idealism. *Transactions of the Charles Sanders Peirce Society, 20*(4), 395–434.
Michael, F. (1988). Two forms of scholastic realism in Peirce's philosophy. *Transactions of the Charles S. Peirce Society, 24*(3), 317–348.
Moore, M. E. (2007). The genesis of the Peircean *Continuum*. *Transactions of the Charles Sanders Peirce Society, 43*(3), 425–469.
Nöth, W. (2010). The criterion of habit in Peirce's definitions of the symbol. *Transactions of the Charles S. Peirce Society, 46*(1), 82–93.
Parker, K. (1994). Peirce's semiotic and ontology. *Transactions of the Charles S. Peirce Society, 30*(1), 51–76.
Popper, K. (1972). Of clouds and clocks. In *Objective knowledge* (pp. 206–255). Oxford: Clarendon Press.
Potter, V. G., et al. (1977). Peirce's definitions of continuity. *Transactions of the Charles Sanders Peirce Society, 13*(1), 20–34.
Rorty, R. (1979). *Philosophy and the mirror of nature*. Princeton: Princeton University Press.
Rosa, A. M. (2003). *O Conceito de Continuidade em Charles S. Peirce*. Lisboa: Fundação Calouste Gulbenkian.
Schelling, Friedrich W.J. (1978). System of Transcendental Idealism. Charlottesville: University Press of Virginia; translation by P. Heath.
Short, T. L. (2010a). What was Peirce's objective idealism? *Cognitio, 11*(2), 333–346.
Short, T. L. (2010b). Did Peirce have a cosmology? *Transactions of the Charles S. Peirce Society, 46*(4), 521–43.
Tiercelin, C. (1998). Peirce's objective idealism: A defense. *Transactions of the Charles Sanders Peirce Society, 34*(1), 1–28.

Peircean Semiotic Indeterminacy and Its Relevance for Biosemiotics

Robert Lane

Accurate writers have apparently made a distinction between the ***definite*** and the ***determinate.*** A subject is ***determinate*** in respect to any character which inheres in it or is (universally and affirmatively) predicated of it, as well as in respect to the negative of such character, these being the very same respect. In all other respects it is ***indeterminate.*** The ***definite*** shall be defined presently. A sign (under which designation I place every kind of thought, and not alone external signs), that is in any respect objectively indeterminate (i.e., whose object is undetermined by the sign itself) is objectively ***general*** in so far as it extends to the interpreter the privilege of carrying its determination further. ***Example:*** "Man is mortal." To the question, What man? the reply is that the proposition explicitly leaves it to you to apply its assertion to what man or men you will. A sign that is objectively indeterminate in any respect is objectively *vague* in so far as it reserves further determination to be made in some other conceivable sign, or at least does not appoint the interpreter as its deputy in this office. ***Example:*** "A man whom I could mention seems to be a little conceited." The ***suggestion*** here is that the man in view is the person addressed; but the utterer does not authorize such an interpretation or ***any*** other application of what she says. She can still say, if she likes, that she does ***not*** mean the person addressed. Every utterance naturally leaves the right of further exposition in the utterer; and therefore, in so far as a sign is indeterminate, it is vague, unless it is expressly or by a well-understood convention rendered general. Usually, an affirmative predication covers ***generally*** every essential character of the predicate, while a negative predication ***vaguely*** denies some essential character. In another sense, honest people, when not joking, intend to make the meaning of their words determinate, so that there shall be no latitude of interpretation at all. That is to say, the character of their meaning consists in the implications and non-implications of their words; and they intend to fix what is implied and what is not implied. They believe that they succeed in doing so, and if their chat is about the theory of numbers, perhaps they may. But the further their topics are from such presciss, or "abstract," subjects, the less possibility is there of such precision of speech. In so far as the implication is not determinate, it is usually left vague; but there are cases where an unwillingness to dwell on disagreeable subjects causes the utterer to leave the determination of the implication to the interpreter; as if one says, "That creature is filthy, in every sense of the term." Perhaps a more scientific pair of definitions would be that anything is ***general*** in so far as the principle of excluded middle does not apply to it and is ***vague*** in so far as the principle of contradiction does not apply to it. Thus, although it is true that "Any proposition you please, ***once you have determined its identity,*** is either true or false"; yet ***so long as it remains indeterminate and so without identity,*** it need neither be true that any proposition you please is true, nor that any

R. Lane (✉)
University of West Georgia, Carollton, Georgia, USA
e-mail: rlane@westga.edu

V. Romanini, E. Fernández (eds.), *Peirce and Biosemiotics,* Biosemiotics 11,
DOI 10.1007/978-94-007-7732-3_4,

proposition you please is false. So likewise, while it is false that "A proposition ***whose identity I have determined*** is both true and false," yet until it is determinate, it may be true that a proposition is true and that a proposition is false.

In those respects in which a sign is not vague, it is said to be ***definite,*** and also with a slightly different mode of application, to be ***precise,*** a meaning probably due to ***præecisus*** having been applied to curt denials and refusals. It has been the well established, ordinary sense of ***precise*** since the Plantagenets; and it were much to be desired that this word, with its derivatives ***precision, precisive,*** etc., should, in the dialect of philosophy, be restricted to this sense. To express the act of ***rendering precise*** (though usually only in reference to numbers, dates, and the like), the French have the verb ***préciser,*** which, after the analogy of ***décider,*** should have been ***précider.*** Would it not be a useful addition to our English terminology of logic, to adopt the verb ***to precide,*** to express the general sense, to render precise? Our older logicians with salutary boldness seem to have created for their service the verb ***to prescind,*** the corresponding Latin word meaning only to "cut off at the end," while the English word means to suppose without supposing some more or less determinately indicated accompaniment. In geometry, for example, we "prescind" shape from color, which is precisely the same thing as to "abstract" color from shape, although very many writers employ the verb "to abstract" so as to make it the equivalent of "prescind." But whether it was the invention or the courage of our philosophical ancestors which exhausted itself in the manufacture of the verb "prescind," the curious fact is that instead of forming from it the noun ***prescission,*** they took pattern from the French logicians in putting the word ***precision*** to this second use. About the same time (see Watts, ***Logick,*** 1725, I, vi, 9 ***ad fin.***) the adjective ***precisive*** was introduced to signify what ***prescissive*** would have more unmistakably conveyed. If we desire to rescue the good ship Philosophy for the service of Science from the hands of lawless rovers of the sea of literature, we shall do well to keep prescind, presciss, prescission, and prescissive on the one hand, to refer to dissection in hypothesis, while precide, precise, precision, and precisive are used so as to refer exclusively to an expression of determination which is made either full or free for the interpreter. We shall thus do much to relieve the stem "abstract" from staggering under the double burden of conveying the idea of prescission as well as the unrelated and very important idea of the creation of ***ens rationis*** out of an{epos pteroen}—to filch the phrase to furnish a name for an expression of non-substantive thought—an operation that has been treated as a subject of ridicule—this hypostatic abstraction—but which gives mathematics half its power.

The purely formal conception that the three affections of terms, ***determination, generality,*** and ***vagueness,*** form a group dividing a category of what Kant calls "functions of judgment" will be passed by as unimportant by those who have yet to learn how important a part purely formal conceptions may play in philosophy. Without stopping to discuss this, it may be pointed out that the "quantity" of propositions in logic, that is, the distribution of the ***first*** subject, is either ***singular*** (that is, determinate, which renders it substantially negligible in formal logic), or ***universal*** (that is, general), or ***particular*** (as the mediaeval logicians say, that is, vague or ***indefinite***). It is a curious fact that in the logic of relations it is the first and last quantifiers of a proposition that are of chief importance. To affirm of anything that it is a horse is to yield to it ***every*** essential character of a horse; to deny of anything that it is a horse is vaguely to refuse to it ***some*** one or more of those essential characters of the horse. There are, however, predicates that are unanalyzable in a given state of intelligence and experience. These are, therefore, determinately affirmed or denied. Thus, this same group of concepts reappears. Affirmation and denial are in themselves unaffected by these concepts, but it is to be remarked that there are cases in which we can have an apparently definite idea of a border line between affirmation and negation. Thus, a point of a surface may be in a region of that surface, or out of it, or on its boundary. This gives us an indirect and vague conception of an intermediary between affirmation and denial in general, and consequently of an intermediate, or nascent state, between determination and indetermination. There must be a similar intermediacy between generality and vagueness. (CP 5.447-450, EP 2:350-353, 1905)

Introduction

Charles Peirce's theory of signs grew in sophistication and detail from its origin in the 1860s through the last decade of his life. But both early on and later, Peirce maintained that no sign is completely determinate, or, better put, that the meaning conveyed by a sign is never as determinate as it could be. In 1870 Peirce wrote that

> [w]e can … say, in a general way, that a term, however determinate, may be made more determinate still, but not that it can be made absolutely determinate. Such a term as "the second Philip of Macedon" is still capable of logical division—into Philip drunk and Philip sober, for example … (CP 3.93, W 2:390)

Years later he distinguished two different kinds of semiotic indeterminacy—*generality* and *vagueness*—and he continued to maintain that every sign is indeterminate to at least some degree: "a sign can only escape from being either vague or general by not being indeterminate. But … no sign can be absolutely and completely [] determinate …" (R 291, CP 5.506, c.1905).[1] So Peirce's considered view was that every sign is either general or vague to some degree. If this is correct, then it is also true of the signs studied by biosemioticians: no instance of biosemiosis is an absolutely determinate sign, and so any instance of biosemiosis is affected by at least some degree of either generality or vagueness.

But caution is needed here. Behind Peirce's view that every sign is indeterminate are subtle and idiosyncratic concepts of generality and vagueness. Care is especially warranted when it comes to Peirce's concept of vagueness, since by "vague" he did not mean exactly what philosophers now call "vague" and which is exemplified by terms and concepts that have borderline cases or "fuzzy" boundaries, like "bald" and "heap."[2] Any exploration of Peirce's work on semiotic indeterminacy needs to take careful account of exactly what Peirce took vagueness and generality to be.

While some biosemioticians have emphasized generality or vagueness in their work, few of them have attempted to exploit or even to explain Peirce's own concepts of generality and vagueness, and this is true even of those who are inspired by Peirce in other ways. For example, Stanley Salthe writes that "[t]he outstanding problem for pansemiosis … is to follow Peirce's suggestion to develop a logic of vagueness," but the concept of vagueness that Salthey employs, as well as that of generality, do not seem to derive from Peirce.[3] Tommi Vehkavaara emphasizes

---

[1] Peirce wrote "indeterminate" rather than "determinate," but as the editors of the *Collected Papers* note (CP 5.506n.1), this was obviously a mistake on his part. As published in the *Collected Papers*, the relevant sentence reads: "But that no sign can be absolutely and completely indeterminate is proved in CP 3.93 where Plutarch's anecdote about appealing from Phillip drunk to Phillip sober is put to use." But at CP 3.93, Peirce argued that no term is completely *determinate*, not that no term is completely indeterminate. (The quoted passage, reprinted at CP 5.506, occurs at R 291:16, but that page is missing from the microfilm copy of the Houghton Library papers, and so I have not been able to check what is published in the *Collected Papers* against a copy of the original MS).

[2] In this contemporary sense of the word, vagueness has been an extremely popular topic among philosophers for the last several years, and the literature on this subject is vast. Two seminal works are Williamson 1994 and Keefe and Smith 1996.

[3] 2007, p. 214.Some of Salthe's comments about vagueness, although not necessarily adhering to a Peircean concept of vagueness, do sound strikingly Peircean. For example:

vagueness, but other than contrasting it with clearness and definiteness, he does not say exactly what he takes it to be, nor does he address Peirce's own non-standard use of "vague" and its cognates.[4]

The goal of this chapter is to fill this lacuna, i.e., to explain Peirce's concepts of semiotic generality and vagueness and to offer an idea about how those concepts might be applied within biosemiotics. I begin with a brief account of his earlier views on the matter before turning to a lengthy examination of his later views. I conclude by considering the prospects for identifying indeterminate instances of biosemiosis, i.e., general or vague signs below the level of human cognition and communication. As we will see, Peirce's later view of semiotic indeterminacy crucially involves his distinction between the immediate object and the dynamical object of a sign, and the question whether there are indeterminate biosigns will turn on whether instances of semiosis below the level of human sign use ever involve immediate objects.

In referring to levels "below" that of human sign use, I have in mind the following spectrum of possible instances of semiosis, ranging through different levels of biological complexity:

- human-to-animal communication
- animal signals
- animal behaviors that interpret environmental phenomena (other than human signaling and the signaling of other animals)
- sign processes occurring within an organ or organ system or across or between different organs or organ systems
- sign processes occurring within a single cell
- molecular structures that represent.[5]

> [A]ll natural systems and objects are to some degree vague (not fully determined, under construction, plastic), while our scientific discourses about them have been as crisp as possible in the interests of measurement and, ultimately, of mathematical modeling. On this score, we might note that having some physical constant to five decimal points does not preclude getting to a sixth, or seventh. The pragmatics imposed by our own material scale generally obviates the need to pursue this process of refinement further than a few decimal points. Note, however, that Nature is vague enough so that if a system much smaller than ours (or in connection with our own nanotechnology) needed to have more precise values, it (we) could in principle construct them. In this example—since measurement is the essential act of science—we discover the vagueness of Nature from a scientific point of view. (p. 210)

This echoes Peirce's insistence that reality itself is infinitely determinable. (CP 3.93 n.1, W 2:390-391n.8, 1870) No matter how precise a true description of the world might be, it is always possible, at least in principle, to make it even yet more precise without rendering it false. But in contrasting vagueness with generality, Salthey seems to take a position very different than Peirce's when he asserts that "[g]enerality may be synthesized out of particulars" (p. 214). On Peirce's view, the general is antithetical to the particular and the individual, and no general can be synthesized or composed out of any number of particulars.

[4] Vehkavaara 2007.

[5] Kull et al. 2009 describes how the structure of a hemoglobin molecule might represent oxygen.

It will be useful to have a technical term to refer to signs involved in instances of semiosis at these levels, and so for that purpose I will use the term "biosign." Human thought and cognition are, of course, biological occurrences—not necessarily in the sense that they are reducible to, or explainable solely in terms of, the terms and concepts of biology, but certainly in the sense that without living human organisms, there would be no human thought or communication. So from one point of view, any instance of human thought or language-use is a *bio*sign. But here I use "biosign" narrowly, to refer only to instances of semiosis "below" the level of human thought and communication. In this narrower sense, the most sophisticated biosign is a human signal interpreted by an animal, as, for example, when a dog responds to the commands of its owner.

A more fundamental question than whether there are indeterminate biosigns is whether there are signs *at all* below the level of human thought and communication. Obviously, anyone committed to the legitimacy of the field of biosemiotics must answer this more basic question in the affirmative. For the purposes of this chapter, I will remain agnostic about whether any purported instance of semiosis below the level of human thought and communication actually is an instance of semiosis. I will not assume, let alone argue, that, e.g., the alarm call of a vervet monkey, or the human genetic code, is a genuine sign. Instead, I will consider a single example, that of a bear interpreting a scent as a sign of food,[6] and on the assumption that this *is* a genuine instance of the sign relation, I will consider how Peirce's conception of semiotic indeterminacy might be applied to it.

Peirce's Early View: Indeterminacy as Generality.

Both early on and later, Peirce took semiotic indeterminacy to be a matter of *underspecificity*; he held that an indeterminate sign is one that is less than maximally specific. In his earliest writings on the subject, Peirce took indeterminacy to be coextensive with generality. He also explained the general/indeterminate by contrasting it with the individual or singular. But here already we run into a significant complication. Peirce recognized two senses of the words "individual" and "singular," and in only one of those senses does the individual/singular stand in contrast to the general/indeterminate. In 1868, he wrote:

> "Singular" and "individual" are equivocal terms. A singular may mean that which can be but in one place at one time. In this sense it is not opposed to general. The sun is a singular in this sense, but, as is explained in every good treatise on logic, it is a general term. I may have a very general conception of Hermolaus Barbarus, but still I conceive him only as able to be in one place at one time. When an image is said to be singular, it is meant that it is absolutely determinate in all respects. Every possible character, or the negative thereof, must be true of such an image. (CP 5.299, W 2:233, EP 1:47)

[6] I borrow this example from Short 2007b.

Here Peirce was articulating a distinction that appears throughout his writings, that between *concrete individuals* and *strict individuals*.[7]

A concrete individual is that which can be in only one place at one time. So any physical object, including the body of a human being, is a concrete individual. Concrete individuality is *not* opposed to generality/indeterminacy; one can have a quite general/indeterminate concept of something that can be in only one place at a time. Hermolaus Barbarus was a concrete individual, yet my own concept of Hermolaus Barbarus is quite general, in that it is not nearly as detailed, and thus not as specific, as it might be. It is the concept of an Italian Renaissance scholar who translated several works of Aristotle, but it does not have much content beyond that, and so it leaves any number of questions about Barbarus unanswered, e.g., whether he was married and whether he had an interest in bird-watching. Others might have a concept of Hermolaus Barbarus that is even more general than mine, e.g., the concept belonging to someone who knows of him only that he was a man who translated the works of Aristotle. As this example illustrates, Peirce conceived of indeterminacy as coming in degrees. On his view, a given concept can be more or less indeterminate, even if that concept is of something that is a concrete individual.

It is the other sort of individuality—*strict* individuality—that is the opposite of indeterminacy. That which is strictly individual is "absolutely determinate in all respects" (CP 5.299, W 2:233, EP 1:47).[8] To understand what Peirce had in mind, consider his 1868 definition of "determined" as "fixed to be *this* (or *thus*), in contradistinction to being this, that or the other (or in some way or other)" (CP 6.625, W 2:155-56, emphases in original), as well as the first definition in his *Century Dictionary* (1889) entry for "determinate": "Having defined limits; fixed; definite; clearly defined or definable; particular: as, a *determinate* quantity of matter." Roughly, for something to be determinate is for it to be limited to be the way that it actually is rather than being some other way.[9] This is in harmony with Peirce's 1868 statement that "all determination is by negation" and so "whatever is absolutely universal [i.e., absolutely general] is devoid of all content or determination." (CP 5.223n.2, W 2:200n.4, EP 1:18n).[10] On the other hand, something that is completely determinate would be subject to no further limiting or specification; it would be maximally specific.[11] The strictly individual is completely determinate and so

[7] I adopt "concrete" as a description of the first kind of individual from Peirce's use at W 3:235, EP 1:107, 1877. I take "strict individual" as a description of the second sort of individual from R 300, CP 4.651n1, c.1907-08; see also W 1:461, 1866, and W 3:93, 1873.

[8] In his *Century Dictionary* (1889) entry for "singular," Peirce noted that "Scotus and others define the singular as that which is here and now—that is, only in one place at one time. The Leibnitzian school defines the singular as that which is determinate in every respect." (CD 5648) That is, Scotus et al. defined the singular as the concretely individual, while Leibniz et al. defined it as the strictly individual.

[9] T. L. Short has proposed that the meaning of "determine" "is that of 'to limit,' as in, 'The water's edge determines where your property ends.'" (2007b, p. 167) See Short, ibid., p.168n.9 for references to the secondary literature on the question of what Peirce meant by "determine."

[10] Peirce also asserted that "[a]ll determination is by negation" at CP 5.294, W 2:231, EP 1:45, 1868.

[11] As we saw above, Peirce indicated that this is what the word "singular" means when images are described as singular. Elsewhere Peirce gave the example of the retinal image of an object,

must be distinguished from that which is general to any degree at all. Conversely, the absolutely indeterminate would not be limited even to the slightest degree.

Since Peirce maintained that every sign is indeterminate to some degree, and since he held that "[a]ll thought … must … be in signs" (CP 5.251, W 2:207, EP 1:24, 1868),[12] he was also committed to what the conjunction of these views implies about human thought, viz., that "no cognition of ours is absolutely determinate." (CP 5.312,, W 2:239, EP 1:53, 1868) One of Peirce's favorite illustrations of this view is that one may think of Philip of Macedon without thinking about him as being drunk or thinking about him as being sober.[13] Thus, Philip "is logically divisible into Philip drunk and Philip sober; and so on; and you do not get down to anything completely determinate till you specify an indivisible instant of time" (W 3:235, EP 1:107, 1877). This is because anything that exists for any duration of time, no matter how short the duration, "will undergo some change in its relations" during that time. (CP 3.93n1, W 2:390n, 1870) For any real thing, S, and any duration, D, through which S exists, there will always be some property, P, that S will have during one portion of D and that S will lack during another portion of D, and so S will be "logically divisible" into S with P and S without P. The upshot is that

> [w]e can only say, in a general way, that a term, however determinate, may be made more determinate still, but not that it can be made absolutely determinate. Such a term as "the second Philip of Macedon" is still capable of logical division—into Philip drunk and Philip sober, for example; but we call it individual because that which is denoted by it is in only one place at one time. It is a term not *absolutely* indivisible, but indivisible as long as we neglect differences of time and the differences which accompany them. (CP 3.93, W 2:390, 1870)

So even a proper name is not a *strictly* individual term; it can always be made more specific and therefore always has some degree of indeterminacy. However, as long as we do not intend for a proper name to refer to a thing that remains completely static in all of its properties as it persists through time, it can refer to a *concrete* individual, something that can be in only one place at one time.

To sum up Peirce's early views:

1. Semiotic indeterminacy (a.k.a. generality) is a matter of underspecificity, i.e., of a sign having a meaning that is not maximally specific.
2. All signs are indeterminate to some degree. A maximally determinate sign is an impossibility.

---

which image does not inform the viewer whether or not the object is sweet. (CP 3.93, W 2:389-90, 1870) It fails to inform the viewer of this because it is general to some degree, i.e., because it is not maximally determinate.

[12] Peirce continued to maintain this view throughout his later life. See, e.g., CP 1.191, 1903, and CP 4.551, 1906.

[13] For uses of the Philip example, see W 3:84-85, 1873; W 3:235, EP 1:107, 1877; CP 1.494, c.1896; R 515:25, n.d.; and R 291, CP 5.506, c.1905. Peirce adopted the example from a story told by Plutarch. "When a wife petitioned Philip of Macedon on behalf of her husband, he happened to be in his cups and dismissed her. 'I shall appeal' she told him. 'To whom?' he asked, confident that there was no higher authority than himself. 'From Philip drunk to Philip sober' she replied, and in due course her appeal succeeded." Wilkinson 1993, p. 394.

3. Since all human thinking occurs in signs, every instance of human thought is indeterminate to some degree.
4. There are concrete individuals in the world, entities that can be in only one place at a time, and we can succeed in thinking about those individuals even though all signs, and thus all thinking, is indeterminate to some degree.

The Mature Account of Semiotic Indeterminacy

A central difference between Peirce's earlier and later views is the latter's recognition of two types of indeterminacy. Peirce called them *generality* and *vagueness*. But the change that Peirce's view underwent was not the addition of a second kind of indeterminacy (vagueness) to a kind already recognized (generality). Rather, Peirce came to believe that there are two different ways in which a sign can be indeterminate. In the vocabulary of the earlier view, he came to believe that there are two ways for a sign to be *general*, i.e., less than maximally determinate. However, in the course of evolving toward this later view, he retained "indeterminate" as a term for any sign that is less than maximally specific, and he began to use "general" as a term for only one such type of sign and to use "individual" for the opposite of the general.[14]

Peirce distinguished between semiotic generality and vagueness in numerous articles and manuscripts, but 1905's "Issues of Pragmaticism" (IP) contains one of the most detailed articulations of the distinction, one couched in terms of the communicational exchanges between the "utterer" of a sign and the "interpreter" to whom the sign is addressed. This characterization of generality and vagueness anticipates the game-theory semantics for quantifier expressions pioneered by Jaakko Hintikka,[15] and for that reason I will refer to it as Peirce's Game-Theoretical Semantics (GTS) for indeterminate signs.[16] In this section I explain the GTS for general and vague signs, as well as for "singular" signs, those that are individual (non-general) and definite (non-vague). As we will see, it is in virtue of a sign's immedi-

---

[14] Peirce complained that "[l]ogicians have been at fault in giving Vagueness the go-by, so far as not even to analyze it" (CP 5.446, EP 2:350) Since Peirce himself had for years failed to recognize vagueness as a separate sort of indeterminacy, it is tempting to think that one of the logicians that Peirce had in mind was himself.

[15] See Brock 1982 and Hilpinen 1982.

[16] Peirce also explained the distinction in terms of the logical principles that do not apply to the two sorts of indeterminate sign: "anything is *general* in so far as the principle of excluded middle does not apply to it and is *vague* in so far as the principle of contradiction does not apply to it." (IP, CP 5.448, EP 2:351) There is a lot to be said about this logical semantics for indeterminate signs, but here my attention is limited to the GTS. But I will note briefly that by saying that "the principle of excluded middle does not apply to" general signs, Peirce did not mean that general signs, including propositions that incorporate general terms, are neither true nor false, and in saying that "the principle of contradiction does not apply to" vague signs, he did not mean that vague signs, including propositions that incorporate vague terms, are both true and false. For more on this point see Lane 1997.

ate object (rather than its dynamical object) that Peirce would classify it as being either general, vague, or singular. A consideration of how Peirce's GTS intersects with the immediate/dynamical object distinction will prove essential to the work of this chapter's third and final section, in which I consider how Peirce viewed cases of semiosis involving neither minded utterers nor minded interpreters.

The Game-Theoretical Semantics for General and Vague Signs

Again, Peirce's illustrations of the GTS for general and vague signs involve an "utterer" and an "interpreter." (IP, CP 5.447, EP 2:350-351). Given his dialogical conception of thinking, according to which the thinking of a single individual is structured as a dialogue, these parties need not be two different people; one and the same person can act as the interpreter of her own earlier thoughts, written or spoken communiqués, or other signs that she herself has produced. Furthermore, the sort of sign involved in nearly all of Peirce's illustrations of his GTS is a proposition. So before we examine the GTS itself, a bit of background is in order on Peirce's theory of the proposition, beginning with his concept of the *dicisign*.

Peirce coined the term "dicisign" (or "dicent sign") to refer to any "sign that *conveys* information" and that "is either true or false … [without] directly furnish[ing] reasons for its being so." (CP 2.309-310, EP 2:275, 1903)[17] A dicisign need not be composed entirely of words; a portrait with a name beneath it conveys information, viz. that the person named looks as he is portrayed in the portrait, and this may be true, if the person in question actually does look like that, or false, if he does not. Further, a dicisign need not be composed of any linguistic signs at all; a weathervane pointing northwest conveys the information that the wind is blowing in that direction, and this, too, may be true or false. Propositions are a species of dicisign; they are *symbolic* dicisigns. (CP 2.320, EP 2:282, 1903)[18]

If a dicisign is "to be understood, [it must] be considered as containing two parts": one or more subject-terms, and a predicate.[19] (CP 2.312, EP 2:27, 1903) Each subject-term "partakes of the nature of an index, in that its function is the

---

[17] This concept is part of the classificatory triad *rheme*, *dicisign* and *argument*. For one explanation of this triad, see CP 2.250 ff., EP 2:292 ff., 1903.

[18] Sometimes Peirce identified dicisigns with propositions (CP 2.357, 1902), so that the aforementioned portrait and weathervane are propositions *simpliciter*. But at other times he restricted "proposition" so that it referred only to symbolic dicisigns. He also sometimes used the term "*quasi*-proposition" (CP 2.250, EP 2:292, 1903), by which I believe he meant non-symbolic dicisigns, and the term "*ordinary* proposition" (CP 2.262 and 265, EP 2:295 and 297, 1903; CP 2.315, EP 2:278, 1903), by which I believe he meant symbolic dicisigns.

[19] I use "subject-term(s)" rather than "subject(s)," since the former is less likely to be mistaken for a term for the thing(s) that the proposition is saying something about. But note that in doing so, I am using "term" in a broad sense, so that it applies not just to words or other symbols but to any sign or aspect of a sign whatsoever that, as a component or aspect of a dicisign, serves to refer to

characteristic function of an index, that of forcing the attention upon its object." (CP 2.357, 1902) It thus "serves to identify something which the predicate represents." (R 280:32, c.1905) A predicate functions like an icon, in that it "excites something like an image or dream in the mind of its interpreter" (ibid.). In essence, a dicisign is a sign that its constituent subject-term(s) and predicate are signs of the same thing. The dicisign composed of the portrait and the name is a sign that the person to whom the name refers resembles the image of the portrait. The weather-vane, as it is pointing to the northwest, is a sign that its object—the wind—has the property of blowing toward the northwest.[20] In the proposition "Silvio Berlusconi is wealthy," the subject-term "Silvio Berlusconi" directs the interpreter's attention to that which is its object, viz. the man Berlusconi, while the predicate, "is wealthy," represents a property. (CP 2.317, EP 2:281, 1903) Rightly interpreted, a dicisign, and thus a proposition, conveys the information that the object of its subject-term possesses the property that is represented by its predicate.

When illustrating the GTS for indeterminate signs, Peirce almost invariably used dicisigns. But in each of those illustrations, it is the subject-term, or the predicate, or both, that is indeterminate, rather than the dicisign considered as a whole. In IP, Peirce began his explanation of the GTS for general signs by considering the norms that govern an ordinary exchange of information between human interlocutors, one of whom has asserted a proposition in which "man" occurs as a subject-term.

> A sign … that is in any respect objectively indeterminate (i.e., whose object is undetermined by the sign itself) is objectively *general* in so far as it extends to the interpreter the privilege of carrying its determination further. *Example*: "Man is mortal." To the question, What man? the reply is that the proposition explicitly leaves it to you to apply its assertion to what man or men you will. (IP, CP 5.447, EP 2:350-351)[21]

If you and I are in conversation and I assert that man is mortal, i.e., that all humans are mortal, I do not mean to limit my claim of mortality to some specific group of human beings. One of the things that I am implying is that you, my listener, may apply the predicate "is mortal" to any human being you like, and I am committing myself to the view that no matter what human being you choose, he or she will be mortal. I am extending to you the "privilege" (or, as Peirce sometimes said, the "right") of interpreting the proposition as if its subject-term, "man," were more

---

the object of that proposition, e.g., the pointing arm of a baby directed toward a flower as the baby says "Pretty." (CP 2.357, 1902).

[20] This example illustrates Peirce's view that a dicisign need not be decomposable into two parts, one a sign of its object and the other a sign of the property that the object is purported, by the dicisign as a whole, to possess. Again, the sign need not actually *have* a part serving as the subject-term and another serving as the predicate; rather, "[i]t must, in order to be understood, *be considered as* containing two parts." (emphasis added).

[21] Peirce provided nearly equivalent definitions of "objective generality" in R 291 ("A sign is objectively *general*, in so far as, leaving its effective interpretation indeterminate, it surrenders to the interpreter the right of completing the determination for himself. 'Man is mortal.' 'What man?' 'Any man you like.'" CP 5.505, c.1905) and in a letter to William James ("The general term leaves the object partly indeterminate, and leaves the person addressed… to make the further determination at his pleasure. 'Men are mortal' = 'Any man *you please* is mortal.'" NEM 3:812, 1905).

specific, i.e., of applying the predicate "mortal" to anything to which the subject-term "man" applies.[22] Just as on the earlier account of indeterminacy, a sign's being indeterminate is a matter of its being less than maximally specific; but here, Peirce explained this in terms of the party in a communicational exchange who may legitimately increase the specificity of the sign. Signs that are typically general when used as subject-terms of propositions include common nouns in the plural ("humans," "rainbows," "possums") and universal quantifier phrases ("Every human," "Any rainbow," "All possums") used as subject terms of propositions. No matter who the utterer of the sign is, an interpreter is free to make the sign's reference more specific without thereby violating the relevant norms of communication.

The "antithetical analogue of generality" (R 291, CP 5.505, c.1905) is vagueness, or as Peirce also frequently wrote, indefiniteness. Rather than extend to its interpreter the right to make its subject-term more specific, a sign that represents its object vaguely withholds that right.

> A sign that is objectively indeterminate in any respect is objectively *vague* in so far as it reserves further determination to be made in some other conceivable sign, or at least does not appoint the interpreter as its deputy in this office. *Example*: "A man whom I could mention seems to be a little conceited." The *suggestion* here is that the man in view is the person addressed; but the utterer does not authorize such an interpretation or *any* other application of what she says. She can still say, if she likes, that she does *not* mean the person addressed. (CP 5.447, EP 2:351)

In this example, it seems as if the utterer alone has the right further to determine the reference of the subject-term of the proposition that she has asserted, and elsewhere Peirce said as much: "insofar as the right of determination is not distinctly extended to the interpreter it remains the right of the utterer." (R 291, CP 5.506, c.1905)[23] But in that same manuscript, Peirce provided an example in which it does not seem that that right is in fact reserved for the utterer of the sign:

> A sign is objectively *vague*, in so far as, leaving its interpretation more or less indeterminate, it reserves for some other possible sign or experience the function of completing the determination. "This month," says the almanac-oracle, "a great event is to happen." "What event?" "Oh, we shall see. The almanac doesn't tell that." (R 291, CP 5.505)

The almanac's claim is *vague*, in that it is indeterminate (unspecific to some degree) and we, its interpreters, are not currently free further to specify its meaning—only some future experience will show more specifically to what event the claim refers. So in at least some moods, Peirce held that it is not necessary that the right to make the subject-term of a sign more determinate be reserved exclusively to the sign's

---

[22] Peirce maintained that any proposition can be analyzed in a number of different ways. In all of the examples of propositions I consider here, I analyze them such that their logical subject-terms, the aspects of the propositions that refer to the propositions' objects, are their grammatical subjects. But for any proposition considered, Peirce would insist that alternative analyses are possible.

[23] Peirce made the same point in R 516:39-40, n.d.: "if a term is indefinite, it is because the utterer reserves a latitude of choice as to what singular it shall be taken to denote"; and at R 9:2, c.1903: "If a sign is apt to represent many things, the option as to what single thing it shall be taken to represent may be reserved by the utterer of it, to whom it naturally belongs; in which case it may be said to be used *vaguely*, or *not definitely*."

utterer. The characteristic mark of a vague sign is that its meaning, which is not wholly determinate, is not given over to its interpreter to further determine.

Peirce also illustrated his GTS with general and vague propositional predicates.

> Suppose that after a person had said that something was much, reserving, of course, his natural right to understand 'much' in any sense the word would bear and that he might choose, a second person were to declare that he was in the wrong. This would be tantamount to declaring the thing not to be much, while renouncing the right to take 'much' in a sense to suit himself, but, on the contrary, allowing the interpreter of his speech to take the word in any legitimate sense that *he* might choose. Such an assertion would be general; for *generality* consists in liberty of choice allowed in the interpretation of the general proposition, term, or other sign. (R 530, pp. 14–15, second pagination, c.1903)

In this example, it is the second speaker, the person who says that the object in question is *not* much, who is speaking generally. But it is not the object of the proposition that has been represented less than determinately; it is, rather, what is being said about that object that is indeterminate. It is as if the second speaker is saying, "That is not much, in any sense of 'much' you please." He is turning over to his interpreter the right to choose from among legitimate interpretations of the word "much" and to understand him as asserting that the subject of their exchange is not much on any of those interpretations. This is analogous to the way in which the utterer of "Man is mortal" turns over to his interpreter the right to choose whichever human being she pleases and to apply the predicate "mortal" to that human.[24] A proposition in which the predicate is being used generally is like a covert universal quantification over admissible interpretations of that predicate. Peirce also suggested that predicates in imperative sentences tend to be general, e.g., the predicate of "paint this in a dark color" "would be *general*, since it leaves the interpreter free to give the other two coefficients of the color [viz., its hue and chroma] any determination that may suit him." (EP 2:395, 1906).

On the other hand, a speaker may reserve for herself the right further to determine the meaning of a predicate within a proposition that she has asserted. In such a case, the predicate is not general, but vague:

> Had I said that [a friend's] hair was red, that would … be … a little "vague," since there are shades of hair between sandy and red which I might one day call red while on another day I might say, "No, that is reddish, but not red." Had I said that his hair was *not* red, that would be vague in the extreme, since it might be black, white, golden, or chestnut. (R 48:8, n.d.)

[24] Elsewhere Peirce suggested that a general predicate should be understood to extend to its interpreter, not a choice from among the various interpretations associated with the terms in question, but a choice from among other objects that are agreed to have the relevant property: "[T]he… judgment 'This chair appears yellow' has vaguely in mind a whole lot of yellow things, of which some have been seen, and no end of others may be or might be seen; and what it means to say is, 'Take any yellow thing you like, and you will find, on comparing it with this chair, that they agree pretty well in color.' It thus directly invites the exercise of a freedom of choice on the part of the interpreter (any one yellow thing answering as well as any other) …" (CP 7.632, 1903) N.b., in this passage Peirce used "vaguely" when he was clearly talking about the sort of indeterminacy that he later called "generality."

In this example, the speaker does not commit herself to the claim that her friend's hair is red on *every* legitimate interpretation of "red." Instead she withholds the right to be more specific about what she means, to make what she means by "red" more determinate by making clear exactly which shade or shades she has in mind. Were she to have asserted that her friend's hair is not red, her claim would have been vague to an even greater degree, since in that case she would be committing herself to even less, i.e., she would be leaving open a broader range of possible determinations of the predicate from which she could later choose. So a proposition in which the predicate term is being used vaguely is like a covert existential quantification over legitimate interpretations of that predicate.[25, 26]

It should be obvious by this point that as Peirce used "vague" and its cognates, he did not have in mind exactly the same semantic property that gives rise to borderline cases or "fuzzy" boundaries. A paradigm case of the latter sort of vagueness, e.g., the predicate "bald," need not always be vague in Peirce's sense. Rather, a given use of that predicate in a proposition could be either vague or general, depending on the circumstances in which the proposition is asserted. I might insist that Michel Foucault was bald with the intention of extending to my interlocutor the right to understand "bald" in any of its legitimate interpretations. On the other hand, I might assert that a recently-shorn David Beckham is bald while meaning to reserve the right to further specify what I mean by "bald."

---

[25] Bergman notes an important difference between subject-term vagueness (which, following Peirce's usage on a variant page of R 283, he calls *indesignance*) and predicate vagueness:

> In the case of indesignance, the utterer can typically designate what the object is; it is only the degree of precision of the designation that might be open to discussion. Normally, a sufficiently precise designation leads to dynamical object determination, assuming that the experiential and discursive requirements are met; ... However, [predicate] vagueness is often more contentious, as it concerns how habitual concepts might or ought to be applied. (2009, p. 154)

As an illustration, Bergman suggests that the indesignance of the subject-term of the proposition "a certain king was immoral" is easily eliminated by the utterer, who can simply say more precisely what king he has in mind, while the vagueness of the predicate could not be so easily eliminated, since there are "indeterminate habits of sign use" regarding "immorality" that do not affect "some king."

[26] There is an additional way in which predicates may be general or vague. "Usually, an affirmative predication covers *generally* every essential character of the predicate, while a negative predication *vaguely* denies some essential character." (IP, CP 5.447, EP 2:351) So a typical assertion of "Murphy is a cattle dog" will attribute *all* essential properties of cattle dogs to Murphy, as if the utterer were saying: choose from among all of the essential properties of cattle dogs any property you please, and you will find that Murphy has that property. Conversely, a normal use of "Murphy is not a poodle" will deny *some* essential property or properties of poodles to Murphy, as if the utterer is saying that there is some unspecified essential property of poodles that Murphy lacks.

Singular Signs

Since the only kinds of semiotic indeterminacy are generality and vagueness, any sign that is individual (non-general) and definite (non- vague) must be completely determinate. Peirce called such signs *singular*, and he described what a singular sign would be, in, for example, this statement of his pragmatic conception of truth:

> [E]very proposition has other subject-terms than which are explicitly set forth as such, and with them corresponding dimensions of the universe. Such are often dates of time, states of things, possibilities, logical, metaphysical, physical, etc. … But in every case, whatsoever, there is a primary universe and subject which is the Truth. Every metaphysician will have his way of describing this. I should say that it is that ideal state of ultimately settled opinion about the matter in hand, which we *hope* will be realized. If so, it plainly allows no liberty of choice about it, either to utterer or interpreter; … it is *singular.* (R 690, 1901)[27]

Peirce frequently acknowledged signs, propositional and otherwise, that are singular. (e.g., R 9:2-3, c.1903; R 530:16-17, c.1903) But as I mentioned at the beginning, Peirce's view, both early and late, was that no sign is *absolutely* determinate in meaning. On his later view, every sign is either general (non-individual) or vague (indefinite) to at least some degree.[28] So how can there be singular signs if in fact every sign is either general or vague?

Peirce's view seems to have been as follows. Although no sign is completely determinate in the sense of allowing for no possible specification on the part of either utterer or interpreter, very often signs are sufficiently determinate to allow for successful human communication and thus might as well be singular for the communicational

[27] This passage is from a handwritten insertion ("A Note on Collections") that Peirce added to his typescript for "On the Logic of Drawing History from Ancient Documents, Especially from Testimonies." That article, without the insertion, was published at CP 7.164-231, and, with the insertion, in HP 2:705-762. The insertion is available online at http://www.iupui.edu/~peirce/ep/ep2/headers/notes/note08.htm, accessed June 28, 2012. Later statements about semiotic determination in terms of the GTS include the following: "[H]onest people, when not joking, intend to make the meaning of their words determinate, so that there shall be no latitude of interpretation at all." (IP, CP 5.447, EP 2:351, 1905); "[W]ords whose meaning should be determinate would leave 'no latitude of interpretation' … either for the interpreter or for the utterer." (R 283, CP 5.448n, EP 2:392, 1906).

As the passage quoted from R 690 illustrates, Peirce held that even those propositions that have indeterminate subject-terms "refer[] to something singular," viz. "the entire universe of all things." (R 530:18-19, c.1903; see also R 291, CP 5.506, c.1905). The point is this: in order for two parties to communicate using a proposition, it cannot be left up to the interpreter to choose whatever universe he likes and understand what's being said in the proposition to apply to that universe. Any successful conversational exchange must assume that the subject of conversation is some one specific universe or some limited range of universes. But this does not mean that neither the subject-term nor the predicate of a proposition used in such an exchange can be indeterminate.

[28] Peirce also maintained that "[e]verything must be, in each respect, either definite or individual." (R 516, n.d.) As we have seen, Peirce used "definite" to mean non-vague, and he used "individual" to mean non-general. So this claim means that everything, including every sign, must be either non-vague or non-general. This is consistent with the claim that every sign it either general or vague to some degree, since every sign can be either general and thus definite (non-vague) or vague and thus individual (non-general) in any given respect.

purpose at hand. Even proper names, which may seem to allow no latitude of interpretation to either utterer or interpreter, are not absolutely free of generality or vagueness: "A proper name is set down as singular, although in absolute strictness, of course it is not so." (NEM 3:812, 1905) As we have already seen, even the name "Philip of Macedon" can be made more determinate, since it does not distinguish between Philip drunk and Philip sober. Peirce continued to use "Philip" as an example of indeterminacy even after he distinguished between generality and vagueness:

> … a sign can only escape from being either vague or general by not being indeterminate. But that no sign can be absolutely and completely determinate is proved in [1870's "Description of a Notation for a Logic of Relatives"] where Plutarch's anecdote about appealing from Phillip drunk to Phillip sober is put to use. (R 291, CP 5.506, c.1905)

If we are having a conversation about Philip and I explicitly assert that I intend all of my claims about him to be true of him at every stage of his life, then, if I assert that Philip was human, I am licensing you to narrow the meaning of "Philip" to, e.g., Philip as a two-year-old, or to Philip as a 20-year-old. But a given proper name need not be general in every instance of semiosis in which it takes part. The name "Philip of Macedon" can also signify the concrete individual Philip vaguely. If in a conversation with a fluent English speaker who knows who Philip of Macedon was, I say "Philip was a drunkard," my listener will probably understand that I do not mean to assert that Philip was a drunkard during every moment of his life. In this conversational exchange, I have reserved the right to narrow the reference of "Philip," to follow my original assertion by saying more specifically, "Of course, I mean only when he was an adult." This example illustrates Peirce's claim that "[e]very utterance naturally leaves the right of further exposition in the utterer; and therefore, in so far as a sign is indeterminate, it is vague, unless it is expressly or by a well-understood convention rendered general." (IP, CP 5.447, EP 2:351) In nearly any ordinary conversation about Philip or any other actual (or even fictional) human being, the speaker would have the right to make the meaning of a given utterance about that person more specific, unless he explicitly gives up that right by saying something like "From now on, everything I say about Philip I mean to be true of him at every stage of his life" or there is some relevant convention according to which that right is extended to an interpreter of his words.

Our intention is usually to make our meanings so determinate that our interlocutors have no leeway to interpret what we are saying in one way rather than another. Peirce's view was that, when it comes to ordinary discourse, we are doomed to fail in this aim. (IP, CP 5.447, EP 2:351) But this is not a hindrance to human communication. Peirce imagined two Englishmen having a conversation after meeting on a train in Europe. When one of them mentions Charles the Second,

> the other need not consider what possible Charles the Second is meant. It is no doubt the English Charles Second. Charles the Second of England was quite a different man on different days; and it might be said that without further specification the subject is not identified. But the two Englishmen have no purpose of splitting hairs in their talk; and the latitude of interpretation which constitutes the indeterminacy of a sign must be understood as a latitude which *might affect the achievement of a purpose*. For two signs whose meanings are for all possible purposes equivalent are absolutely equivalent. (CP 5.448 n.1, EP 2:393, 1906; emphasis added)

If distinguishing between Charles at noon on his 23rd birthday and Charles exactly one minute later is irrelevant to the purpose of the Englishmen's conversational exchange, then the fact that they *could* use a more specific sign than the name "Charles II," one that would call their attention to Charles at one of those times, is also irrelevant. So far as their purpose is concerned, the subject-term that they do in fact use, while subject to further specification and thus indeterminate, may as well be absolutely singular. What is true of subject-terms is true of predicates: strictly speaking, there is no such thing as a determinate predicate, but it is possible to communicate with predicates that are so determinate as to be indistinguishable, for the practical purpose at hand, from predicates that *are* determinate.

> Suppose the chat of our pair of Englishmen had fallen upon the color of Charles II's hair. Now that colors are seen quite differently by different retinas is known. That the chromatic sense is much more varied than it is positively known to be is quite likely. It is very unlikely that either of the travelers is trained to observe colors or is a master of their nomenclature. But if one says that Charles II had dark auburn hair, the other will understand him quite precisely enough for all their possible purposes; and it will be a determinate predication. (CP 5.448 n.1, EP 2:393, 1906)

The point has been well stated by Jarrett Brock: "A sign may be regarded as sufficiently determinate as long as whatever latitude of interpretation it allows does not affect our purposes one way or the other."[29]

It is worth noting that on Peirce's view, a sign itself, e.g., the spoken words "Charles the Second," cannot, on its own, convey to its interpreter that it signifies the English Charles the Second. A sign's object "cannot be fully revealed or brought to light by any study of the sign alone, as such. Knowledge of it must come from some previous or collateral source." (EP 2:404, 1907) In this example, when the utterer refers to Charles the Second, his interlocutor, in interpreting those words to be about the English Charles the Second, is relying on knowledge gained from collateral experience, e.g., knowledge that the utterer is English. Had the utterer been Spanish rather than English, then his listener might have had some doubt as to whether his words had as their object Charles the Second of England or Charles the Second of Spain.[30]

Dynamical and Immediate Objects

So far I have been speaking simply of the *object* of a sign. But in doing so, I have been ignoring an important distinction from Peirce's later semiotics, that between an object's *dynamical object* and its *immediate object*. But I cannot continue to ignore this distinction, since it is in virtue of a sign's immediate rather than its dynamical object that Peirce would classify that sign as being general, vague or (relatively) singular.

---

[29] 1969, p. 345.

[30] For more on the role of an interpreter's collateral experience in determining the correct interpretation of a given sign, see 8.178 and 183, EP 2:493 and 495-496, NEM 3:840 and 843-844, 1909; and CP 8. 314, EP 2:498, 1909.

A sign's dynamical object (or as Peirce also wrote, its *dynamic*, or *genuine*, or *real object*)—henceforth "DO"—is "the Reality which by some means contrives to determine the Sign to its Representation" (CP 4.536, 1906), "the Object as it is regardless of any particular aspect of it, the Object in such relations as unlimited and final study would show it to be" (CP 8.183, EP 2:495, 1909). If one of our traveling Englishmen asserts that Charles II was a drunkard, the DO of that proposition is Charles II himself, the formerly living, breathing, human being.[31]

Not every sign has a DO, but every sign does have an immediate object (henceforth "IO"):[32]

> [C]ollateral observation, aided by imagination and thought, will usually result in some idea, though this need not be particularly determinate; but may be indefinite in some regards and general in others. Such an apprehension, approaching, however distantly, that of the Object strictly so called, ought to be, and usually is, termed the "immediate object" of the sign in the intention of its utterer. (R 318, EP 2:409, 1907)

A sign's IO is a "mental representation" (R 318, CP 5.473, 1907), "the Object as cognized in the Sign and therefore *an Idea*" (CP 8.183, EP 2:495, 1909), "the idea which the sign is built upon" (R 318, EP 2:407, 1907.) It is "the Object as the Sign itself represents it, and whose Being is thus dependent upon the Representation of it in the Sign" (CP 4.536, 1906.) When one Englishman says that Charles II was a drunkard, the IO of his words is not Charles himself; it is an idea of Charles, an idea that is conveyed to his interpreter if their communication is successful.[33]

As this example suggests, the IO is not an idea that is unique to one person, in the sense of being in his mind and his mind only. Rather, it is an idea that can be shared by both utterer and interpreter, as it is in Peirce's example of saying to his wife, "It is a stormy day." The IO of that sign is "the notion of the present weather *so far as this is common to her mind and mine*" (CP 8.314, EP 2:498, 1909, emphasis added.) When Peirce asserts that it is a stormy day, he has in his mind the idea of the current weather conditions; this is the IO of the proposition that he has asserted, and if his wife successfully interprets what he has said, then that idea is conveyed to her. Each of them undergoes a separate cognitive event—a distinct, person-specific mental process. But the IO of the sign is not person-specific. It is an idea that can be *shared* between the utterer and interpreter of that sign.

Elsewhere Peirce made this same distinction, between ideas as cognitive processes unique to specific persons and ideas as sharable among two or more persons. He distinguished *thinking*, actual cognitive processes that occur in the minds of individual beings, from *thoughts*, "the objects which thinking enables us to

[31] At least once, in a 1909 letter to William James, Peirce wrote that a sign's DO may be "altogether fictive" (CP 8.314, EP 2:498, 1909). This may be why he, on at least one occasion, referred to the DO of a sign as its *quasi-real* object (R 145s:28, n.d.).

[32] Peirce wrote in an April 1907 letter to Papini that "all signs necessarily have Immediate Objects, but not all have Real Objects." (Ketner 1995, p. 287).

[33] On my interpretation, the IO of a sign and the DO of that same sign are numerically distinct entities. On this point I agree with Hilpinen 2007, but my understanding of the IO/DO distinction is very different from those presented in Ransdell 1976, Short 2007a and 2007b, and Bergman 2009.

know." (CP 1.27, 1909) While instances of thinking are *internal*, i.e., dependent on what someone or other thinks, thoughts are *external*, i.e., independent of what anyone—i.e., of any actual being—happens to think.[34] As Max Fisch succinctly put it, "Thinking is a matter for psychology, thought for logic. Thought is type; thinking is token."[35] While thinking is person-specific, thoughts are sharable among persons, such that when you and I both think that pomegranates are delicious, we are thinking the same thought, and that same thought can be shared in common among different sign tokens, be those tokens written, spoken, or mental. However, despite being external, thoughts are not independent of all *possible* mental activity. Peirce rejected the view that, e.g., the thought embodied in an argument "is only realized when [it] is understood," but he did not make the being of that thought independent of all possible understanding: "the logical form [i.e., thought] is already realized in the symbol itself", "they [i.e., thoughts] only are by virtue of possible thought [i.e., thinking]." (W 1:165-166, 1865) External thoughts are dependent for their being on the *possibility* of interpretive action. Without that possibility, there would be no external thoughts, no content that can be brought before a human mind in thinking.

Peirce used the term "idea" to discuss the same sort of sharable content in a manuscript from the mid-1890s, a manuscript in which he also indicated the role that that sort of content plays in semiosis:

> A sign, or *representamen* … stands for something, its *object*. It stands for that object, not in all respects, but in reference to a sort of idea, which I have sometimes called the *ground* of the representamen. "Idea" is here to be understood in a sort of Platonic sense, very familiar in everyday talk; I mean in that sense in which we say that one man catches another man's idea, in which we say that when a man recalls what he was thinking of at some previous time, he recalls the same idea, and in which when a man continues to think anything, say for a tenth of a second, in so far as the thought continues to agree with itself during that time, that is to have a *like* content, it is the same idea, and is not at each instant of the interval a new idea. (CP 2.228, c.1897)

The "idea" that Peirce mentioned here is, I believe, what he later dubbed the "immediate object" of the sign, the idea that the utterer of the sign has in mind as the

[34] To get a bit clearer on the distinction between the external and the internal, we need to distinguish among some related concepts. While the external is that which is independent of anyone's actual thinking, the *real* is that which is independent of what anyone actually thinks *about it* (e.g., CP 7.339, W 3:29, 1873). Thus, everything that is external is *ipso facto* real, but not everything real is external. Some things are internal but nonetheless real, i.e., some things depend upon what someone thinks but not upon what anyone thinks *about them*. Last night I dreamed that I was a famous musician, and this is a real fact about me: I actually did have such a dream. And since dreaming is a form of thinking, the fact that I had that dream is a fact about my thinking, and thus is dependent on my thinking; so that dream was *internal* (to my mind). But the fact that I dreamed about being a famous musician does not depend on whether I, or anyone else, believes that I dreamed about being a famous musician. So it is a real fact about me that I had that dream. And the content of the dream was, of course, unreal, or *fictional*, since I am not in fact a famous musician. Generally speaking, individual instances of thinking are internal, while sharable thought types are external, in that they have being apart from any *actual* instance of thinking. For more on Peirce's distinction between external and internal thoughts, see Lane 2009, pp. 4 ff. Peirce sometimes called the DO of the sign its *external object* and called its IO its *internal object* (R 145s:28-30, n.d.; see also CP 8.354, EP 2:485, 1908). But we should not assume that Peirce was here employing the technical meanings of "external" and "internal" described above.

[35] 1986, p. 360.

meaning that she wishes to convey in using the sign. But as that idea is thought by its utterer, it is a token of a type. Peirce's thinking about the current weather conditions—the cognitive process he was undergoing when he asserted that it is a stormy day—is a token of the public, sharable thought-type. If his wife has interpreted his words correctly, then there will be a separate token of that same type in her mind as an interpretant of that sign.[36]

In summary: An utterer of a sign forms an idea—the IO of that sign, an internal, mental representation, an instance of "thinking"—that is a token of an external, general thought-type; that token may, but need not, correspond to a real DO. In successful communication, an interpreter forms a mental token of the same external thought-type, which serves as (at least one of the) interpretants in that instance of semiosis. So in successful communication involving an utterer and an interpreter, there are at least two distinct tokens of the same external, sharable thought-type: the IO (an idea in the mind of the utterer) and the interpretant in the mind of the interpreter.

Immediate Objects and Indeterminacy

As I mentioned above, Peirce held that it is in virtue of a sign's IO that the sign should be classified as either general, vague (indefinite), or singular. He classified signs in this way in R 284, a manuscript of 1905, as well as on pages written in his Logic Notebook (R 339) in that same year:

> [A] sign may … represent[] its *immediate* object as Priman, that is as *indefinite*, for the indefinite is, in so far, priman, and the priman is always indefinite …
> [A] *Vague Sign* ... represents its Immediate Object in the logically formal character of the Priman, which is Indefiniteness. (R 284:60-61 recto, 61 verso, 1905)
> There are signs of each of which the immediate object is only a possible presentment of a dynamic object, a fragment of it, the rest being held in reserve, so that there is nothing in the immediate object to prevent contradictory attributes being separately possible of it. Thus "A certain man" may turn out to be rich. He may turn out and be poor. Such a sign may be termed an *indefinite sign*. (R 339:256 *recto*, October 10, 1905)[37]

[36] Peirce distinguished a number of different types of interpretant that may be associated with a given sign. Here I am claiming that in successful communication of the sort just described, there is, in the interpreting mind, as *one* of the relevant sign's interpretants, a token of the same thought-type that is tokened in the utterer's mind. One candidate for this token idea in the mind of an interpreter is the *immediate interpretant*, which Peirce described as

> the immediate pertinent possible effect in its unanalyzed primitive entirety. It is for instance in the case of a sign interpreted by a mind, *that idea (in a very exterior sense)* which must be apprehended in order that the sign should at all fulfill its function, this idea being presented whole and unanalyzed. It may be a quality of feeling more or less vague or an idea of an effort or experience awaked by the air of previous experience and may be the idea of a form or anything of a general type. (R 399:288 recto, October 23, 1906, emphasis added).

[37] This reflects Peirce's Logical Semantics for indeterminate signs, according to which a vague sign is one to which "the principle of contradiction" does not apply; see note 16.

The vagueness of a sign is a matter of how it represents its IO to its interpretant. Suppose you and I are in conversation, and I say, "A certain man is rich." In this proposition, the subject-term, "a certain man," is vague, in that the idea that is its IO does not "present" a definite man. "A certain man" is the sign of a "*possible* presentment of a dynamic object" (emphasis added). I may well have in mind a definite wealthy person; that possibility is not ruled out by the fact that the IO of the words that I have spoken is the idea of a *possible* choice or selection of some definite man. If our communication is successful, then the idea of such a possibility is at least part of what is conveyed to you by my words. But you will also understand that I do not extend to you the right to further determine the subject-term, i.e., the right to choose a definite man to whom the predicate is to be applied. If there is something real that counts as the DO of "a certain man," it is not a Peircean Second (existent) or Third (general); it is a Peircean First ("Priman"), the real *possibility* of choosing a definite human to whom the predicate can be applied.

Peirce next explained singular signs in terms of their IOs:

> [A] sign may … represent[] its *immediate* object … as Secundan, that is as if a definite individual existent …
> [A] *Singular Sign* … represents its Immediate Object in the logically formal character of the Secondan, which is Definite Individuality. (R 284:60-61 recto, 61 verso, 1905)
> There are signs of each of which the immediate object holds nothing in reserve, by supplying which the utterer can afterward limit it, nor allows any freedom of interpretation, the immediate object precisely denoting the dynamical object. Such a sign is called a *Singular Sign*, a term in the use of which a certain latitude must be allowed, however; or else there will be no occasion on which it can be applied. (R 339:256 *recto*, October 10, 1905)

The subject-term of the proposition "Silvio Berlusconi is wealthy" has as its IO an idea that "precisely denot[es]" the man Berlusconi, who is a definite (non-vague) individual (non-general), a Peircean Second (or "Secondan"), and the term's DO. When I assert that Berlusconi is wealthy, I have in mind the idea of the man Berlusconi, and that idea—the sign's IO—is a token of an external thought. The name's IO—an idea in my mind—is, as it serves to determine that sign, above a certain threshold of determinacy: it is so determinate that it "holds nothing in reserve" that I can later cite in order to make my meaning even more specific. The thought-type of which my idea is a token, although not completely determinate, is, in a case of successful communication, sufficiently determinate that a token of it, as an interpretant in the mind of the interpreter, will specify the man Berlusconi, and that will be specific enough for most conversational purposes.[38]

[38] Note Peirce's claim that a singular term represents its IO—e.g., the utterer's token of the thought of Berlusconi—"as if a definite individual existent." The point cannot be that the name "Silvio Berlusconi" represents *the utterer's idea of Berlusconi* as being a singular existent thing; the name is a sign of the man Berlusconi, not a sign of the utterer's idea of Berlusconi. We should read the claim that a singular sign "represents its *immediate* object … as if a definite individual existent" to mean that a singular sign is one that, in its connection with its IO, comes to represent a definite individual existent.

Finally, Peirce described how a sign's IO might be such as to render the sign general, or, as Peirce sometimes wrote, distributive [39]:

> [A] sign may … represent[] its *immediate* object … as Tertian, that is as distributively general. …
> [A] *General Sign* ... represents its Immediate Object in the logically formal character of the Tertian, which is Distributive Generality. (R 284:60-61 recto, 61 verso, 1905)
> There are signs of each of which the immediate object is represented as exchangeable for any existent within specified or understood limits. Such may be termed a *distributive sign.* (R 339:256 *recto*, October 10, 1905)

Suppose that I assert that "All humans are mortal" in an ordinary conversational exchange in which the subject-term "all humans" is general. That subject-term represents its IO—a thought—"as distributively general." If we are communicating successfully, it conveys to you, the interpreter, the idea that what I have in mind is "exchangeable" for any definite, individual human being, or as we have seen Peirce say elsewhere, the idea that you may select any human being you like and that I am committed to the claim that whichever one you select is mortal. If the subject-term of "All humans are mortal" has a DO, it is a Peircean general or Third ("Tertian"), something like a rule or general principle governing an unlimited number of potential selections that the interpreter might make from among the collection of all human beings.

Indeterminate Biosigns

As we have just seen, Peirce held that a sign's indeterminacy depends on its IO. It follows that if biosigns do not have IOs, then, at least on the Peircean conception of semiotic indeterminacy, there will be no indeterminate biosigns. So the question before us is whether biosigns have IOs.

In each of the examples we have considered so far, the IO has been an external thought a token of which is in the mind of the sign's utterer. But no biosign below the level of human-to-animal communication has an utterer, a "mind[], or theatre[] of consciousness … that *utters* the sign (whether acoustically, optically, or otherwise)." (R 318, EP 2:403) However, this does not mean that such biosigns cannot be genuine signs by Peirce's lights. Peirce held that some genuine signs lack utterers, e.g., "symptoms of disease [and] signs of the weather"; a sign can "convey intelligence" even without a mind to "utter" it. (R 318, EP 2:404) Paired with his view that every instance of semiosis whatsoever involves an IO, whether or not it involves a DO, what follows is that even signs without utterers must have IOs. And since Peirce classified signs as general, vague or singular based upon their IOs, this

[39] Elsewhere he made clear that general terms, such as the subject-terms of "Men are mortal" and "Pigs have great pointed snouts," refer to "one collective object distributively taken." (NEM 3:812, 1905), and he defined a general sign as one that "represents its Immediate Object in the logically formal character of the Tertian, which is Distributive Generality." (R 284:61 *verso*, 1905).

opens the possibility that signs without utterers, including biosigns, may be general or vague.

Peirce explained this in R 318 (1907).[40] Before distinguishing between IOs and DOs or even using the word "object" to denote that which a sign represents, he set out to find that which is an "essential" "ingredient" of the utterer of an uttered sign, his assumption being that that ingredient will be present in every case of semiosis, whether or not an actual utterer is involved. When there is no actual utterer, this ingredient "will function as a sort of substitute for an utterer … or at any rate fulfill [] nearly the same, but a more essential, function" as an utterer. (R 318, EP 2:404) This ingredient turned out to be "the *Object* of the sign;—the *immediate* object, if it be the idea which the sign is built upon, the *real* [i.e., dynamical] object, if it be that real thing or circumstance upon which that idea is founded, as on bedrock." (R 318, EP 2:407) Although Peirce did not say so explicitly, it is only the IO that can be an "essential ingredient" of a sign's utterer; the "essential ingredient" characterization would make no sense applied to a sign's DO,[41] and as I mentioned earlier, Peirce held that *every* sign has an IO, whether or not it has a DO.

As we saw earlier, Peirce held that no sign, all on its own, can convey to an interpreter what its object is. A correct interpretation of a sign depends on its interpreter having some collateral experience that enables her correctly to identify the sign's object. After giving several examples to illustrate this point, Peirce drew a conclusion about signs that have both an actual utterer and an actual interpreter and that "stand alone without any context": the object of such a sign is what the utterer "has in mind, but which it does not occur to him to express, because he well knows that the interpreter will understand that he refers to that, without his saying so." (Ibid.) His example is an utterer saying "Fine day!" to an interpreter without saying explicitly what day he means. In ordinary circumstances, no such explicit specification is necessary; the utterer

> does not dream of any possibility of the interpreter's thinking of any mere *desire* for a fine day that a Finn of the North Cape might have entertained on April 19, 1776. He means, of course, to refer to the actual weather, then and there, where he and the interpreter are alike influenced by the fine weather, and have it near the surface of their common consciousness. (Ibid)

In this example, the IO is an external thought a token of which is in the utterer's mind: the thought of the current weather conditions in the local environment. The utterer's words have that thought as their IO, even though the utterer does not say, "The current weather conditions in the local environment are fine!"[42]

---

[40] R 318 is a lengthy manuscript containing several variants of a "letter to the editor" that was rejected by the *Nation* and by the *Atlantic Monthly*. Overlapping portions of this MS are published in the *Collected Papers* (CP 1.560-562, CP 5.11-13, and CP 5.464-496) and in EP 2:398-433.

[41] Unless, of course, the sign in question is the sign of a mental state, like the subject-term of the proposition, "The immediate object of Smith's assertion that Berlusconi is rich is an idea in Smith's mind." That subject-term has as its DO an idea in Smith's mind, which is an "essential" part of Smith *qua* utterer of the proposition that Berlusconi is rich.

[42] Even in this sort of case, where the sign itself says explicitly that it is the current weather conditions in the local environment that are fine, an interpreter still needs collateral experience correctly

The notion of what an utterer "has in mind, but which it does not occur to him to express" is essential to Peirce's explanation of how a sign without an utterer can involve an IO—or at least, it is essential when the sign in question "stands alone without any context." In order to identify the IO of a sign that lacks an actual utterer, we are to consider what a *hypothetical* utterer would have had in mind but not expressed in the sign itself. To illustrate this strategy, Peirce gave the following example:

> Marine fossils found on a mountain, considered as a sign of the sea level having been higher than the level of deposit of those fossils, refer to a distant but indefinite date. Here, there is no utterer; but this [i.e., the indefinite date] is what might have remained unexpressed in the mind of the utterer, though essential to the significance of the sign, if that sign had been devised and constructed to give the human race a first lesson in geology. (Ibid.)

On the analysis Peirce suggested, the sign's IO is the thought "of a distant but indefinite date"—a thought that a hypothetical utterer *would have had* had the fossil been designed to convey information about past sea levels, e.g., were it placed on earth by some divine being as an indicator of the earth's past state. Since there is no actual utterer, there is no token of the relevant thought in the mind of an utterer; still, there is the external thought, the general thought-type, which, if the sign is correctly interpreted, is apprehended by its interpreter and a token of which is thus an interpretant of that sign in the interpreter's mind.

Peirce's fossil example is also an illustration of semiotic vagueness (indefiniteness). To see this, notice that the sign in question is a dicisign—it does not simply indicate an object but says something about that object and is thus capable of being either true or false. It is interpreted to mean that at some indefinite time in the distant past, the sea level was higher than that of the fossil itself.[43] The IO is the thought of an *indefinite* date; applying the GTS for vague signs, we can say that the IO is the thought of a possible choice from among various times in the distant past, but a choice that an interpreter is not licensed to make for herself. The IO is "only a possible presentment of a dynamic object, a fragment of it, the rest being held in reserve" (R 339:256 *recto*, October 10, 1905). A present-day interpreter would be justified in taking the fossil as a sign that the sea level was higher at *some* point in the distant past, but not in "further determining" the object of that sign by inferring, e.g., that it was higher exactly 10,000 years ago.

Peirce's analysis of a sign that lacks an actual utterer suggests how we might go about identifying the IOs of biosigns that lack utterers and thus how we might classify those signs as either general, vague, or singular. But biosigns below the level of human-to-animal communication do not just lack utterers; at least arguably, they also lack interpreters, in Peirce's sense of "mind that *interprets* the sign." (R 318, EP 2:403) If Peirce were to maintain that any genuine sign must involve a minded

---

to interpret the sign, specifically, to know that it is *these* weather conditions at *this* location that the sign has as its object.

[43] Since the object of a dicisign is identical to the object of the subject-term (or of that which, after analysis, is identified as serving the function of subject-term) of that same dicisign, the IO of the representamen fossil is the same as the IO of the "subject-term" of the fossil (i.e., of that aspect of the fossil that serves the logical function of a subject-term).

interpreter, then he would be committed, not simply to the claim that no biosign is semiotically indeterminate, but also to the claim that no biosign is a genuine sign at all. So before considering how the approach just sketched might be applied to biosigns, we should consider, at least briefly, how Peirce addressed the question of signs lacking actual interpreters. Succinctly put, he addressed that question in the much same way as he addressed that of signs lacking actual utterers: "If a sign has no interpreter, its interpretant is a 'would be,' i.e., is what it *would* determine in the interpreter if there were one." (R 318, EP 2:409) To identify the interpretant(s) of a sign lacking a minded interpreter, we are to consider what interpretant(s) a *hypothetical* interpreter would undergo were he to interpret the sign. The fossil is a sign even if not interpreted by any human being or other intelligent interpreter; it is so because it is *interpretable* as such a sign, and the lack of any actual interpreter, and thus of any actual interpretants, does not undermine this interpretability.

To illustrate how Peirce's hypothetical-utterer-and-interpreter approach might be applied to a biosign lacking both a minded utterer and minded interpreter, I will adapt an example from T. L. Short.[44] Consider a bear that, upon detecting the scent of rot in a log, begins clawing at the log in order to reveal and eat the grubs within. Here the scent is a sign of grubs, and on Short's view it is interpreted as such by the bear's clawing—the clawing is an interpretant of the smell of rot.[45] Here I do not wish to raise the question whether Short's analysis is the best that can be given of this instance of semiosis or the question whether Peirce himself would have accepted all of the details of Short's analysis. Instead, I wish simply to use the bear example as a case of semiosis in which there is no actual utterer and no actual *minded* interpreter.[46] The questions at hand are these: If there is no minded interpreter interpreting the scent of rot, can that scent actually function as a sign? And if so, in the absence of an actual utterer, can the scent, considered as a sign, have an IO and therefore be indeterminate?

The approach taken by Peirce in R 318 indicates that he would answer the first question in the affirmative. To identify the interpretant(s) of the scent, we are to consider a hypothetical interpreter and ask what interpretants there *would have been* had that interpreter taken the scent to be a sign of grubs. Those "would-be's," including, perhaps, the thought (i.e., the external thought-type) of grubs, are the interpretants of the sign, even if the sign itself is never actually interpreted. So even if we decline to follow Short in analyzing this case such that the bear's clawing

[44] 2007b, at, for example, pp. 153, 156, 159, 160, 172, 301, and 302.Short is skeptical toward much of the biosemiotic enterprise; see, e.g., 2007 p. xiv. On Short's reconstruction of Peirce's theory of signs, genuine semiosis requires, not just purpose, but purposefulness, and purposefulness requires an entity that is capable of self-correction. Some non-human animals are capable of this, but not all are, and there is no purposefulness "below" the level of the behavior of non-human animals.

[45] Peirce recognized interpretants other than mental states, e.g., the action that a solder performs upon hearing his commanding officer yell "Ground arms!" is an *energetic* interpretant of that commander's words. (R 318, CP 5.475) So the clawing behavior of the bear may be an energetic interpretant of the scent of rot.

[46] As Short states, the bear has no intentional states upon which the instance of semiosis in question depends. (2007b, p. 301)

turns out to be an interpretant of the scent, we can still adopt Peirce's approach and maintain that the scent is a sign, or, at least, that it does not fail to be sign for lack of an interpretant.

In order for the scent to be a genuine sign, it needs to be related not just to an interpretant but also to an object. If there is such an object, and in particular, if there is an *immediate* object, then, from Peirce's perspective, the scent is not only a genuine sign, but one that can be categorized as either general, vague or singular. Now obviously, the scent has not been "uttered" by any mind. But as we have seen, Peirce's approach was to say that there is nonetheless an IO, viz. the thought that a hypothetical utterer *would have had* had the sign been created in order to convey meaning. Even in the absence of an utterer who is thinking the relevant thought, there is still the thought-type a token of which an utterer *would* have been thinking had the sign actually been uttered. In the present example, rather than imagining a divine being fashioning a fossil to convey information about the earth's history, Peirce would have us imagine someone placing the scent on the log to serve as a sign.

But as a sign of what? Up to now, I have been describing the object of the scent-sign as simply *grubs*. But that is too simple, even on an analysis that yields the bear's clawing as the sole interpretant of the scent.[47] That clawing does not interpret the scent as a sign of grubs (or insects, or protein, or food) only. The scent is more informative than that, even if the clawing is its only interpretant. It conveys the information that there are grubs *here*, *in this log*. The clawing, after all, is not a general, undirected flailing-about of paws; it is targeted at the log. So I think the better analysis of this instance of semiosis will understand the scent to be a dicisign, one that does not simply represent part of the world, but that represents part of the world as being a certain way and is thus capable of being true or false. And as Peirce understood, this sort of sign can be a dicisign without being composed of separate or separable parts, one serving as a subject-term and the other serving as a predicate. What is important for the odor's status as a dicisign is that "[i]t must, in order to be understood, *be considered as* containing two parts." (CP 2.312, EP 2:277, 1903)

Applying Peirce's hypothetical utterer approach to the scent understood as a dicisign, we can imagine an utterer who brings it about that the log smells like rot in order to signal the presence of grubs, i.e., in order to represent the fact that there are grubs in the log. Our hypothetical utterer would have had the idea that there are grubs in the log, so the relevant IO is the external thought that grubs are present in the log, the thought that English speakers might convey by speaking the words "There are grubs in that log" or by pointing at the log and saying "Grubs." On this analysis, the scent indicates its object determinately: *this* log. Better put, it represents its object *relatively* determinately, since even "this log," as a sign of a specific log, could be more determinate, by, e.g., specifying one area of the log as opposed to others, or specifying a span of time during which the relevant predicate is supposed to apply to the log.

---

[47] As Short presents the example, the object of the scent is simply *grubs*. (2007b, p. 159)

The aspect of the dicisign that functions as a predicate is a more promising source of indeterminacy. The scent represents the log as containing grubs, so the aspect that functions as a predicate represents the property of containing grubs. Is this aspect of the sign either general or vague? Again, Peirce's GTS explains a general sign as one that extends the right of making the sign more specific to its interpreter and a vague sign one as that reserves as that right for the utterer. So to see whether the predicate-aspect of the scent is either general or vague, it seems that we need to think in terms of such rights. How might this "rights" talk get cashed out in the case of the scent? In many of the examples of general or vague signs that Peirce himself provided, the rights referred to were rights that obtain under the general norms of human communication, norms that grow out of actual human communicative practice. Pretty clearly, there is no analogous system of norms governing bear-log interaction, and to say that a "right of interpretation" has been either extended to or withheld from our bear friend sounds odd at best.

So rather than think in terms of rights that the bear may have with regard to the scent, I suggest that we extend Peirce's hypothetical utterer-and-interpreter approach by understanding the norms that are appealed to in his GTS as those that *would* govern communicational exchanges between two hypothetical interlocutors. To apply this strategy to the present example, we can consider the norms under which hypothetical interlocutors would be operating were the scent a sign in a communicational exchange between them. Imagine an utterer who attempts to convey the idea that there are grubs in the log, not by saying "This log has grubs in it" but by bringing it about that there is a scent of rot on the log.[48] Of course, no norms currently obtain that govern the placement of scents on objects so as to indicate the contents of those objects—humans do not regularly communicate with each other that way. But to identify the norms relevant to this example, we can consider a hypothetical utterer uttering a proposition, one expressed in a language that *is* governed by familiar norms. In following this approach, we imagine one person asserting a proposition that conveys the relevant information, e.g., "This log contains grubs." Here the predicate would be indeterminate, since it would not be maximally specific; for example, it does not specify what species of grub the log contains and is thus like a quantifier phrase ranging over various acceptable interpretations of the word "grub." And further, it would be vague, since it would

[48] To interpret the sign correctly, an interpreter would need to know that that sort of scent is frequently associated with grubs. But this knowledge is not the result of the kind of collateral experience that is necessary for an interpreter correctly to identify the object of a sign; it is more similar to the knowledge of a natural language that an interpreter must possess in order to understand words in that language at all: "I do not mean by 'collateral observation' acquaintance with the system of signs. What is so gathered is *not* COLLATERAL. It is on the contrary the prerequisite for getting any idea signified by the sign. But by collateral observation, I mean previous acquaintance with what the sign denotes." (CP 8.179, EP 2:494, 1909) In the present example, the collateral experience necessary for identifying the log as the object of the scent-sign would include the experiences of the sort by which humans learn to tell the direction or object from which a scent is emanating.

not extend to its (hypothetical) interpreter the right to make it more determinate by choosing from among those acceptable meanings.[49]

This strategy seems to involve a kind of anthropomorphism: it requires that we understand a sign as being meaningful only in virtue of the fact that it could be interpreted by us, and it requires that we understand semiotic indeterminacy as deriving from the fact that, in a communicational exchange between two of us, the right to further specify the sign's meaning belongs to one or the other participant in that exchange. This will probably be objectionable to many biosemioticians. But within the broader framework of Peirce's philosophical system, this apparent anthropomorphism is no such thing—or, at least, it is not a matter of projecting a uniquely *human* way of understanding signs onto the non-human world. On Peirce's view, signs are essentially cognizable, not necessarily by members of our species, but by some or other "scientific intelligences," intelligent beings capable of "learning by experience." (CP 2.227, c.1897)[50] This is just another way of putting the point that even a sign without an utterer has an IO, an external thought-type, something that is essentially *thinkable* whether or not any actual being ever thinks it. As Max Fisch put it,

> how extremely rare it is for an ill human being or other animal to be observed at all by a trained and skilled diagnostician, and how much escapes even the most skilled! But the symptoms and other signs are there, and so are the interpretants to which they *would* lead an ideally qualified observer and interpreter. The thought is "there," though there be no thinker of it.[51]

As this suggests, the scent of the log is a sign that there are grubs present, *even if that sign is never actually interpreted*, either by a minded interpreter (like an entomologist specializing in the study of grubs) or a bear whose clawing may

[49] This strategy suggests a way of responding to Vehkavaara's skepticism about the idea of an IO being present in the purported biosign involved when a bacterium follows a chemical gradient:

> I have no doubt that the behavior of *E. coli* is purpose-oriented (self-functional) and sign-directed, but certain non-desirable conclusions will follow if we think that the nutrition gradient is the object of a chemical sign — a sign interpreted by the bacteria concluding appropriate flagellar movements. The problem is that there is no immediate object, no ground of representation *for the bacterium*. (2002, p. 10, emphasis in original)

On Peirce's hypothetical-utterer-and-interpreter strategy, if the chemical gradient is a sign of nutrition, it is not a sign of nutrition *to the bacterium*, but rather to a hypothetical minded interpreter.

[50] Peirce held that humans are not the only scientific intelligence in the cosmos and not the only ones capable of intellectual understanding:

> We may take it as certain that the human race will ultimately be extirpated; because there is a certain chance of it every year, and in an indefinitely long time the chance of survival compounds itself nearer and nearer zero. But, on the other hand, we may take it as certain that other intellectual races exist on other planets, —if not of our solar system, then of others; and also that innumerable new intellectual races have yet to be developed; so that on the whole, it may be regarded as most certain that intellectual life in the universe will never finally cease. (CP 8.43, EP 1:235, W 5:227, 1885).

[51] 1986, p. 360.

count as a sort of interpretant. Succinctly put, the upshot of Peirce's approach to the issue of the meaning of unuttered, uninterpreted signs is this: if a tree rots in the forest and there is no one there to smell it, it may nonetheless make an indeterminate sign.

References

Bergman, M. (2009). *Peirce's philosophy of communication*. New York: Continuum.

Brock, J. (1969). *C. S. Peirce's logic of vagueness*. Unpublished doctoral dissertation, University of Michigan, Ann Arbor.

Brock, J. (1982). Peirce's anticipation of game theoretic logic and semantics. In M. Herzfeld & M. Lenhart (Eds.), *Semiotics 1980, proceedings of the fifth annual meeting of the semiotics society of America, held October 16-19, 1980, in Lubbock, Texas* (pp. 55–64). New York: Plenum Press.

Fisch, M. (1986). *Peirce, semeiotic and pragmatism: Essays by Max H. Fisch, Ketner, K., and Kloesel, C., eds*. Bloomington: Indiana University Press.

Hilpinen, R. (1982). On C. S. Peirce's theory of the proposition: Peirce as a precursor of game-theoretical semantics. *The Monist, 65*(2), 182–188; reprinted in Eugene Freeman. (Ed.). (1983) *The relevance of Charles Peirce* (pp. 264–270). La Salle: The Hegeler Institute.

Hilpinen, R. (2007). On the Objects and Interpretants of Signs: Comments on T. L. Short's Peirce's Theory of Signs. Transactions of the Charles S. Peirce Society, 43(4), 610–618.

Keefe, R., & Smith P. (1996). *Vagueness: A reader*. Cambridge: MIT Press.

Ketner, K. (1995). *A thief of Peirce: The letters of Kenneth Laine Ketner and Walker Percy*. In P. Samway (Ed.). Jackson: University Press of Mississippi.

Kull, K., et al. (2009). Theses on biosemiotics: Prolegomena to a theoretical biology. *Biological Theory, 4*(2), 167–173.

Lane, R. (1997). Peirce's 'Entanglement' with the principles of excluded middle and contradiction. *Transactions of the Charles S. Peirce Society, 33*(3), 680–703.

Lane, R. (2009). Persons, signs, animals: A Peircean account of personhood. *Transactions of the Charles S. Peirce Society, 45*(1), 1–26.

Ransdell, J. M. (1976). Another interpretation of Peirce's semiotic. *Transactions of the Charles S. Peirce Society, 12*(2), 97–110.

Salthe, S. (2007). Meaning in nature: Placing biosemiotics with pansemiotics. In M. Barbieri (Ed.), *Biosemiotics: Information, codes and signs in living systems* (pp. 207–217). New York: Nova Science Publishers.

Short, T. (2007a). "Response" to contributions to a symposium on Short 2007. *Transactions of the Charles S. Peirce Society, 43*(4), 663–693.

Short, T. L. (2007b). *Peirce's theory of signs*. Cambridge: Cambridge University Press.

Vehkavaara, T. (2002). Why and how to naturalize semiotic concepts for biosemiotics. *Sign Systems Studies, 30*(1), 293–313. The page reference is to the version of the paper posted at http://utafi.academia.edu/TommiVehkavaara/Papers/267417/Why_and_how_to_naturalize_semiotic_concepts_for_biosemiotics. Accessed 27 July 2012.

Vehkavaara, T. (2007). From the logic of science to the logic of the living: The relevance of Charles Peirce to biosemiotics. In M. Barbieri (Ed.), *Introduction to biosemiotics* (pp. 257–282). Dordrecht: Springer.

Wilkinson, P. R. (1993). *A thesaurus of traditional english metaphors* (2nd ed.). New York: Routledge.

Williamson, T. (1994). *Vagueness*. New York: Routledge.

Peircean Habits, Broken Symmetries, and Biosemiotics

Eliseo Fernández

> ...We place ourselves, then, at the beginning of time. Qualities are already possible. Actual existence has begun. Accidental reactions are taking place. Several continua are established. A tendency toward generalization is operative.
>
> But as yet no *thing* can be said to exist; much less any personal consciousness. The accidental reactions are purely accidental, unregulated in any degree by law, the work of blind and brutal chance.
>
> But now the tendency toward generalization which is already operative, and which indeed is more ancient than actual existence itself, begins to group the accidental reactions into fragmentary continua. Into continua, because such is the logical nature of generalization. Into fragmentary continua because the tendency to generalization has to fight the lawless brutality of chance with its youthful freakishness, and ebullient vivacity. At first, during the early eternities, those generalizations and continua are smashed as soon as they appear. For they are them selves of a haphazard kind with no vitality. But in this endless haphazard shindy between generalization and chance this generalization happens to come about, namely a limited but still a general tendency toward the formation of habits, toward repeating reactions that had already taken place under like circumstances. Now the difference between this generalization, this tendency toward law, and the rest was that this was one which by its own law was always tending to grow stronger. Therefore, although this was doubtless smashed like the others billions upon billions of times, to use a hyperbole of stating matters infinitely weaker than I really mean, yet still, it was often springing from its ashes, and on the whole was tiring out the lawlessness, until at length, —of course after an infinite lapse of time subsequent to the first moment, although infinitely long ago, there came to be a decided and so to say a sensible degree of tendency in nature to take habits.
>
> This was the earliest of the laws of nature and was and still is continually strengthening itself. A habit of acquiring habits began to be established, and a habit of strengthening the habit of acquiring habits, and a habit of strengthening that habit, and so on a*d infinitum.*
>
> The acquiring [of] a habit is nothing but an objective generalization taking place in time. It is the fundamental logical law in course of realization. When I call it objective, I do not mean to say that there really is any difference between the objective and the subjective, except that the subjective is less developed and as yet less generalized. It is only a false word which I insert because after all we cannot make ourselves understood if we merely say what we mean. (Peirce, MS 942, 1898)

E. Fernández (✉)
Linda Hall Library of Science and Technology, Kansas City, USA
e-mail: fernande@lindahall.org

V. Romanini, E. Fernández (eds.), *Peirce and Biosemiotics,* Biosemiotics 11,
DOI 10.1007/978-94-007-7732-3_5,

Introduction

With the exception of the incipient discipline of biosemiotics, the influence of Peircean ideas upon science in general—and biology in particular—has been rather oblique and peculiar. In most cases this influence has not arisen from familiarity with his writings, but quite indirectly, through rediscovery of those ideas by the scientists themselves. Moreover, these have often been reluctant rediscoveries, motivated by theoretical impasses that proved insurmountable short of painfully rejecting well-entrenched, even cherished, philosophical stances and presuppositions.

Among these unsavory notions we count, for instance, Peirce's rejection of absolute determinism, his advocacy of objective randomness, and his conception of the evolutionary emergence of the laws of physics.[1] There are also other originally Peircean ideas of greater interdisciplinary scope, whose ongoing recognition may influence the future orientation of biosemiotic research. Among these we encounter a relational turn in the life sciences, best exemplified by the rise of systems biology, indicative of a tendency to countenance a relational view of reality. This stance is firmly in keeping with Peirce's rejection of the nominalistic bias that he so much deplored in modern philosophy and science, and is auspicious to the integration of biosemiotic ideas in biology.

The deep, direct influence of Peircean ideas on biosemiotics aside, other related notions are being rediscovered, notably Peirce's advocacy of various forms of causation that had been barred from scientific orthodoxy until recently. Somewhat reluctantly, biologists are beginning to abandon their predilection for the narrow form of efficient causation that characterized physical theories before the rise of quantum physics, nonlinear dynamics and complexity theories.

This chapter concentrates on one central idea that Peirce was in the process of extending and generalizing towards the end of his career: the paramount notion of *the habit of acquiring habits*. We will see how this conception brings into synthetic conciliation other convergent Peircean ideas and how it might further be developed in light of recently introduced conceptions in science and philosophy, such as the role of symmetry breaking in the emergence of evolutionary novelties.

Habits and Cosmological Speculation

Some of the ideas we shall consider are hard to grasp because of multiple difficulties. One of these impediments is the need to be conversant with specialized concepts within several diverse disciplines. To lessen this difficulty, technicalities

[1] Here and elsewhere the noun "emergence," the verb "to emerge," and derived expressions, refer to the rise of novel forms of order, variety and complexity through the breaking of symmetries (or equivalently, as intimated further on, by the acquisition or breaking of habits). There is at present a vast and complex philosophical bibliography on emergence, often concerned with metaphysical issues (such as the relations between mind and matter) not necessarily pertinent to the ideas discussed here. The acceptation adopted herein is increasingly encountered in the scientific literature and in many cases can be traced to the very influential paper "More is different" (Anderson 1972).

will be relegated to footnotes to avoid breaking the continuity of the exposition. Other obstacles include a high level of abstraction, and the fact that Peirce's accounts of his cosmology are often scattered, extremely sketchy or condensed, and far from transparent.

A further hindrance is that these ideas tend to remain inaccessible unless we break loose from the grip of tacit, nominalistic assumptions imbued in our educational formation. Perhaps the most inimical to our aims is the prevailing conflation of generality and abstraction. For Peirce the generality of abstract concepts is nothing but a logical analog of the real, *concrete generality* at work in the evolution of the world.

A recent notable book bears the title *Biology's First Law: The Tendency for Diversity and Complexity to Increase in Evolutionary Systems* (McShea and Brandon 2010). This title would announce most aptly the ideas discussed here if it were only to read ***Nature's*** *First Law*. Of course, its authors focus on biological evolutionary systems characterized by heredity and variation. Yet Peirce's conception of evolution is of a much wider scope. He applied his notion of concrete generalization to account for the tendency of diversity and complexity to increase within one all-embracing evolutionary system: *the entire physical universe.*

Peirce himself gave the tersest possible account of his speculative, evolutionary cosmology:

> The state of things in the infinite past is chaos, tohu bohu, the nothingness of which consists in the total absence of regularity. The state of things in the infinite future is death, the nothingness of which consists in the complete triumph of law and absence of all spontaneity. Between these, we have on our side a state of things in which there is some absolute spontaneity counter to all law, and some degree of conformity to law, which is constantly on the increase owing to the growth of habit. The tendency to form habits or tendency to generalize is something which grows by its own action, by the habit of taking habits itself growing. (CP 8.317, 1891)

Let us try to unpack and expand some relevant points of this overly concise statement, assisted by other pronouncements scattered throughout Peirce's writings. In his view our world, with all its immense complexity and variety, emerged gradually from an original firstness, a realm of pure spontaneity and undetermined, undifferentiated potentiality, through the occurrence of a radically fortuitous, uncaused event that Peirce named a *flash*:

> Out of the womb of indeterminacy we must say that there would have come something, by the principle of Firstness, which we may call a flash. Then by the principle of habit there would have been a second flash. Though time would not yet have been, this second flash was in some sense after the first, because resulting from it. Then there would have come other successions ever more and more closely connected, the habits and the tendency to take them ever strengthening themselves, until the events would have been bound together into something like a continuous flow. (A Guess at the Riddle, EP1, 278, 1888)

Peirce propounded an original version of *creatio ex nihilo*:

> The initial condition, before the universe existed, was not a state of pure abstract being. On the contrary it was a state of just nothing at all, not even a state of emptiness, for even emptiness is something. (CP 6. 216, 1898)
> But this is not the nothing of negation. For *not* means *other than*, and *other* is merely a synonym for the ordinal numeral *second*. As such it implies a first; while the present pure zero is prior to every first. The nothing of negation is the nothing of death, which comes

> *second* to, or after, everything. But this pure zero is the nothing of not having been born. There is no individual thing, no compulsion, outward nor inward, no law. It is the germinal nothing, in which the whole universe is involved or foreshadowed. As such it is absolutely undefined and unlimited possibility—boundless possibility. There is no compulsion and no law. It is boundless freedom. So of *potential* being there was in that initial state no lack. (CP 6: 217, 1898)

Out of the original nothing, the emergence of increasingly complex processes and things is driven by the progressive action of a generalizing tendency—one that constrains the original spontaneity and in so doing generates regularity, i.e., compliance with law. This generalizing tendency takes hold in the world of processes and things through the formation of habits. From this perspective, the fundamental laws of physics are conceived as primordially established habits of unmatched generality and scope.

Peirce's conception of inquiry demands the rejection of any appeal to ultimate and inexplicable things. Because we tend to explain all else by an appeal to laws, he saw that the only explanation for the laws themselves is their evolution out of a state of lawlessness[2], i.e., absolute chance:

> Now the only possible way of accounting for the laws of nature and for uniformity in general is to suppose them results of evolution. This supposes them not to be absolute, not to be obeyed precisely. It makes an element of indeterminacy, spontaneity, or absolute chance in nature. (CP 6.13, 1891)
>
> Mr. Herbert Spencer wishes to explain evolution upon mechanical principles. This is illogical, for four reasons. First, because the principle of evolution requires no extraneous cause, since the tendency to growth can be supposed itself to have grown from an infinitesimal germ accidentally started. Second, because law ought more than anything else to be supposed a result of evolution. Third, because exact law obviously never can produce heterogeneity out of homogeneity; and arbitrary heterogeneity is the feature of the universe the most manifest and characteristic. Fourth, because the law of the conservation of energy is equivalent to the proposition that all operations governed by mechanical laws are reversible; so that an immediate corollary from it is that growth is not explicable by those laws, even if they be not violated in the process of growth. (CP 6.14, 1891)

Habits and "Concrete Generalization"

At times Peirce uses "tendency" and "habit" interchangeably, yet they are distinguishable. A tendency is a simple "would-be": its reality is merely potential and it consists in the fact that under circumstances of a characteristic type a certain change, describable in general terms, would occur more often than not. A habit, on the other hand, is a second-order tendency involving a relational complex of tendencies. It is a tendency to repeat past changes and a tendency to prompt similar changes in the

[2] During the twentieth century several eminent physicists separately rediscovered Peirce's conception of the evolutionary rise of the laws of nature from former, less exacting laws. Among them we count Dirac, Feynman, Nambu, Thirring, and others. Lee Smolin has repeatedly given credit to Peirce for this idea, which he has further developed (see e.g. Smolin 2009, 2012a, 2012b).

future *each time a triggering combination of similar circumstances is itself repeated*. It is also a self-referential generalizing tendency that further generalizes itself:

> .... all things have tendency to take habits. For atoms and their parts, molecules and groups of molecules, and in short every conceivable real object, there is a greater probability of acting as on a former like occasion than otherwise. This tendency itself constitutes regularity, and it is continually on the increase [...] It is a generalizing tendency; it causes actions in the future to follow some generalization of past actions; and this tendency is itself something capable of similar generalizations; and thus is self-generative. (CP 1.409, 1890)

The generalization effected by habit is not an abstract idea. It is a concrete (Peirce says "objective") process, taking place in the physical world. At the creation of the universe,

> [a] habit of acquiring habits began to be established, and a habit of strengthening the habit of acquiring habits, and a habit of strengthening that habit, and so on ad infinitum. The acquiring [of] a habit is nothing but an objective generalization taking place in time. (NEM IV: 140, no date)

Through the concept of the habit of acquiring habits Peirce could envision a simultaneous account of both the creation of the universe, from a primeval nothingness of undifferentiated possibilities, as well as of the emergence of the laws of nature, from a self-generalizing tendency towards increasing regularity[3].

Habits and Symbols

In Peirce's mature thought, the concept of habit—the key notion of his evolutionary cosmology—becomes a central conception of his expanded system of semiotics, which by this time subsumes his entire system of logic. [4] *Habit* and *law* have become practically synonymous and lay the ground for an improved definition of the most developed kind of sign—the *symbol*:

> A Symbol is a law, or regularity of the indefinite future. Its Interpretant must be of the same description; and so must be also the complete immediate Object, or meaning. But a law necessarily governs, or "is embodied in" individuals, and prescribes some of their qualities. ('A Syllabus of Certain Topics of Logic', EP 2:274, 1903)

---

[3] Our brief exposition cannot do justice to the complex network of ideas behind Peirce's cosmological speculations. As is often the case these ideas have deep roots in his mathematical thought. For an excellent analysis of their connection to the mathematical notion of the Absolute and to time and geometry see Kalil 2011.

[4] As Winfried Nöth remarks: "...Peirce's much broader perspective of the symbol as a sign guided by onto- and phylogenetic habit serves as a synechistic bridge to overcome two dualisms which have prevailed in the history of semiotics, the dualism of culture vs. nature and the dualism of the conventional vs. the innate, i.e., between signs culturally transmitted by teaching and learning and signs genetically inherited and interpreted by instinctive dispositions. Against the dualism culture vs. nature, Peirce proposes that the habit by which symbols are interpreted is conventional or natural. Against the dualism of the conventional vs. the innate, Peirce postulates that the habit which determines the symbol is an "acquired or inborn" disposition (Nöth 2010, p. 84–85).

The fairly contemporaneous definition he offered in Baldwin's dictionary also refers to *habit*:

> Symbol. A Sign (q.v.) which is constituted a sign merely or mainly by the fact that it is used and understood as such, whether the habit is natural or conventional, and without regard to the motives which originally governed its selection. (Dictionary of Philosophy & Psychology, vol. 2; CP 2.307, 1902)

By following Peirce's applications of the notion of habit, we ascend through the cosmic evolutionary scale toward increasingly higher levels of organization and variety. We trace the formation of stars, the advent and evolution of organisms and their complex associations, and finally reach that peculiar kind of organism that writes and reads pages such as this. Through the agency of language, our use of a refined kind of symbol allows us in a sense to transcend the biological world. We are organisms immersed in human culture, a realm of historically constituted symbols and artifacts in accelerated evolution. Animals communicate by means of rudimentary types of symbols to a limited extent, [5] but we are uniquely historical animals with a collective capacity for creating symbols that are vehicles of *intellectual concepts*.

In 1906 or thereafter, Peirce came to conceive of intellectual concepts as specially developed symbols or symbolic actions. He articulated these conceptions through the introduction of different kinds of symbolic interpretants. The logical interpretant in particular is not a sign as previously defined by Peirce; it is the acquisition of a habit. The final interpretant of a symbol is a change of habit. About the symbol that is a vehicle of intellectual concepts he says:

> [...] if this sign be of an intellectual kind—as it would have to be—it must itself have a logical interpretant; so that it cannot be the *ultimate* logical interpretant of the concept. It can be proved that *the only mental effect that can be so produced and that is not a sign but is of a general application is a habit-change*; meaning by a habit-change a modification of a person's tendencies toward action, resulting from previous experiences or from previous exertions of his will or acts, or from a complexus of both kinds of cause. ("Pragmatism", CP 5.475, 1907, *emphasis* added, variants in EP2: 398-433)

In the ideas of habit and habit change we can discern a unifying thread that runs though both Peirce's cosmology and his semiotic system, and finally ties them together to the pragmatic maxim, the "kernel of pragmatism":

> Intellectual concepts [...] essentially carry some implication concerning the general behaviour either of some conscious being or of some inanimate object, and so convey more, not merely than any feeling, but more, too, than any existential fact, namely, the 'would-acts' of habitual behaviour; and no agglomeration of actual happenings can ever completely fill up the meaning of a 'would-be.' But that the total meaning of the predication of an intellectual concept consists in affirming that, under all conceivable circumstances of a given kind, the subject of the predication would (or would not) behave in a certain way,—that is, that it either would, or would not, be true that under given experiential circumstances (or under a given proportion of them, taken as they would occur in experience) certain facts would exist,—that proposition I take to be the kernel of pragmatism. More simply stated,

[5] The study of symbols in animal communication is an expanding field of research in biosemiotics. See, e.g. Ribeiro et al. 2007, Slobodchikoff et al. 2009, and Queiroz 2012.

> the whole meaning of an intellectual predicate is that certain kinds of events would happen, once in so often, in the course of experience, under certain kinds of existential circumstances. ("Pragmatism," EP 2:401-402, 1907)

As Lucia Santaella has cogently remarked:

> When he discovered the role of the logical interpretant in habit and of the ultimate interpretant in the change of habit, Peirce amalgamated the processual nature of semiosis with pragmatism. The evolutionist character of his pragmatism resulted from this synthesis. (Santaella 2004, p. 75)

Symmetry and the Breaking of Symmetry

Peirce was not able to develop his speculative views on the origin of the universe and its laws into an experimentally testable theory. It was impossible—he died before the emergence of some early twentieth century discoveries that made possible for the first time the creation of a genuinely scientific cosmology.[6] Among these findings there are two that are especially important; one empirical and the other theoretical.

The first all-important discovery was an astronomical one, made more than a decade after Peirce's death. This was Edwin Hubble's demonstration of the on-going metric expansion of the universe, through the observation that distant galaxies are moving away from us at speeds approximately proportional to their distance to the earth. [7] This empirically established fact was to become the cornerstone of scientific cosmology since the third decade of the past century. A similar, crucial empirical discovery (not far from the first one in significance) was the detection of the cosmic microwave background radiation in the 1960's.

The second paramount discovery, arguably the deepest theoretical insight of twentieth-century physics, was the realization of *the intimate relationship between the laws of physics and the notion of symmetry*. This recognition is an essential grounding element of relativity theory, quantum physics and condensed matter physics, i.e., of most of current physics. The term *symmetry* characterizes a process or law that remains unchanged (invariant) after physical or mathematical operations

---

[6] The history of the rise and consolidation of scientific cosmology through impassionate debates in the third and fourth decades of the last century is ably and vividly depicted in Gale 2011. An eminent Peircean scholar, Thomas Short, is quite unsympathetic to Peirce's cosmological speculations (see Short 2011), which he declares a failure, adding that Peirce did not have a cosmology because he was unable to make experimentally testable predictions. This is an idiosyncratic acceptation of the term "cosmology," contrary to standard uses (e.g. Aristotelian cosmologies), and in most cases if we were to adopt it the term would apply only metaphorically to ideas developed before roughly 1930.

[7] Extrapolating backwards in time, Hubble's discovery indicates that some 14 billion years ago all energy (before the formation of matter) was concentrated in an extremely small and dense region. The release of that energy represents what is known as the "Big Bang." This fact remarkably agrees with Peirce's intuition that the universe came into existence in a way that logically "precedes" the origination of time itself, at a finite temporal distance from our present.

are performed upon it. These operations must combine according to some simple mathematical rules spelled out by group theory.

Twentieth-century physics went beyond the traditional task of finding invariances in natural phenomena through the discovery of laws. It moved into deeper theories of higher generality by examining the *invariances manifested by the laws themselves* under suitable transformations. The special theory of relativity, for example, is based on the invariance of the laws of physics after changing from one inertial frame of reference to another.

Long before gaining supreme conceptual status through its formalization in group theory, the idea of symmetry—and the closely related notions of stability, equilibrium, reversibility, continuity and determinism—had informally guided the development of physical theory. In the twentieth century a complementary and equally central conception, *symmetry breaking*, arose in the explanation of phenomena involving the antithetical notions of instability, disequilibrium, irreversibility, discontinuity and indeterminism.

In physics textbooks symmetry breakings are frequently illustrated through events known as *critical phase transitions*. The most familiar example is that of the abrupt qualitative changes that water undergoes at normal atmospheric pressure when its temperature is raised or lowered continuously. Water in the liquid phase displays sudden density changes at the freezing and boiling points, at which the temperature goes through a sharply defined critical value. The driving mechanism of critical phase transitions is a *breaking of symmetry* in the medium undergoing the changes. When water freezes, for instance, the rotational symmetry (indifference to direction) in the motion of the molecules is broken. Before the transition the molecules are free to move randomly in all possible directions, but after going through the critical temperature threshold they suddenly become arranged along a single, *randomly selected* direction, forming an ice crystal.[8]

The mechanism of symmetry breaking is currently invoked in the explanation of an extraordinarily wide range of phenomena spanning all levels and scales of nature, from the origin and formation of the elementary particles to the emergence and evolution of the entire universe. Most importantly for our purposes, as outlined in the following sections, they also allow us to show how Peirce's prescient—albeit necessarily vague—speculations can be given a more precise rendering in terms of contemporary scientific ideas and a suitably generalized notion of habit change.

---

[8] Critical phenomena are pervasive in nature. Typical examples are phase transitions, as in the example of the change of liquid to solid water, the spontaneous magnetization of an iron bar when the temperature goes trough the critical Curie point, the onset of turbulence in fluids, superconductivity, etc. These transitions are characterized by the emergence of new forms of behavior through the appearance of new structures, and a transition from a disorderly state to a more orderly one. Phillip Anderson's seminal 1972 paper, "More is different," initiated an ongoing program to account for the emergence of novelty and complexity through the phenomenon of symmetry breaking. See e.g., Anderson 1972; Fernández 2012a.

Cosmic Evolution and the Breaking of Symmetry

Peirce anticipated in a qualitative way several key discoveries of contemporary physics and cosmology. With hindsight we now recognize in the original quantum vacuum that void of actualities and plenum of possibilities that he identified as the primordial nothingness. We can also identify quantum fluctuations (the spontaneous creations and annihilations of "virtual particles" out of vacuum energy) with the randomly created "flashes" that Peirce envisioned as randomly emerging from the primeval vacuum.

The genesis of the universe from a quantum fluctuation is a widely held explanation in contemporary cosmology. Frank Wilczek states this pithily: "The answer to the ancient question 'Why is there something rather than nothing?' would then be that 'nothing' is unstable" (quoted in Stenger 2006). This is more than a witty jest: the instability of the quantum vacuum is a deep, unifying theoretical link between contemporary cosmology and quantum physics. The original vacuum must be conceived in the analogy of a "space" preceding our ordinary tridimensional space and endowed with an ideal maximum of symmetry.

According to contemporary theory the tiny domain of super-concentrated energy that initiated the big bang emerged through a primordial phase transition. It led to an exponentially rapid expansion of the universe, known as "cosmic inflation."[9] This enormous expansion took place in an exceedingly short fraction of a second, and after this stage the universe continued expanding, albeit at a much slower rate.

Current cosmology, in conjunction with the standard model of particle physics, explains the genesis of the first elementary particles and the formation of structures of increasing complexity and regularity as originating from quantum fluctuations during the inflationary period. New entities and processes continued developing through a series of phase transitions as the universe "cooled" (its energy concentration decreased) as a result of the cosmic expansion. As the energy level decreased, a postulated supersymmetry was successively broken into more restricted symmetry groups and three of the fundamental forces of nature (i.e., the strong and weak nuclear forces and the electromagnetic interaction) split in succession from an initial unified force that reigned at higher energy levels. Processes analogous to this cosmic genesis, driven by symmetry breakings, are discernibly at work in the continuous generation of novelties, order and complexity that characterizes the evolution of the universe from the formation of stars and galaxies to the creation of living systems.

---

[9] Alan Guth, the original proponent of "inflation," explains this difficult idea in very simple terms: "Inflationary theory takes advantage of results from modern particle physics, which predicts that at very high energies there should exist peculiar kinds of substances which actually turn gravity on its head and produce repulsive gravitational forces. The inflationary explanation is the idea that the early universe contains at least a patch of this peculiar substance. It turns out that all you need is a patch; it can actually be more than a billion times smaller than a proton. But once such a patch exists, its own gravitational repulsion causes it to grow, rapidly becoming large enough to encompass the entire observed universe" (Guth 2001).

The following section introduces some tentative ways of developing Peirce's conception of habit and habit changing in light of present knowledge, with the hope of applying Peirce's thinking in the explanation of evolutionary novelties and the role of semiosis in living systems.

Habits, Causation and Complexity

Physics concentrated on time-reversible, linear and deterministic processes through most of its history. These processes characterize non-complex systems, usually modeled in terms of a limited set of simple basic entities interacting according to computationally simple rules. In the nineteenth century thermodynamics introduced the concept of entropy to deal with phenomena, such as energy dissipation and irreversibility, which were not amenable to explanation under the former scheme. But it was thought that these phenomena, manifest at the macrolevel, could in principle be explained statistically in terms of a reversible, deterministic dynamics involving large assemblies of basic, identical components, at the microlevel.

Something approaching a revolution started to take form in the 1960s—a transformation of the scope and methods of scientific explanation, driven in part by the increasing availability of powerful and inexpensive programmable computers. It became clear that protracted iteration of very simple rules over small sets of data could produce structures of unforeseen complexity (e.g., fractals). Work on nonlinear dynamics disclosed the characteristics and ubiquity of chaos: a deterministic dynamics where tiny differences in initial conditions lead to diverging and unpredictable trajectories. The study of thermodynamic systems far from equilibrium opened new vistas on self-organizing systems, such as living beings, that display novel forms of laws and causal operation. All of these developments fall under the scope of *complexity theory*, a not-yet-unified and growing framework of partially overlapping concerns, concepts and methodologies. In what follows we explore some possible applications of the Peircean concept of habit to the new forms of causation displayed by complex systems.

In physics the expression "laws of nature" is often reserved for the fundamental laws that universally constrain all natural phenomena, following an historically entrenched belief that no special laws or forms of causation emerge at different levels of scale or complexity. In this perspective the so-called laws invoked in the so-called "special sciences" are just locally constrained manifestations of the same fundamental laws. The study of complex systems suggests that complex phenomena, over and beyond their compliance with the fundamental laws, display kinds of action and passion that betray the dictates of new kinds of laws and causal determination.[10]

[10] Causation (and therefore explanation) in self-organizing complex systems takes new forms as a consequence of features, such as historicality, contextuality, and hierarchical organization, which are not apparent in simpler, linear systems. The emergence of new and locally manifest laws is

Peirce insightfully anticipated to some extent this discrepancy between the fundamental laws of dynamics and other special, experimentally discovered laws:

> ...In the first place, people are persuaded that everything that happens in the material universe is a motion completely determined by inviolable laws of dynamics; and that, they think, leaves no room for any other influence. But the laws of dynamics stand on quite a different footing from the laws of gravitation, elasticity, electricity, and the like. The laws of dynamics are very much like logical principles, if they are not precisely that. (CP 1.348, 1903)

As we know now, the dynamical laws that Peirce viewed as quasi-tautologies are logical consequences of global symmetries that hold at all points of space and time. We can think of them as universal habits. The other laws he mentions, related to the forces of gravity and electromagnetism, are indeed different. They are grounded on principles of local gauge symmetry, i.e., invariant under transformations that change from point to point in space and time. Perhaps we can think of them as localized habits. But in biosemiotics we are interested in new kinds of laws that emerge in organisms as complex systems far from equilibrium, and we ponder on their possible relations to Peircean habits and on the special role of semiosis in biological causation.

The subject of causation in complex systems is too complex (!) and present discussions too fluid and intricate to be explored here, even superficially. Although some of those issues may prove quite relevant to future applications of the Peircean conception of habit, we must be content here with a few remarks and suggestions.

First of all it seems that the mechanism of symmetry breaking may provide *a concrete physical model of habit acquisition and habit change*. The discontinuous passage from liquid water to rigid, crystalline ice in our previous example offers an elemental image of the creation of a new regularity by the spontaneous (random) restriction of previously possible kinds of behavior. As with any other habit the restriction will recur each time similar circumstances (temperature, pressure, etc.) obtain. As we noted earlier, contemporary physics traces the emergence of novel forms of regularity and structure in general to the action of symmetry breakings.[11]

in different ways brought about by the presence of changing dynamic constraints. We cannot go into these vast and contentious topics here. For the role of dynamic constraints on the emergence of new kinds of regularities see e.g., Juarrero 2009. For various forms of non-standard causation in complex systems see e.g., papers by Noble 2012; Ellis 2012; Bishop 2012, and several other authors in the special issue on top-down causation of *Interface*, February 6, 2012.

[11] The rise of novelty through symmetry breaking is virtually coextensive with an increase of information. As John Collier has repeatedly noted, "information originates in symmetry breaking" (see e.g. Collier 1996; Fernández 2011). There are several partially incompatible notions of "information" used at present in biosemiotics, but a consensus seems to be arising that the standard (Shannon's) conception so much valued by communication engineers is insufficient to meet the needs of semiotics. Peirce described the rise of novelty as "... 'organized heterogeneity,' or, better, rationalized variety." (CP 6.101). André de Tienne has brilliantly examined Peirce's evolving notion of information in connection with Peirce's late model of semiosis as a medium of communication, discussed later on (de Tienne 2006). Roy Frieden and Vinicius Romanini have given a new explication of semiotic information in terms of Fisher information (Frieden and Romanini 2008).

Habits and New Forms of Causation

In this section we review some ideas on the possibility of generalizing further the notion of habit through its interpretation in terms of symmetry breaking. We shall now consider some simple habits and some compound habits in which simple component habits reinforce or counteract each other's effects.

Peirce conceived individual things as bundles of habits of reaction:

> [...] nothing can crowd a place for itself but an act of reaction. The mind, by its instinctive adaptation to the Outer World, represents things as being in space, which is its intuitive representation of the clustering of reactions. What we call a Thing is a cluster or habit of reactions, or, to use a more familiar phrase, is a centre of forces. (CP 4.157, c. 1897)

At present we regard ordinary things as objects that behave according to the laws of classical physics. As such they are endowed with individual identities through their sharp locations in space and time. We contrast them to the "particles" of quantum physics, which are not things. They lack the spatial determinations which are the individuating features defining the "classicality" of ordinary, macroscopic objects. Furthermore, their properties are merely potential, and cannot be thought as having been predetermined before becoming actualized through irreversible interactions with classical objects, such as measuring instruments. In present-day physics a randomly driven, irreversible process known as "decoherence" describes how such a system of co-present possibilities coalesces into a spatially individuated object. This happens when some of the possible properties become actualized through the forcible exclusion of others.[12]

In Peircean terms we can say that the deterministic, linear and time-reversible laws encoded in the quantum formalism yield only the probabilities that a particular selection of quantum "would be's" is to become actualized. Decoherence, in a way that remarkably resembles the action of a phase transition, irreversibly and randomly singles out for actualization one possible set of embodied habits from innumerable competing ones.

It would seem that current discussions of unorthodox forms of causation among biologists and systems scientists might open a new field of applicability to the Peircean conception of habit. Discussions of bottom-up, top-down, and circular causation have brought to the fore notable connections with phenomena produced by the breaking of symmetry in unorganized physical systems. One such instance is the emergence of similar or generic types of behavior at higher levels in systems that at their lowest, microscopic levels show totally different structures and behaviors. In living systems another striking similarity with ordinary critical phenomena

[12] According to the principle of superposition, before a measurement is performed quantum systems are in all their possible states simultaneously, and this leads to the wave-like interference (coherence) famously exemplified in the two-slit experiment. Decoherence refers to the spontaneous suppression of quantum interference through complex interactions (entanglement) of the quantum entities with their environment. See e.g., Bacciagaluppi 2012.

appears in the transmission of causal determinations across different length scales and levels of organization.[13]

These similarities have led some authors (see e.g., Longo et al. 2011, 2012) to advance the concept of *extended criticality*. In their view organisms can be envisioned as complex, far-from-equilibrium systems permanently undergoing an uninterrupted flow of continuous phase transitions. In living beings the singular critical points that characterize ordinary phase transitions give way to continuous intervals of extended criticality, where old symmetries are constantly broken and new ones are simultaneously created. Extended criticality suggests the existence of an activity through which the acquisition, rejection and alteration of habits becomes a continuous process that emerges at the level of living functionality.

Habits and Semiosis

The action of signs is a very particular kind of causal determination and its role and manifestation in living systems is one of the main concerns of biosemiotics. Symbols are signs in which the object determines its interpretant through the existence of innate or acquired habits. In animals the existence of a nervous system makes possible the triggering of innumerable established habits through the action of fixed neural reflexes. The evolutionary emergence of conditioned reflexes provided efficient means for the individual acquisition of new habits, essential for learning new kinds of behaviors. Eventually they also made possible the use of symbols, such as warning calls and gestures. But biosemiotics needs to address the action of signs in all organisms, from bacteria to rabbits, not just in animals that are proficient in the use of symbols in communication between members of the same or different species. A much more basic function is the action of signs between parts of an organism (e.g., the organelles of a bacterium) and between those parts and the environment.

Towards the end of his career (circa 1906), Peirce started developing a new account of semiosis in terms of acts of communication, and this approach may turn out to be most suitable for the needs of biosemiotics. He attempted to define a sign as a special medium of communication:

> That which is communicated from the Object through the Sign to the Interpretant is a Form. It is not a singular thing; for if a singular thing were first in the Object and afterward in the interpretant outside the Object, it must thereby cease to be in the Object. The Form that is communicated does not necessarily cease to be in one thing when it comes to be in a different thing, because its being is a being of the predicate. The Being of a Form consists in

[13] The term "scaling" refers to a feature of systems that show self-similar properties around their critical points, and "universality" to the fact that otherwise dissimilar systems display similar behavior near those points. A very elementary explanation of the appearance of universality classes and their relations to the phenomenon of scale invariance can be found in Fernández 2012b. Batterman 2011 offers and in-depth, competent, and rather technical treatment of these issues in connection to current discussions on emergence and reductionism.

> the truth of a conditional proposition. Under given circumstances, something would be true. The Form is in the Object, entitatively we may say, meaning that that conditional relation, or following of consequent upon reason, which constitutes the Form, is literally true of the Object. In the Sign the Form may or may not be embodied entitatively, but it must be embodied representatively, that is, in respect to the Form communicated, the Sign produces upon the Interpretant an effect similar to that which the Object itself would under favorable circumstances. (MS 283, partially reprinted in EP2, 371-397, 1906)

Making abstraction of its physical embodiment we can consider this *form* in the analogy to a *disembodied habit*, similar to quantum "particles" before they undergo decoherence. But semiosis takes place in the domain of classical physics and therefore the transmission must occur through the mediation of a classical vehicle, such as e.g., the action of ligands on cellular receptors. The process is nevertheless the result of semiosis rather than ordinary physical causation. This is so because the action is quite independent of the physical nature of the vehicle, i.e., vehicles of totally different physical nature may mediate the same communication. The physical vehicle provides a transitory and ephemeral embodiment to the transmitted habit.

Concluding Remarks

The preceding discussions and observations were aimed at elucidating Peirce's conception of habit and some closely associated ideas (habit acquisition, habit change, and the habit of acquiring habits) in connection with their present and future applications in biosemiotics. Our goal was to achieve a better understanding of these notions by following two often-intercepting routes. We examined the unifying role of these ideas in bringing together apparently separated strands of Peirce's thought in such topics as cosmology, the character and evolution of the laws of nature, and his mature views in semiotics. In parallel discussion we surveyed several developments in science and philosophy during the last 100 years, to explore how they may help deepen, extend, or otherwise modify these conceptions. We hazarded along the way some speculative interpretations of the ideas considered.

In spite of the great achievements of such founding figures as Thomas Sebeok and Jesper Hoffmeyer the budding discipline of biosemiotics is still searching for a unified theoretical framework, and the settling of such basic issues as the nature of semiosis and the semiotic workings of the organic codes is still far from attaining overall consensus. One central question at present is the place of semiosis in the context of various new forms of causal determination disclosed by work in systems biology and complexity theories. It is to be hoped that renewed interest in Peirce's ideas about habits and their possible elaboration and extension, as outlined in this chapter, will contribute to the clarification and resolution of these and other issues of foundational import in biosemiotics.

References

Anderson, P. W. (1972). More is different. *Science, 177,* 393–396.
Bacciagaluppi, G. (2012). The role of decoherence in quantum mechanics, the stanford encyclopedia of philosophy (Summer 2012 ed.), Edward N. Zalta (ed.). *http://plato.stanford.edu/archives/sum2012/entries/qm-decoherence/.*
Batterman, R. W. (2011). Emergence, singularities, and symmetry breaking. *Foundations of Physics, 41*(6), 1031–1050.
Bishop, R. C. (2012). Fluid convection, constraint and causation. *Interface Focus, 2*(1), 4–12.
Collier, J. (1996). Information originates in symmetry breaking. *Symmetry: Culture & Science, 7,* 247–256.
De Tienne, A. (2006). Peirce's logic of information. Seminario del Grupo de Estudios Peirceanos, Universidad de Navarra. www.unav.es/gep/SeminariodeTienne.html.
Ellis, G. F. R. (2012). Top-down causation and emergence: Some comments on mechanisms. *Interface Focus, 2*(1), 126–140.
Fernández, E. (2011). Energy, semiosis and emergence. Eleventh annual international gatherings in Biosemiotics. New York City, USA. http://www.lindahall.org/services/reference/papers/fernandez/Energy_semiosis_and_emergence.pdf. Accessed 21–26 June 2011.
Fernández, E. (2012a). Symmetry breaks out—a fundamental concept jumps over disciplinary barriers. Midwest Junto for the History of Science. University of Missouri of Science and Technology, Rolla, MO, U.S.A. Fifty-fifth annual meeting, March 23–25, 2012.
Fernández, E. (2012b). Habit and generalization, V Jornadas Peirce en Argentina. Academia Nacional de Ciencias de Buenos Aires. http://www.lindahall.org/services/reference/papers/fernandez/habit_and_generalization.pdf. Accessed 23–24 Aug 2012.
Frieden, R. B., & Romanini, V. (2008). Eluding the demon-how extreme physical information applies to semiosis and communication. *Cognitio-Estudos (PUC-SP), 5,* 52–63.
Gale, G. (2011). Cosmology: Methodological debates in the 1930s and 1940s. In N. Z. Edward (Ed.), *The stanford encyclopedia of philosophy* (Winter 2011 ed.). http://plato.stanford.edu/archives/win2011/entries/cosmology-30s/.
Guth, A. (2001). A golden age of cosmology. Edge. http://www.edge.org/conversation/a-golden-age-of-cosmology.
Juarrero, A. (2009). Top-down causation and autonomy in complex systems in Murphy. In E. Nancey (Ed.), *Understanding complex systems, downward causation and the neurobiology of free will* (pp. 83–102). Berlin: Springer.
Kalil, R. (2011). La metafísica del Absoluto, c. 1881–1891. *Cuadernos de sistemática peirceana, 5*(3), 5–51. *http://acervopeirceano.org/wp-content/uploads/2012/11/CSP3.pdf.*
Longo, G., & Montévil, M. (2011). From physics to biology by extending criticality and symmetry breakings. *Progress in Biophysics and Molecular Biology, 106*(2), 340–347.
Longo, G., et al. (2012). No entailing laws, but enablement in the evolution of the biosphere. arXiv 1201.2069.
McShea, D. W., & Brandon, R. N. (2010). *Biology's first law: The tendency for diversity and complexity to increase in evolutionary systems.* Chicago: University of Chicago Press.
Noble, D. (2012). A theory of biological relativity: No privileged level of causation. *Interface Focus, 2*(1), 55–64.
Nöth, W. (2010). The criterion of habit in Peirce's definitions of the symbol, transactions of the Charles S. *Peirce Society, 46*(1), 82–93.
Queiroz, J. (2012). Dicent symbols in non-human semiotic processes. *Biosemiotics, 5,* 319–329.
Ribeiro, S., et al. (2007). Symbols are not uniquely human. *Biosystems, 90,* 263–272.
Santaella, L. (2004). O papel da mudança de hábito no pragmatismo evolucionista de Peirce. *Cognitio,* 5(1), 75–83.
Short, T. L. (2011). Reading peirce differently: A response to David Dilworth. *Cognitio, 12*(2), 257–271.

Slobodchikoff, C. N., et al. (2009). Prairie dog alarm calls encode labels about predator color. *Animal Cognition, 12*, 435–439.
Smolin, L. (2009). The unique universe: Against the timeless multiverse. *Physics World, June issue*, 21–26.
Smolin, L. (2012a). Unification of the state with the dynamical law. arXiv: 1201.2632.
Smolin, L. (2012b). A perspective on the landscape problem. arXiv: 1202.3373.
Stenger, V. (2010). Did the universe come from nothing? *Reality Check, 20*(4). http://www.csicop.org/sb/show/did_the_universe_come_from_nothing.

Semeiotic Causation and the Breath of Life

Menno Hulswit and Vinicius Romanini

A single Assertion has but a single Predicate; but the simplest Assertion has more than one Subject, unless it be such a statement as "It rains," where one of the Subjects is expressed otherwise than in words. But I must explain myself more fully, and in the way which alone will be truly expressive, namely, by examples. I will, however, first remark that the Proposition that embodies an Assertion has the same Subjects and Predicate as the Assertion itself. Take the Proposition "Cain killed Abel." This is identically the same Proposition as "Abel was killed by Cain"; it is only the grammatical dress that is different. Other things being equal, everybody will prefer the former. Why? Because it is simpler, but why is it simpler? Because in putting the cause before the effect, it in that respect diagrammatizes the truth. What are the Subjects of this Proposition. Cain, first: that is not only a Subject of the Proposition, but is the principal Subject of the Assertion which a historian would naturally make. But in the Proposition Cain and Abel are as Subjects on one footing precisely (or almost precisely, for Cain is preponderant in causality). But besides these, "killed" = committed *murder* upon, is a third Subject, since no study of the words alone, without extraneous experience, would enable the Addressee to understand it. What, then, is left to serve as Predicate? Nothing but the *flow of causation*. It is true that we are made acquainted even with that in Experience. When we see a babe, in its cradle bending its arms this way and that, while a smile of exultation plays upon its features, it is making acquaintance with the flow of causation. So acquaintance with the flow of causation so early as to make it familiar before speech is so far acquired that an assertion can be syntactically framed; and it is embodied in the syntax of every tongue. However, it is not because of this physiological fact, that it becomes proper to draw the line between Subjects and Predicate here; neither is it because of the psychical fact that human minds naturally think in a way broadly (i.e. a little) similar to the forms of syntax; nor is it even because of the metaphysical truth, that the order of syntax is the law of Time and of Becoming. This is proved by the facts, first, that it is necessary that Reasonings by which we discover and defend the order of Causation, of human thoughts, of time, and of becoming, themselves presuppose the recognition of the corresponding order in syntax; and secondly, by this, that it has not been Time, or Causation, or the structure of the human mind nor human anatomy and physiology that have, any or all of them, determined that that ought to be the order of syntax that in fact ought to be so, but precisely the contrary, it is the fact that the order of Syntax ought to be as

M. Hulswit (✉)
Radboud University, Nijmegen, The Netherlands
e-mail: m.hulswit@phil.ru.nl

V. Romanini
University of São Paulo, São Paulo, Brazil
e-mail: vinicius.romanini@usp.br

V. Romanini, E. Fernández (eds.), *Peirce and Biosemiotics,* Biosemiotics 11,
DOI 10.1007/978-94-007-7732-3_6, © Springer Science+Business Media Dordrecht 2014

> in fact it ought to be that has determined, first, Real Being and Time to take the same form, and then, that it should become natural to the mind and should be the pattern of physical action. (MS 664, 10-13, 1910)

Introduction

Peirce's theory of signs might be defined as the result of a life-long struggle to describe and understand how mental processes such as consciousness, perception, sensation, emotion, cognition, representation and communication cooperate to produce a general conception of the real, so as to justify our beliefs and guarantee our conduct. During his studies he came to realize that all these mental phenomena are part of or involve genuine triadic relations, which might be generally defined as relations of three correlates welded by a continuous process of information and development. No only intellectual phenomena are based on semeiosis, but also every phenomenon related to life. The conclusion is that there must be a fundamental law underlying mental and biological processes. This law should be explained in terms of a mental causation by which purposeful or quasi purposeful (i.e. non-conservative or "finious", CP.4.741, c. 1898) processes are able to develop towards general final states. This is his "law of mind":

> Mind has its universal mode of action, namely, by final causation. The microscopist looks to see whether the motions of a little creature show any purpose. If so, there is mind there. Passing from the little to the large, natural selection is the theory of how forms come to be adaptive, that is, to be governed by a quasi purpose. It suggests a machinery of efficiency to bring about the end—a machinery inadequate perhaps—yet which must contribute some help toward the result. But the being governed by a purpose or other final cause is the very essence of the psychical phenomenon, in general. (CP 1.269, 1902).

Not only mind has this universal mode of action but, conversely, Peirce advocates that the universe must be defined as partaking of the nature of mental phenomena, for development, growth and increase of complexity are seen in every corner of the real. The schema of time is the logical pattern of causation, and this is the distinctive character of all mental and living phenomena:

> Question any science which deals with the course of time. Consider the life of an individual animal or plant, or of a mind. Glance at the history of states, of institutions, of language, of ideas. Examine the successions of forms shown by paleontology, the history of the globe as set forth in geology, of what the astronomer is able to make out concerning the changes of stellar systems. Everywhere the main fact is growth and increasing complexity. Death and corruption are mere accidents or secondary phenomena. Among some of the lower organisms, it is a moot point with biologists whether there be anything which ought to be called death. Races, at any rate, do not die out except under unfavorable circumstances. From these broad and ubiquitous facts we may fairly infer, by the most unexceptionable logic, that there is probably in nature some agency by which the complexity and diversity of things can be increased; and that consequently the rule of mechanical necessity meets in some way with interference. (EP I: 307-8, 1904)

This general mode of being of reality, which grows in a way very similar to the growth of a ***symbol*** (EP II: 323) drove Peirce to apply his logic, or semeiotic, to cosmological and metaphysical expeditions. The result is his objective idealism in connection with his doctrine of continuum, which he named synechism, intended to be a mathematical and logical theory of reality whose basic premise is to forbid any dry cut between mental and physical processes. On the contrary, mental and physical processes are seen as different stages of evolutionary processes involving habit-taking and habit-change due to the interplay between absolute chance ("pure tychism") and generality.

From this view comes his famous dictum that the universe is perfused with signs, if not composed exclusively of signs (CP 5.448, 1906). On one side, symbols are legisigns grounded by habits of great plasticity and capability to internalize novelty brought by chance (as every mental and biological phenomenon). On the other side, the laws of nature are signs grounded by habits that have lost most of their freshness and originality, becoming inveterate. The mechanical laws of Physics and their subsidiary concepts of force and determinism are only epistemological devices created by scientists to facilitate the manipulation and comprehension of discrete *facts*, while the ontological structure of reality is always non-deterministic and is composed of non-mechanical, continuous *processes* or *events*, a doctrine Peirce named synechism:

> Synechism is founded on the notion that the coalescence, the becoming continuous, the becoming governed by laws, the becoming instinct with general ideas, are but phases of one and the same process of the growth of reasonableness. This is first shown to be true with mathematical exactitude in the field of logic, and is thence inferred to hold good metaphysically. (Peirce 1998, p. 302, c. 1902)

Since synechism must also be applied to dissolve any separation between inorganic and organic reigns, the obvious conclusion is that nothing is completely dead, at least not in the logical or semeiotic sense:

> I am bound to maintain that an idea can only be affected by an idea in continuous connection with it. By anything but an idea, it cannot be affected at all. This obliges me to say, as I do say, on other grounds, that what we call matter is not completely dead, but is merely mind hidebound with habits. It still retains the element of diversification; and in that diversification there is life. (CP 6.158, 1892)

The universe in general and the life-processes in particular, are governed by the same purpose: the growth of reasonableness through the conjoint tendencies of generalization and variation:

> For principle of progress or growth, something must be taken not in the starting-point, but which from infinitesimal beginning will strengthen itself continually. This can only be a principle of growth of principles, a tendency to generalization. Assume, then, that feeling tends to be associated with and assimilated to feeling, action under general formula or habit tending to replace the living freedom and inward intensity of feeling. This tendency to take habits will itself increase by habit. Habit tends to coordinate feelings, which are thus brought into the order of Time, into the order of Space. Feelings coordinated in a certain way, to a certain degree, constitute a person; on their being dissociated (as habits do sometimes get broken up), the personality disappears. Feelings over whose relations to their

> neighbors habit has acquired such an empire that we detect no trace of spontaneity in their actions, are known as dead matter. (CP 6.585)

What we want to stress is that Peirce's cosmological metaphysics, a fundamental piece to understand his contribution to biosemiotics, is supported by logical and mathematical concepts. Indeed, in his many attempts to provide a comprehensive classification of sciences departing from Mathematics and Phenomenology (or Phaneroscopy), such "psychical phenomena, in general" are put under "Nomological Psychognosy." Peirce himself explains that this science coincides roughly with what is called Psychology, but focused only on the study of the laws of mind and not to its application to particular or social cases (CP 7.362, 1902). Here becomes clear where Peirce would classify biosemiotics in his general scheme. He would put it under the science of General Psychonomy:

> I shall, then, recognize that Nomological Psychognosy must separate into two Suborders; the first of General Psychonomy, the second of Special Psychonomy, or Nomological Psychology. The former will study the law of final causation and seek to formulate it with exactitude, while the latter will study subordinate laws of mind, of which that of Association is the first. General Psychonomy will have four Families. Family 1 will define the essence of Mind and the law of final Causation, together with its application to non-biological phenomena. We may term it phenomenalistic Pneumatology. Family 2 will show how final causation works in the development of biological stocks. Family 3 will study the law in its application to biological individuals and to consciousness. Family 4 may be called Demonomy, although it should study the application of the law, not merely to Societies, but also to mere Associations, such as that of a profession. (CP 7. 373-4, 1902)

From the above it becomes clear that a science of biosemiotics inspired by Peirce's philosophy and semeiotic must take into account, as one of its central research programs, his teleological approach of causation in its relation to both synechism and tychism (continuity and chance). It must also acknowledge his semeiotic conception of reality based on a metaphysical defense of objective idealism, which will naturally lead us to explain semeiosis as a formal causation process of transmitting a form, or idea, from the dynamic object to the final interpretant. Life is not an emergent phenomenon waiting for explanation, but a fundamental component of every genuine semeiotic relation.

Peirce on Causality and Causation

Before developing a Peircean semeiotic approach to causation and its relevance to biosemiotics, we must discuss Peirce's conception of cause. Though Peirce never explicitly formulated a theory of causation, there are a number of places (especially in his so-called cosmological papers) where he severely criticized the principle of *causality*. Moreover, in 1902, within the general context of a discussion of natural classes, Peirce gave a highly perceptive analysis of the problem of teleology, in which he exposed some remarkable insights regarding *causation*.

Peirce's Critique of the Principle of Causality

Peirce's sixth and seventh lecture of his Cambridge series on *Reasoning and the Logic of Things* (RLT 1898) are particularly important to the understanding of his critique of the principle of causality. In these papers, Peirce's concern is not an elucidation of causation as such, but rather a clarification of the concept of cause within the context of scientific, causal explanation. The sixth lecture—"Causation and Force"—starts with the following important observation:

> Those who make causality one of the original uralt elements in the universe or one of the fundamental categories of thought,—of whom you will find that I am not one,—have one very awkward fact to explain away. It is that men's conceptions of a cause are in different stages of scientific culture entirely different and inconsistent. The great principle of causation which, we are told, it is absolutely impossible not to believe, has been one proposition at one period in history and an entirely disparate one [at] another and is still a third one for the modern physicist. The only thing about it which has stood [...] is the name of it. (RLT 197, 1898)

Peirce observed that the history of the concept of cause reveals a discrepancy between the constancy in the use of terminology and a gradually increasing equivocal character in the conception itself. At least three different and mutually incompatible meanings may be discerned: (a) the Aristotelian conception (AC), (b) 'the modern physicist's conception' (MPC), and (c) 'the currently accepted view' (CAV). (RLT 197-202, 1898).

Peirce approached the problem from three different angles: (1) a logical analysis of the different concepts of cause, (2) an analysis of scientific knowledge of natural processes, and (3) an analysis of mental processes. Each of these will be discussed separately.

Three Concepts of Cause

The concept of cause was first explicitly formulated by Aristotle. According to Peirce, the original idea of an efficient cause is that of an agent, more or less like a man. It is prior to the effect, in the sense of having come into being before the latter, but it is not transformed into the effect. In this sense, it may happen that an event is a cause of a subsequent event; seldom, however, is it the principal cause. Far less are events the only causes (CP 6.600, 1893).

Moreover,

> It is generally held that the word cause has simply been narrowed to that one of the four Aristotelian causes which was named from the circumstance that it alone produces an effect. But this notion that our conception of cause is that of the Aristotelian efficient cause will hardly bear examination. The efficient cause was in the first place generally a thing not an event, then something which need not do anything; its mere existence might be sufficient. Neither did the effect always necessarily follow. True when it did follow it was said to be compelled. But it was not necessary in our modern sense. That is, it was not invariable. (RLT 198, 1898)

Thus, according to Peirce, Aristotle considered an efficient cause to be (i) *an agent who compelled another agent or a thing* to behave in a certain way, (ii) *prior* to its effects, (iii) *not transformed* into its effects, and (iv) *not necessitating* its effects (in the sense of being a concurrence of antecedents to which a given phenomenon is invariably and unconditionally consequent).

AC differs in at least two respects from "our conception of cause" (CAV and MPC): (1) Aristotelian efficient causes are usually things or substances rather than events, and (2) there is no necessary (invariable) relation between certain types of causes and certain types of effects.

According to CAV, a cause is "an instantaneous state of things perfectly determinative of every subsequent state." (CP 6.600, 1893) More precisely, the commonly accepted view of 'the great principle of causation'—which had its origin in Stoic philosophy (EP I: 299, 1892)—involves three propositions:

1. The state of things at any one instant is completely and exactly determined by the state of things at one other instant.
2. The cause or determining state of things *precedes* the effect or determined state of things in time.
3. No fact determines a fact *preceding* it in time in the same sense in which it determines a fact *following* it in time (RLT 198–199, 1898).

MPC, on the other hand, concerns those phenomena that are governed by the Law of the Conservation of Energy, in which the future determines the past in precisely the same way in which the past determines the future. MPC involves the following three propositions:

1. The state of things at any one instant is completely and exactly determined by the state at *two* other instants.[1]
2. Cause and effect are *simultaneous*.
3. The positions at the two later instants determine the position at the earliest instant *in precisely the same way* in which the two positions at the two earlier instants determine the position at the latest instant (RLT 198–202, 1898).

Thus, according to Peirce, there is a flat contradiction between the basic propositions of CAV and those of MPC. CAV and MPC are therefore irreconcilable concepts of cause. Moreover, both are incompatible with AC.

[1] "According to the dominant mechanical philosophy, nothing is real in the physical universe except particles of matter with their masses, their relative positions in space at different instants of time, and the immutable laws of the relations of those three elements of space, time, and matter. Accordingly, at any one instant all that is real is the masses and their positions, together with the laws of their motion. But according to Newton's second law of motion the positions of the masses at any one instant is not determined by their positions at any other single instant even with the aid of the laws. On the contrary, that which is determined is an acceleration [the effect is the acceleration]. Now an acceleration is the relation of the position at one instant not to the position at another instant, but to the positions at a second and a third instant" (RLT 199, 1898). For further explanation, see RLT 199–201.

Analysis of Natural Processes

In his "The Doctrine of Necessity Examined" (1892) Peirce offered a complex, multi-layered argument against the 'doctrine of necessity,' which was commonly supposed to be a postulate of scientific reasoning. According to Peirce, the doctrine of necessity is the idea that "the state of things at any time, together with certain immutable laws, completely determines the state of things at every other time" (EP I: 299, 1892).

One of Peirce's main arguments against this postulate was that the variety and increasing complexity of nature cannot be explained solely on the basis of "the rule of mechanical necessity." The operation of mechanical law cannot create diversity where there was no diversity before. Under given circumstances, "mechanical law describes *one* determinate result" (CP 1.161. 1897). Since "variety can spring only from spontaneity," the laws of nature must be probabilistic rather than deterministic:

> By thus admitting pure spontaneity or life as a character of the universe, acting always and everywhere though restrained within narrow bounds by law, producing infinitesimal departures from law continually, and great ones with infinite frequency, I account for all the variety and diversity of the universe, in the only sense in which the really sui generis and new can be said to be accounted for. (EP I: 308. 1892)

Peirce insists that the doctrine of necessity is incompatible with the simple fact that there is growth wherever we look: "Now, the essential of growth is that it takes place in one determinate direction, which is *not* reversed. Boys grow into men, but not men into boys. It is thus an immediate corollary from the doctrine of the conservation of energy that growth is not an effect of force alone" (CP 6.555, 1887).

Since each event is characterized by an aspect of irreducible novelty or objective chance, natural processes are always irreversible. And because natural processes consist of causally linked events, cause-effect relationships too must *be irreversible*. Thus Peirce accepts the third proposition of CAV. Further on we will see that (on the basis of an analysis of mental phenomena) Peirce also accepts CAV's second proposition: causes precede their effects in time.

On the other hand, Peirce rejects all three propositions of MPC on account of the irreversibility of all natural processes. This irreversibility extends not only to processes that are inexplicable by the Law of the Conservation of Energy (such as birth, growth, life, conduction of heat, combustion) but even to physical processes that appear to be reversible. Though so-called mechanical processes approach the laws of mechanics to a degree, they never do so perfectly.

Thus, *Peirce held that our scientific study of natural processes forces us to accept the view that (i) cause-effect relationships are irreversible, (ii) causes only partially determine their effects (there is always an aspect of indeterminateness or chance involved), (iii) cause-effect relationships are determined by probabilistic laws.*

Analysis of Mental Processes

According to Peirce, causation, as distinct from the action of force, is not only "a real, and fundamental, and vital element [...] in the outer world," but it is also at work in the inner world. (RLT 220, 1898). Indeed, "the very conception of causality has its origin in our tendency to seek relations in nature analogous to intellectual relations" (MS 963, 2, c.1893). Given Peirce's attention to the original meanings of concepts[2], and given his view that causality is an anthropomorphic conception, he may have held the view that the analysis of mental processes is the very best way to study causality. Whether or not he actually did so, he surely accepted all the conclusions regarding causality of his analysis of mental processes.

For Peirce, there is not the slightest evidence that the doctrine of necessity holds in the domain of the mental. That is to say, "from the state of feeling of any instant, there is no reason to suppose the states of feeling at all other instants are thus exactly calculable" (EP I: 309, 1892). Ideas seem to suggest other ideas, rather than to necessitate them:

> ... no mental action seems to be necessary or invariable in its character. In whatever manner the mind has reacted under a given sensation, in that manner it is the more likely to react again; were this, however, an absolute necessity, habits would become wooden and ineradicable, and no room being left for the formation of new habits, intellectual life would come to a speedy close. Thus, the uncertainty of the mental law is no mere defect of it, but is on the contrary of its essence. The truth is that mind is not subject to "law," in the same rigid sense that matter is. It only experiences gentle forces which merely render it more likely to act in a given way than it otherwise would be. There always remains a certain amount of arbitrary spontaneity in its action, without which it would be dead. (EP I: 329, 1893)

Thus, Peirce concludes that the attempt to explain away the variety and apparent arbitrariness of mental action in favor of absolute determinism, does not agree with the observed facts (EP I: 329-30). Moreover, none of the three propositions of MPC applies to the realm of our mental experience. Here it is appropriate to quote Peirce at length:

> ... when from the world of physical force we turn to the psychical world all is entirely different. Here we find no evident trace of any state of mind depending in opposite ways upon two previous states of mind. Every state of mind acting under an overruling association produces another state of mind. Or if different states of mind contribute to producing another, they simply act concurrently, and not in opposite ways, as the two earlier positions of a particle of matter do, in determining a third position. I come down in the morning; and the sight of the newspaper makes me think of the Maine, the breakfast is brought in, and the sight of something I like puts me into a state of cheerful appetite; and so it goes all day long. Moreover, the effect is not simultaneous with the cause. 1 do not think of the explosion of the Maine simultaneously with seeing the newspaper, but after seeing it, though the interval be but a thirtieth of a second. Furthermore, the relations of the present to the past and to the future, instead of being the same, as in the domain of the Law of Energy, are utterly unlike. I remember the past, but I have absolutely no slightest approach to such knowledge of the future. On the other hand 1 have considerable power over the future, but nobody except the Parisian mob imagines that they can change the past by much or by little. Thus all three propositions of the law of causation [MPC] are here fully borne out. (RLT 201-02, 1898)

Thus, *Peirce's analysis of our mental experience confirms his analysis of our scientific experience of the outer world in all respects:* (*i*) *cause-effect relationships are irreversible,* (*ii*) *causes only partially determine their effects, and* (*iii*) *cause-effect*

[2] See Peirce's "The Ethics of Terminology" (EP II, item 19).

relationships are determined by the laws of association, which are in some sense analogous to the probabilistic laws of nature.[3] *Moreover, (iv) on the basis of his analysis of mental phenomena Peirce concluded that causes precede their effects in time.*

Though Peirce preferred to use the term 'force' within the realm of pure physics, and to restrict the term 'cause' to pure psychics, he felt that Occam's razor compels us to *hope* that the one mode of action be somehow reducible to the other:

> We have, then, these two modes of action, the conservative and the causational, the former rather the dominant one in pure physics, the latter dominant in psychics. Our logical impulse, which prompts us to try to understand the universe, and as an essential condition of doing so to bring all its action under a single principle, this impulse, I say, compels us to ***hope*** that it may in some way be shown either that all causational action is conservative at bottom or that all conservative action is causational at bottom. (RLT 237-38, CP. 7.505, 1898)

Peirce's solution is that, both from an ontological and from an evolutionary point of view, causal action is primary, not only in the inner world but also in the outer world.

Causal action is ontologically more fundamental than force, since all natural processes are irreversible, and therefore causal. Thus so-called purely mechanical processes too must be causal. Moreover, according to Peirce's grand cosmological hypothesis, causal action is also primary from an evolutionary point of view. For cosmic evolution is characterized by a creative advance from pure arbitrariness or chaos toward "an absolutely perfect, rational and symmetrical system" in the infinitely distant future, though at any time in this asymptotic approach toward complete order "an element of pure chance survives" (EPI: 297, 1891).

It may seem odd against this background that Peirce refused to reckon cause as a philosophical word. He thought it was too ambiguous or equivocal. Thus, when he did use it, he tried to reduce its equivocal character by the distinction between the terms *cause*, *force*, and *explanation*. *Cause* was to be taken in its original, albeit crude, Aristotelian sense, while he reserved *force* for the context of dynamics, and *explanation* for a more general, logical context (CP 6.600, 1893).

However, despite Peirce's original intention to stick to the Aristotelian conception of efficient cause, some ten years later, in 1902, Peirce developed what might be called a proto-theory of causation, which—though it was Aristotelian in its reliance on the mutual interdependence of final and efficient causation—deviated in some significant respects from Aristotle's conception. Undoubtedly the most important difference is that it was closer to Peirce's categoreal scheme, which involves an event ontology, than to Aristotle's categoreal scheme, which involves a substance ontology. In the next subsection we will briefly explain Peirce's 1902 conception of causation.

[3] According to Peirce, association is "a habit or disposition of mind in consequence of which an idea of on description is likely to bring into comparative vividness of consciousness an idea of another description" (RLT 232, 1898).

Peirce's Conception of Causation

In his 1902 paper "On Science and Natural Classes" (EP II, item 9) and in some closely related papers[4], Peirce developed the highly original view that each act of causation involves an efficient component, a final component, and a chance component. The efficient aspect of causation is that each event or fact is produced by a previous event or fact (the efficient cause). The teleological aspect is that each event is part of a chain of events with a definite tendency. The chance component is that each event has some aspect that is determined neither by the efficient nor by the final cause.

According to Peirce, *final causes* are general types that tend to realize themselves by (teleologically) determining processes of *efficient causation*. Final causes are basically habits: they ('habitually') direct processes toward an end state. Like human habits, habits of nature (laws of nature) too are final causes because they display tendencies toward an end state. Moreover, habits are not static 'entities' for they may evolve in the course of time. Peirce called the possible evolution of final causes "developmental teleology."

Thus final causes are not future events, but general possibilities. The end state of the process to which the act of causation belongs can be reached in different ways. Moreover, the process involved is irreversible.

The idea that *efficient causation* can only be understood within the context of final causation is absolutely central to Peirce's conception of causation. According to Peirce:

> Efficient causation [...] is a compulsion determined by the particular condition of things, and is a compulsion acting to make that situation *begin* to change in a perfectly determinate way; and what the general character of the result may be in no way concerns the efficient causation. (CP 1.212, 1902; italics ours)

Thus, efficient causation, considered apart from its final causational component, is not directed toward an end in any way; it is *blind compulsion*. We understand better now why, contrary to the mechanical conception according to which causes are the inactive nodes in a law-like implication chain, Peircean efficient causes are the *active initiators of a change*. Thus, 'A is the cause of B' means that B results partly from some activity or influence originating in A. Efficient causation thus considered is a dyadic relation between two concrete individual events. Final causation, on the other hand, is a triadic relation between the general final cause, the concrete efficient cause, and its concrete effect. The *general* final cause (C') determines or mediates the production of the *individual* effect (B) by the *individual* efficient cause (A). The produced effect (B) is in its turn a novel efficient cause, which functions as a *means* for the attainment of the *end* (C). Schematically this may be represented as follows:

[4] Peircean processes are creative in a triple sense: (1) each event involved in the process contains an element of irreducible novelty; (2) the end state of a process can be reached in different ways; whenever one way or line of causation be blocked, it may originate new lines; and (3) the end state toward which a process tends, may evolve spontaneously.

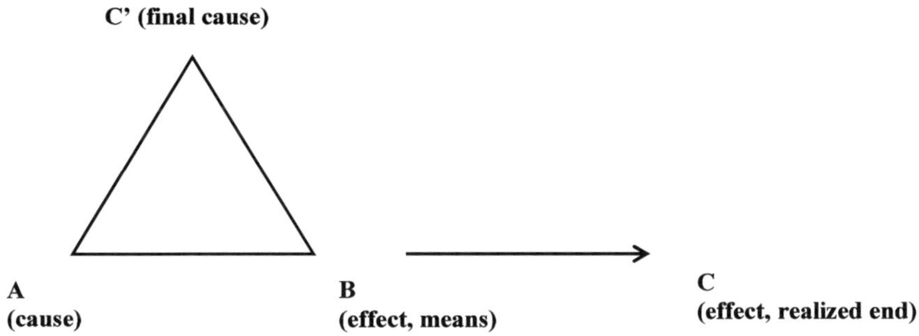

Thus, final causation and efficient causation are just two aspects of one and the same process, which involves a triadic relationship between A, B and C'. At the beginning of the causal process, the *concrete* realized end C does not yet exist. It therefore cannot in any way influence the causation of B by A. Final causation does not entail backward causation. Thus, taking an often-quoted example given by Uexküll (1934), the individual event of a tick jumping from the bush after a long period of inactivity and trying to reach the source of butyric acid it perceives in the environment in order to get the food it needs to survive (C) cannot be fully understood by just referring to the relationship between the high concentration of butyric acid in its surroundings (A) and the movements of its legs as it jumps (B), while leaving aside the instinctual habit of feeding itself that governs its actions (C'). Its jump from the bush wasn't just an accident; it was really driven by purpose: to feed itself and survive by sucking the blood of a mammal. The purpose of the tick is a conditional disposition, or habit, to act accordingly, which could have been realized in countless different events.

Moreover, according to Peirce, every event (as part of a process) is characterized not only by an aspect of final causation and an aspect of efficient causation, but also by an aspect of *objective chance*. Each process involves an aspect of irreducible novelty or objective chance at every stage of the process, which cannot be reduced to efficient or final causation. In the above given diagram, each stage of the causal process (level A-B-C) involves an aspect of objective chance.

Peirce's conception of causation is therefore characterized by a triple interdependence of final causation, efficient causation, and chance. Keeping in mind that in our brief historical survey we distinguished two mutually incompatible conceptions of cause—the *Aristotelian-scholastic Conception*, according to which causes are the active initiators of a change, and the *Scientific Conception*, according to which causes are the inactive nodes in a law-like implication chain—we may conclude now that Peirce's conception of efficient cause forms, in some sense, an ingenious middle way between these two conceptions. On the one hand, Peircean efficient causes are the active initiators of a change (rather than the inactive nodes in a law-like implication chain); 'A is the cause of B' means that B results partly from some activity or influence originating in A. On the other hand, however, Peirce held that "the action of a *cause* is essentially a case of the operation of a law, and implies a law." (MS 318: 20, 1907) In this context, the term 'law' must be understood in the broad sense of habit, final cause, or general disposition; it involves natural laws as well as personal predispositions to act in a certain way.

Finally, Peircean laws are probabilistic rather than deterministic, cause-effect relationships are irreversible, and efficient causes precede their effects. Furthermore, depending on the context, efficient causes are either events or facts. The latter claim, however, needs to be qualified. This will be done in the next subsection.

Causality and Causation: Facts Versus Events

One of Peirce's main points of critique of the principle of causality concerns the problem of the causal *relata*. We decided to discuss it separately, not only because the issue is very important for the development of a Peircean approach to causation, but also because Peirce's observations are at first glance ambiguous, and might therefore easily be misunderstood.

The later Peirce (from 1893 onwards) insisted that the causal relata are *facts* rather than events. He went as far as stating that no one will ever understand the problem of causality unless he sees that both the cause and its effect are *facts*. (MS 647:10, 1910) In a discussion of the nature of the causal relata, Peirce gave the following description of fact:

> A *Fact* […] is so much of the Real Universe as can be represented in a Proposition, and instead of being, like an Occurrence, a *slice* of the Universe, it is rather to be compared to a chemical principle *extracted* therefrom by the power of thought; and though it is or may be Real, yet, in its Real Existence it is inseparably combined with an infinite swarm of circumstances, which make no part of the Fact itself. (MS 647: 00010, 1910)

Thus, we must sharply distinguish events from facts, which are only such abstracted parts of an event as are expressible in a proposition (NEM IV: 252, 1904). In the following passage, a *cause* is described as the minor premise of a deductive argument, of which the major premise is the relevant law of nature and the conclusion the effect. Just like the effect, the cause is a *fact*:

> So far as the conception of cause has any validity, […] the cause and its effect are two facts. Now, Mill seems to have thoughtlessly or nominalistically assumed that a fact is the very objective history of the universe for a short time, in its objective state of existence in itself. But that is not what a fact is. A fact is an abstracted element of that. A fact is so much of the reality as is represented in a single *proposition*. If a proposition is true, that which it represents is a *fact*. If according to a true law of nature as major premise it syllogistically follows from the truth of one proposition that another is true, then that abstracted part of the reality which the former proposition represents is the *cause* of the corresponding element of reality represented by the latter proposition. Thus, the fact that a body is moving over a rough surface is the cause of its coming to rest. It is absurd to say that its color is any part of the cause or of the effect. The color is a part of the reality; but it does not belong to those parts of the reality which constitute the two *facts* in question. (RLT 198, 1898)

Accordingly, causal propositions never refer to events in their full particularity; they refer only to such abstracted elements of events called facts. For instance, the question "What was the cause of the eruption of that mountain?" means "What is the fact from which [,] according to the principles of physics, necessarily resulted the *fact* that the mountain suddenly burst?" (MS 478:155-56, 1903)

By thus insisting that the causal relata are *facts*, Peirce makes clear that the general context of his discussion of the causal relata is *epistemological* rather than ontological. For, since a real fact is defined as the ground of a true proposition, there are no facts in and by themselves, independent of propositions. The context of facts is inherently epistemological.

Contrary to causes and effects, which are *facts* (abstracted elements of experience) reality is purely a matter of *events*:

> What is reality? Perhaps there isn't any such thing at all. As I have repeatedly insisted, it is but a retroduction, a working hypothesis [...]. But if there is any reality, then, so far as there is any reality, what that reality consists in is this: that there is in the being of things something which corresponds to the process of reasoning, that the world lives, and **moves**, and **HAS ITS BEING**, in [a] logic of events. (RLT 161, 1898)

Thus, while cause and effect are abstracted elements of concrete events, *events* rather than facts constitute 'the basic furniture of the world.' But contrary to the received view, Peircean events are neither changes in substances, nor do they presuppose the existence of substances. Instead, by saying that existing things are 'laws' unifying series of events, Peirce suggested that events are ontologically prior to substances;

> ... individual existence, in the strictest sense, only belongs to a single event which happens when and where it does and has no other being. For though we speak, for example, of Philip of Macedon as an individual, yet 'Philip drunk' and Philip sober' were different. The 'existing' thing is only individual in the sense of being a continuous law regulating and unifying events of a series of instants. (MS 478:47-48, 1903)

Hence, it would appear that Peirce's conception of reality involves an event ontology in the strictest sense of the word. Concrete things and concrete persons owe their genetic identity to some 'law' or final cause, which regulates and unifies a series of events. Peirce in the end dismisses the idea that an event is an adjectival, an abstracted element of something more concrete. *The most concrete or most determinate level of reality is not a substance or an enduring individual, but a momentary state or event in an ongoing process.*

Whereas events are continuous (inasmuch as they do not have a definite beginning and a definite end (EP I: 314-15, 1892), only those abstracted aspects of them called facts are trivially discrete. For, while events are temporal particulars, facts are neither temporal nor particular. Only facts, therefore, can be expressed in a proposition, and only facts can be components of a deductive argument. Events in their full particularity never can.

Though Peirce nowhere explicitly made the distinction between causality and causation, he clearly did so implicitly. Causality involves a relationship between facts, but causation is purely a matter of events. Talking about the relationship between discrete causal facts implies that one abstracts from a continuous process of causation. Semeiotically, this abstraction involves the creation of a diagram embodying the abstracted relations, or facts, so as to allow deductive reasoning, an intellectual work that Peirce calls "collateral experience," and without which no sign can convey information about its object. The fact, then, is a proposition considered as a first, or its syntax (CP 2.230).

Given that, according to Peirce, causation is a productive event that it part of a teleological process, we will now pursue the problem of the relationship between events and processes.

Events and Processes

In the previous subsection we described Peircean events as definite temporal units, or cross sections of some actual processes. We also said that the most concrete level of reality is a momentary state or *event* in an ongoing *process*. This raises the question, "How are Peircean events related to Peircean processes?" In order to answer this question, we must first clarify what Peirce meant by *event* and by *process*.

1. Though Peirce would not deny the reality of macro-events, such as witnessing a concert, 'Peircean events,' as we defined them, are micro-events, that is to say, minimal temporal 'slices' of an ongoing process. If our interpretation so far is correct, *Peircean events share the following characteristics: (i) temporal extension, (ii) absence of definite temporal limits, and (iii) an infinitesimal duration.* Moreover, *(iv) they are ontologically prior to substances.*
2. Since Peirce never explicitly discussed the concept of process, the best we can do is to consider what he said about natural (that is to say, irreversible, teleological, evolutionary, etc.) processes.

   In the previous section, it was shown that Peirce held the view that each process consists of a continuous, causal sequence of events, which is regulated by some final cause by virtue of which there is order in the world. This view may be clarified by making use of Dorothy Emmet's definition of process:

   > ... processes are changes with an internal order which distinguishes what is happening within the process from forces acting on it from outside. [...] [They have] a direction, though not necessarily a *terminus ad quem*. (Emmet 1992, p. 35)

   Indeed, Peircean processes too are characterized by an internal order and by directionality. In fact, the principle of the internal order of a Peircean process *is* its directionality, which in turn is determined by its final cause. The final cause sets the boundary conditions of the change toward an end state, which itself is, within certain limits, independent of any "forces acting on it from outside."

   Thus, we propose to give the following characterization of a "Peircean process:" *It is a continuous sequence of events that derives its unity or internal order (distinguishing it from other processes) from a final cause, which directs the sequence to some end state which itself may evolve.*
3. We are ready now to have a closer look *at the relationship between processes and events*. Because Peircean processes and Peircean events are intimately related inasmuch as each event is part of a process, while each process consists of a continuous sequence of events, processes and events
   differ mainly in respect to *complexity, teleology, and coherence*.

Consider the example of someone who, while driving a car from A to B, crashes the car before reaching B. While driving a car from A to B is a good example of a process, the several phases of the change involved count as an event. Driving a car is a complex action, while crashing a car is (comparatively) simple. Hence, it would appear that, *while processes are characterized by complexity, events are simple*.

Moreover, while driving a car from A to B is a teleological process, the crashing of the car (*qua* crashing) is bereft of any teleology. Thus, it would appear that, contrary to processes, events (*qua* events) are not immediately characterized by any teleology. An event just happens. The teleology only comes in if we consider the event as a part of an ongoing process. Thus, *processes are characterized by teleology, which events lack.*

Closely related to the previous point is that processes, such as driving a car from A to B, are characterized by a certain internal order or coherence. Events, however, being simple in themselves, lack this characteristic.

Thus, Peircean processes differ from events inasmuch as only the former are characterized by complexity, teleology, and coherence.

It is important, however, to realize that the suggested distinction is to some extent arbitrary. It owes its justification primarily to our (physical) inability to observe any teleology during an extremely short time span (events are characterized by an infinitesimal duration). But, as it will be shown further on, though events as such lack any recognizable teleology, as part of a teleological process, they are nevertheless teleologically determined by the final cause of the process. Thus, eventually events are just micro-processes.

A Semeiotic Approach to Causation

The close relationship between causation and signs is easily shown, for signs play a major part in our acquaintance with causes. Smoke is a sign of fire. In medicine, symptoms are signs that point to causes. Nevertheless, the apparently intimate relationship between causation and signs does not by itself imply that the relationship is transparent. It is all the more surprising that, as far as we know, it has never received any notable attention in contemporary discussions of causation. In this section we will not only pay attention to this relationship, but we will make it the cornerstone of our theory of causation.

Some Formal Characteristics of Semeiosis

Peirce used the term 'semeiotic' for the scientific study of semeiosis, or the systematic study of the formal structure of "the general conditions of signs being signs" (CP 1.444, c. 1896). The most important formal characteristic of a sign is that it involves a three-term relationship between a *sign*, its *object*, and its *interpretant*. This relationship is irreducibly triadic, that is to say, it is not reducible to a summation of dyadic relations (CP 2.274, c.1902).

There are dozens of tentative definitions of sign given by Peirce, but we chose one that is more directly related to what we have been discussing so far:

> (…) a sign endeavours to represent, in part at least, an Object, which is therefore in a sense the cause, or determinant, of the sign even if the sign represents its object falsely. But to say that it represents its Object implies that it affects a mind, and so affects it as, in some respect, to determine in that mind something that is mediately due to the Object. That determination of which the immediate cause, or determinant, is the Sign, and of which the mediate cause is the Object may be termed the ***Interpretant***… (CP 6.347 c. 1909)

Thus, a sign is not a thing as such, but *anything* may be a sign by virtue of being triadically related to an object and an interpretant. The interpretant is the *respect* in which sign and object are related; it must be understood, not as an act of interpretation, but as "the sign's proper significate effect" (CP 5.473, 1907). Thus, if clouds are a sign of rain to come, the rain to come is the object of the sign, and the expectancy of rain is the interpretant.

Peirce's semeiotic is closely related to his categoreal scheme. The basic idea behind this scheme is that all phenomena are infected by Firstness, Secondness, and Thirdness. Whatever it may be we are describing, be it theories, arguments, or events, there is always an element that does not refer to anything beyond itself (First), an element that is related to this first element but not to anything else (Second), and an element that relates or mediates between the first and the second element (Third) (e.g. CP 8.328, 1904). In other words: Firstness is a monadic relation, Secondness a dyadic relation, and Thirdness a triadic relation. Signs are paramount examples of genuine triadic relations:

> Now in genuine Thirdness, the first, the second, and the third are all three of the nature of thirds, or thought, while in respect to one another they are first, second, and third. The first is thought in its capacity as mere possibility; that is, mere mind capable of thinking, or a mere vague idea. The second is thought playing the role of a Secondness, or event. That is, it is of the general nature of experience or information. The third is thought in its role as governing Secondness. It brings the information into the mind, or determines the idea and gives it body. It is informing thought, or cognition. **But take away the psychological or accidental human element, and in this genuine Thirdness we see the operation of a sign.** (CP 1.537, 1903, emphasis added)

Moreover, the triadic relation involved in a sign is eminently dynamic. Signs do something:

> All dynamical action, or action of brute force, physical or psychical, either takes place between two subjects (whether they react equally upon each other, or one is the agent and the other patient, entirely or partially) or at any rate is a resultant of such actions between pairs. But by 'semiosis' I mean, on the contrary, an action, or influence, which is, or involves, a cooperation of *three* subjects, such as a sign, its object, and its interpretant, this tri-relative influence not being in any way resolvable into actions between pairs. (CP 5.484, 1906)

Peirce defines semeiosis in an explicitly causal terminology and denies that the triadic action constituting semeiosis could be reduced to dyadic cause-effect influences. Semeiosis is irreducibly triadic. Thus, if someone beats his fist on the Table (sign) to express his anger (object), the expression of fear in a witness's face (interpretant) is triadically produced, and is therefore a semeiotic effect.

Semeiosis Provides the Formal Structure of Causation

In the first place, the hypothesis that semeiosis provides the formal structure of causation is supported by *Peirce's hierarchical classification of the sciences* (c.1900), which arranges the abstract sciences in a ladder such that each science derives its fundamental principles from the findings of the more abstract science that occupies the rung above. In Peirce's own words: "I would classify the sciences upon the general principle set forth by August Comte, that is, in the order of abstractness of their objects, so that each science may largely[5] rest for its principles upon those above it in the scale while drawing its data in part from those below it" (RLT 114, 1898).

Thus, according to Peirce's Cambridge lectures of 1898, *Reasoning and the Logic of Things*,[6] there is a hierarchical relationship between Logic (which he defined as Semeiotic) and Metaphysics, such that Logic is one scale higher on the ladder of abstractions. Hence metaphysics must draw its principles from semeiotic: it "must take as the guide of its every step the theory of logic [semeiotic]" (RLT 116; also 123, 1898).

The idea that semeiotic is a more general discipline than metaphysics entails that metaphysical concepts presuppose semeiotic principles, while semeiotic concepts do not presuppose metaphysical principles. Thus, the concept of causation (which is a metaphysical concept) presupposes semeiotic principles.

Our second argument concerns the relationship between semeiosis, Thirdness, and causation. We already established that the most important formal characteristic of semeiosis is its irreducible triadicity. There are, however, some indications that Peirce expanded the view that semeiosis involves Thirdness into the much stronger thesis that semeiosis is the *genuine* form of Thirdness:

---

[5] In the explanation of his final scheme of classification (1904), Peirce had changed the word 'largely' to 'exclusively.' "This classification [...] is to be regarded as simply Comte's classification, corrected. That is to say, the endeavor has been so to arrange the scheme that each science ought to make appeal, for its general principles, exclusively to the sciences placed above it, while for instances and special facts, it will find the sciences below it more serviceable" (MS L 107, 1904).

[6] Peirce's Classification of the Sciences according to his 1898 Cambridge Lectures is as follows;

A. THEORETICAL SCIENCE
Mathematics
Philosophy
a. Logic
b. Metaphysics
3. The Special Sciences
a. Psychical
b. Physical
B. PRACTICAL SCIENCE

For Peirce's final scheme, see his "A Brief Intellectual Autobiography" (MS L 107, 1904).

> In its genuine form, Thirdness is the triadic relation existing between a sign, its object, and the interpreting thought, itself a sign, considered as constituting the mode of being a sign. A sign mediates between the *interpretant* sign and its object. (LW 31, 1904; cf. CP 1.537, 1903)

Thus, it would appear that each *genuine* triadic relation is basically a relation between a sign, an object, and an interpretant. Thus semeiosis is not only irreducibly triadic, but each triadic relation in the strongest sense is also formally semeiotic.

Peirce makes a distinction between genuine and (two forms of) degenerate triadic relations.[7] Only the former cannot be reduced to a summation of Seconds. Thus "[t]he genuine third category is where there are three objects each having a character which essentially supposes the other two." (MS 304: 41, 1903) Thus, while 'A gives B to C' is an example of genuine Thirdness, 'A putting down B, which is picked up by C' is an example of degenerate Thirdness (LW 29-30, 1904).

Now, we know that causation involves a *genuine* triadic relation between an efficient cause, a final cause, and an effect, for each of these terms "essentially supposes the other two." And if causation involves *genuine* Thirdness, and if *genuine* Thirdness is formally semeiotic, then it would appear that causation is formally semeiotic. In other words: *semeiosis provides the format structure of causation.*

Though these two arguments do not unconditionally prove our hypothesis that causation presupposes semeiosis, rather than the converse, they make clear that our hypothesis fits Peirce's theoretical framework.

A semeiotic Approach to Causation

It may be helpful to point out again that the semeiotic approach to causation that will be defended, is Peircean in the sense that it is primarily based upon Peirce's semeiotic and upon his conception of causation. In order to see what new light Peirce's semeiotic may throw on the concept of causation, it is helpful to recall some formal characteristics of Peirce's conception of causation. This will allow us to elaborate an anatomy of (Peircean) events, and finally to discuss some more formal characteristics of signs.

Some Formal Characteristics of Causation

So far it was established (see A Semeiotic Approach to Causation) that a coherent theory of the relationship between events and processes is one of the primary requirements for any theory of causation. Peircean *events* have been characterized by (i) temporal extension, (ii) the absence of definite temporal limits, (iii) an infinitesimal duration, and (iv) ontological priority in respect of substances. Moreover, a Peircean *process* was defined as "a continuous sequence of events that derives its

[7] Peirce borrowed the terms 'genuine' and 'degenerate' from the geometry of plane curves, where they refer to the irreducibility or reducibility of a figure to simpler figures. (MS 304: 35, 1903). For a thorough analysis of "Genuineness and Degeneracy in Peirce's Categories," see Kruse 1991.

unity or internal order (distinguishing it from other processes) from a final cause, which directs the sequence to some end state, which itself may evolve" (3.4). Peircean processes differ from Peircean events inasmuch as only the former are characterized by complexity, teleology, and coherence.

Moreover, each productive event or act of causation was seen to involve a dynamic relation between a general final cause, a concrete efficient cause, and its concrete effect. Thus, each act of causation involves two aspects that must be sharply distinguished: the fact *that* something is caused, and *what* it is that is caused. While the former is explained by the efficient cause, the latter is explained by the final cause. Thus, the final cause does not determine *that* an event occurs; it only determines *what* type of event it will be, if the event occurs. In other words: it does not determine the action qua action, but it determines the general type or *form* of the action.

In Peirce's Conception of Causation, we gave the schematic representation of Peirce's conception of causation. In that schema, the sequence A → B → C represents a *continuous* sequence of events. Thus, between any event A, apparently causing another event B, there is an innumerable series of events affecting B, which may therefore as well be called the cause of B. Describing A as *the* cause of B abstracts from the aspect of continuity involved in causation; in other words, one changes his perspective from causation toward causality (and thereby from events to facts).

From the dynamic point of view, the scheme represents the *production* of a sequence of *events*, in which A-B (as mediated by C') must be thought of as *one* concrete, productive event. This event is part of the continuous sequence A→ B→ C, which derives its unity teleologically from one and the same final cause (C'). It is only after the *one* concrete, productive event A-B has occurred, that abstraction yields A as the cause and B as its effect; for it is only after an effect has been produced, that one can speak of 'causes' and 'effects.'

In view of this brief survey of Peirce's conception of causation, we may now provide an anatomy of Peircean events.

The Anatomy of Events

According to Peirce, each event has an aspect of Firstness, Secondness, and Thirdness. The aspect of Firstness concerns the particular qualities—which as such are mere possibles—that are involved in the event. Secondness concerns the aspect of action *hic et nunc*, independently of any uniformity or law. Thirdness concerns the aspect of habit, or continuity of form; Thirdness "determines the suchness of that which may come into existence, when it does come into existence" (EP II: 269, 1903). Thus, each event owes its definiteness or *form* to Thirdness, which is of the nature of habit or final cause.

Though Firstness, Secondness, and Thirdness seem to reflect the traditional modal categories of, successively, *possibility, actuality*, and *necessity*, the necessity of Thirdness is not absolute, but it is *conditional necessity* of the form*: if A, there will be a tendency toward B* (e.g. EP II: 271, 1903).

Peirce held the view that modal distinctions correspond to temporal ones; thus *possibility, actuality*, and *conditional necessity* (or potentiality) are the modes of, respectively, present, past, and future, (CP 5.459, 1905). Accordingly, in each Peircean event three distinct temporal phases may be recognized: (1) reception of causal influence from the past, (2) (present) self-determining activity, and (3) influence upon subsequent events.

Thus, every *present event* is conditioned by *the past* and conditions the future. Though the present event conforms to the past, it nevertheless contains an element of irreducible novelty, for causal conditioning always leaves a range of open possibilities, however small it may be sometimes. Every present event is self-actualizing; it is an act of decision that selects and actualizes one possibility among various possibilities. Which of the causally possible events occurs is decided in the event itself, rather than beforehand.

Though future events are not necessitated in their full particularity, certain abstract, more or less *general*, features are determined in advance. Hence, given a certain event, it is necessary that a certain *kind* of event will occur, while it is not necessary that this rather than some other event will occur. For, *it belongs to the nature of an event, not only* ***that*** *it creates a new event (Secondness), but that it creates a new event* ***of a certain type or natural class*** *(Thirdness)*. To what type a new event will belong, is determined by a (cluster of) previous event(s), and by the relevant law or final cause. For instance, an egg produced by a turtle will always be a turtle egg, partly because the event emerges from biological processes that occur in the mother turtle, and partly because of the 'law' of the development of a turtle egg.

Thus, each event:

1. is *causa sui*, that is, it involves a decision selecting one possibility among various possibilities,
2. is conditioned by a cluster of previous events, and by the relevant final cause,
3. conditions and limits the range of possibilities for subsequent events.

This view, which we consider to be totally compatible to Peirce's approach, entails that the relevant final cause is the *perspective* by virtue of which the efficient cause and its effect are related. This claim needs some clarification. Consider the macro-event of 20 million people watching a live speech by the new president of the United States on some controversial political subject. No doubt, there will be certain recognizable patterns in the reactions of the audience. Democrats will tend to react differently from Republicans. But on a closer view it will appear that the speech will yield 20 million different reactions, for each observer watches the speech event in the light of his own expectations, predispositions, feelings, etc. In other words, it is by virtue of the fact that each spectator observes the event from his own *perspective*, that the very same event yields such different reactions.

Such perspectives are basically habits, which help determine the future reactions of a person to specific events. For instance, if a person's perspective involves an abhorrence of discriminatory rhetoric, each utterance that smacks of discrimination may cause him to be angry. *Thus each perspective reflects a conditional necessity, which has the structure of a material implication.* The probable or potential reaction of, say, John to an event of the type X may therefore be described as follows: If an event of the sort X occurs, then John will have a strong tendency to react in such and such a way (way Y).

Schematically, the occurrence of John's fury may be represented as follows:

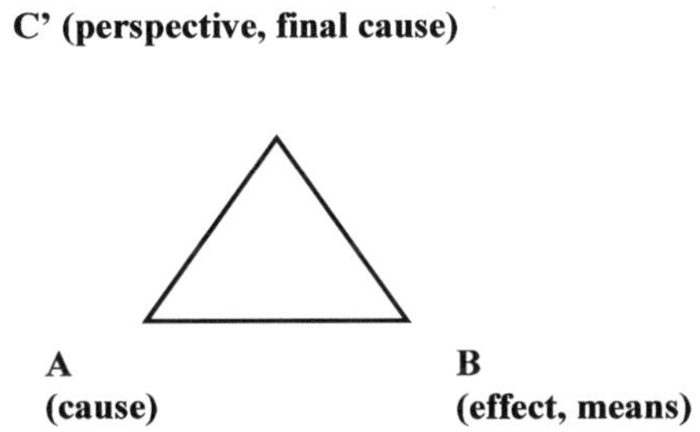

While A is John's perception of discriminatory talk, B is John's angry reaction, and C is John's abhorrence of discrimination, A-B-C is the one productive event of John's becoming angry.

It would appear that each new event assimilates the past event from a definite perspective. The past event is related to the present event as an efficient cause to its effect. This relationship is mediated by the final cause, which is the perspective.

Whereas each present event assimilates the past event from one definite perspective, each past event may be assimilated from innumerable different perspectives. For instance, the just-mentioned speech-event of the new president was perceived from 20 million different perspectives. It would seem that we have an important clue here that may be useful to our semeiotic approach to causation. For the very same speech-event functioned as 20 million different signs, because there were 20 million different interpretants. Hence, it would appear that, from a semeiotic point of view, *the perspective from which the previous event is assimilated is precisely the interpretant.*

If the latter hypothesis is correct, and if causation does indeed presuppose semeiosis, then, in order to better understand causation, we must now try to obtain an even more precise idea of what semeiosis involves. We will see that it involves the transmission of forms from causes unto effects. Because the expression "transmission of forms" may reek of Aristotelianism, it is of the utmost importance to spell out what it refers to.

Signs as Media for the Transmission of Forms

Peirce translated the concept of 'form' in terms of his own framework. This view is most clearly expressed in the following passage in which he explains that forms must be seen as general rules:

> The being of a Form consists in the truth of a conditional proposition. Under given circumstances, something would be true.[8] (EP II: 544, n. 22, 1906)

This description of form, though referring to the Aristotelian terminology, cannot be understood without the *pragmatic context* that is basic to Peirce's thought. Forms are not related to properties that are inherent to things, because our belief in properties is really our belief in the conditional fact that, given certain interactions, such and such effect would produce itself. In other words, our very reference to 'properties' of things already presupposes a possible immanent relation, which is the hallmark of all interaction.

But in order to understand the relationship between forms as general rules and causation, two intermediary steps are required. It has been repeatedly pointed out that the logical structure of thought, rules and generals is the same; it is the structure that is expressed in the material implication: if *p*, then *q*. This is the logical equivalent of what Peirce called Thirdness. If such is the case, and if forms are equivalent to rules, clearly the transmission of forms involves the transmission of Thirdness. Thus, a different perspective on the same past event leads to a different type of recollection.

Thus, *within Peirce's framework the burning question of causation must be understood in terms of the question of how Thirdness may be transmitted. And in this respect there is no doubt that the answer must be sought in Peirce's conception of semeiosis*. For indeed, the preceding quote regarding forms, which are to be understood as rules, is part of a larger context in which Peirce explains what he meant by semeiosis. (EP II: 544, n.22, 1906).

The premise of Peirce's argument is that the object of a sign is always a rule, and inasmuch as a rule has the same logical structure as a thought, the object of the sign may be considered to be a thought. The sign represents this rule or thought, and by doing so, yields an interpretant which the object itself would have yielded, were it not that the relation between the object and the interpretant must be mediated through the sign. When smoke acts as a sign of fire, it represents the rule "if you get close to fire, you will get burnt," and by so representing that rule it yields the interpretant "et away unless you want to get burnt," which is exactly the rule which fire by itself would yield without the smoke if that were possible. *The function of the sign therefore is to mediate in the transmission of a rule from the object to the interpretant.*

The only problem remaining is to show how the transmission of forms within semeiosis is related to the transmission of forms within causation. At the outset an important point that—though obvious—is often overlooked, is that the question of causation always enters only after a particular effect has struck our attention. The obviousness of this is echoed in the principle that "every event has a cause." Put in another way, this amounts to saying that every effect is the 'sign' of some event that is considered to be its cause. Thus causation always implicitly involves semeiosis.

According to the received view, causation requires two and only two elements: a cause and its effect. Semeiosis on the other hand, inherently involves three aspects:

---

[8] See page 92 for a full quotation on Peirce's view of the being of a form.

the sign, its object and its interpretant. It should be clear from the given analysis *that the effect is the sign, while the cause is the object. The paramount issue then is*, "What sense can be made of the interpretant within the context of causation?" The answer should by now be rather simple and straightforward. An effect can be the sign of a cause only from a certain perspective. It is only because the death of a person is important to us that we look after its cause. However, it is not only the effect that is selected by our perspective: the very same event (effect) may be said to have different causes, depending on one's perspective. Thus, for example, while the cause of the death may be the deprivation of oxygen for a pathologist, it may be the act of a criminal who deliberately deprived the victim of oxygen for a policeman, whereas it may be some traumatic experience of the criminal in the view of a psychiatrist. Speaking of cause and effect without at least implicitly indicating the perspective from which an effect is produced by its cause is as vacuous as speaking of natural classes without specifying the perspective from which certain properties are considered important. This perspective is precisely where interpretant and final causation coincide.

Although a reference to final causes has become a virtual taboo in modern philosophy, it should be emphasized again that, within the perspective of Peirce, final causes are not events in the future bringing about events happening now; final causation merely refers to that sort of general rule in virtue of which events that follow upon each other may be thought of in relation to one another. Thus, though an apple falling to the earth undeniably involves the mass of the apple, the mass of the earth, and the distance between those masses, there can be causation only inasmuch as the behavior of the earth and the apple in respect to each other happen within the boundaries set forth by the law of gravity; this law states no less and no more than that, given the masses of the apple and the earth, and given the distance between them, the apple and the earth will behave in such determinate way. This conditional relation constitutes on the causal level the final cause, and on the semeiotic level the interpretant.

Apparently, Peirce's concept of *Form* has roughly the same function as the Aristotelian formal cause: it explains the apparent stability of the world. However, it radically differs in at least one respect from the Aristotelian formal cause. Whereas the formal cause was thought to explain the stability of the world by explaining the structure of *things*, Peircean Forms explain the stability of the world by explaining the dynamic *relations* between events. In other words: *the structure of the Forms themselves is relational, rather than something embodied in a substance.*

If our hypothesis is correct that semeiosis concerns the aspect of the transmission of forms from causes into effects, then, in order to better understand causation, we must now have a more precise look at what goes on in semeiosis.

Signs Conveying the Form-Aspect of Causation

We established that Peirce held the view that signs convey significate forms, which are of the general nature of thought. We established, furthermore, that a productive event assimilates a past event from a definite perspective, which is the final cause, and that, on the semeiotic level, this perspective or final cause is the interpretant.

Peirce defined the *interpretant* as "all that is explicit in the sign itself apart from its context and circumstances of utterance" (CP 5.473, 1907). For instance, the interpretant of the symptoms of measles is the real law-like relationship between measles and its symptoms. The symptoms are there, and so is the interpretant to which they *would* lead a qualified doctor.[9] Thus, the thought is 'there,'even in case there be no thinker of it. Peirce provides an interesting example from Paleontology:

> (...) if, for example, there be a certain fossil fish, certain observations upon which, made by a skilled paleontologist, and taken in connection with chemical analyses of the bones and of the rock in which they were embedded, will one day furnish that paleontologist with the keystone of an argumentative arch upon which he will securely erect a solid proof of a conclusion of great importance, then, in our view, in the true logical sense, that thought has already all the reality it ever will have, although as yet the quarries have not been opened that will enable human minds to perform that reasoning. For the fish is there, and the actual composition of the stone already in fact determines what the chemist and the paleontologists will one day read in them. (...) It is, therefore, true, in the logician's sense of the words, although not in that of the psychologist's, that the thought is already expressed there. (EP II: 455)

Hypothetical thinking may be regarded as that sort of thinking in which we ascribe to a particular event the status of being the sign of an object. And since "reasoning from effect to cause" is an instance of hypothesis (EP I: 194, 1878), it would appear that *causal explanations are abductions or hypotheses in which the effects are conceived as functioning as signs which point to their causes. The relevant' law' or habit is the interpretant by virtue of which the effect is related to its cause.* For instance, the symptoms of a disease point to its cause. The relevant medical law is the interpretant by virtue of which the symptom (sign) is related to its cause (object). Considered from the perspective of causation as a productive event, the law determines that the symptoms caused by the particular disease belong to a certain type or natural class. Measles, for instance, causes red spots rather than tooth aches.

We may now consider the following schematic representation of a productive event in semeiotic terms:

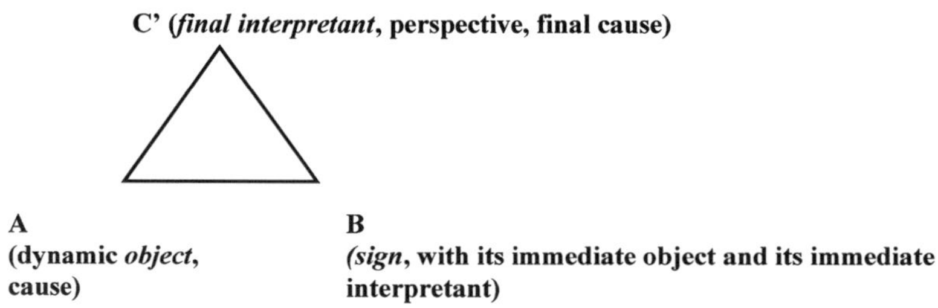

[9] Consider, for example, the following passage: "If a sign has no interpreter, its interpretant is a 'would be,' i.e., is what it would determine in the interpreter if there were one." (EP II; 409, 1907)

Thus, while A stands for measles, B stands for red spots, and C stands for the law that measles causes red spots, A-B-C' represents the structure of the single productive event of the emergence of red spots. Thus the emergence of red spots is a new event, which assimilates the past event—some functional disorder called measles—from a definite perspective. This perspective is the law that measles causes red spots. This law is the *Form*, which is conveyed in the sign. The *Form*, which is really embodied in the object, is embodied in the sign only in a representative sense. Thus, the red spots represent the embodied Form of measles.

If all this seems fairly abstract, a reference to some insights gained from modern psychological research may illustrate the point. Modern scientific research on memory confirms that recollection is a productive event, an activity here and now, rather than some passive effect of a past cause. Moreover, if it is right that in memory we have concrete examples of the interconnectedness of events, then this modern research on memory confirms our hypothesis that each productive event involves an assimilation of a past event from a definite perspective. For recent research on memory has shown that the way in which people remember past events depends upon their intentions and objectives at the moment they try to remember the event (see, for instance, Schaeter 1996, Chap. 2). In terms of our scheme of productive events, this means that the present recollection event is not only determined by the past, remembered event, but also by the present perspective on the past event. While the past event functions as the efficient cause of our recollection experience, the perspective is the final cause. Thus, a different perspective on the same past event leads to a different type of recollection. Apparently, each recollection is an interpretation of the past event in a sort of constructive narrative.

The above-given examples suggest the paradoxical conclusion that causes are both objective and perspectival; not only are causes part of what objectively goes on, but they are also determined by the perspective on what goes on. They are facts rather than events, because they are abstractly derived aspects from the concrete, objective stream of events. As such they are partially determined by concrete reality and partly by the abstractive power of our thought. (Though in the example of the measles, the objective law was emphasized, it is only by virtue of the modern scientific theoretical perspective that the disease can be explained).

We are ready now to understand our guess at the riddle of how forms can be transmitted from a cause into its effect. There is an obvious structural similarity between the assimilation of a past event by a present productive event, and the representation of a past object in a present sign.

This structural similarity can only be explained if we assume semeiotic principles to be really operative in each act of causation. *The in itself baffling transmission of forms in causation is explained in terms of the transmission of forms in semeiosis, simply because the transmission of forms in causation is basically a semeiotic process*. Thus Peirce's contention that "general principles are really operative in nature" (EP II: 183, 1903) is translated into our hypothesis *that 'semeiotic principles are really operative in each productive event.'* [10]

---

[10] This is our interpretation of the meaning of Peirce's "broader conception" of sign. (LW 81, 1908)

The *prediction of future events*, which is, in some sense, the counterpart of causal explanation, also implicitly refers to semeiosis. In this case, the sign stands in relation to its object as an efficient cause to its effect. Thus, clouds are a sign of rain to come (object) because clouds (condensed vapor) may cause rain. The interpretant—clouds perceived as signifying possible rain—reflects the law-like relationship between clouds and rain under certain conditions. Indeed, the interpretant expresses the law-like relationship between cloud-signs and possible rain (object).

Reasoning from an event ('cause') to its possible effect is, just like reasoning from effect to cause, an instance of hypothesis. Thus, it would appear that *predictions of future events are hypotheses in which certain events ('causes') are conceived as functioning as signs that point to their would-be effects. The relevant 'law' is the interpretant by virtue of which the cause is related to its effect.*

So far we have explained the role of signs in single productive events, in causal explanations of past events, and in our predictions of future events. There remains, however, one more problem to be discussed: *the relationship between signs and causal chains*. We established that causes and effects are abstractly derived aspects from a continuous chain of events, and that whatever we call the cause of a certain event depends on a perspective in view of which the continuous sequence is observed. But if our conclusion is correct that effects function as signs pointing to their causes (causal explanations), or that causes function as signs pointing to their would-be effects (predictions), then signs cannot be discrete entities, but must be abstractions of a continuous process of semeiosis. Fortunately, the objection is not serious, for the idea that (Peircean) signs are discrete involves a complete misunderstanding of Peirce's concept of sign for two reasons. First, the objection presumes that signs are a species of entities, while in fact they constitute a certain aspect of events. Furthermore, as Joseph Ransdell has pointed out, atomicity and signs are mutually exclusive concepts:

> … the individual 'links' in the chain [of semeiosis] are always potentially analyzable into innumerably many possible sub-links and sub-sub-links […], since a sign is not a 'logical atom' but only something which it happens to be intellectually profitable to treat as a unit at a certain point in an analysis, though always analyzable in principle into sub-signs if or when the analytic project should require it, and which may itself be a sub-sign within some more exclusive sign. (Ransdell 1986, p. 677)

Thus it is because both semeiosis and causation involve real continuity, that any of the innumerable events of which a causal chain consists, may function as a sign which points to its cause (causal explanation) or to its would-be effect (prediction).

In the next section, we will see whether, and to what extent, our semeiotic approach to causation can handle the main requirements for a theory of causation as put forward in A Semeiotic Approach to Causation.

General Requirements for a Semeiotic Theory of Causation

Two generally accepted requirements for a theory of causation are: it should be coherent, and it should involve a coherent theory of events and processes.

The theory presented meets the first requirement inasmuch as it is based on one, and only one, categoreal framework. Contrary to the received view, which is caught between a substance ontology and a fact ontology, the theory presented is based upon an event ontology in the strictest sense of the word (that is to say, one that is not in any way based upon a substance ontology). The theory meets the second requirement by conceiving the relationship between processes and events as expressed in our definition of a (Peircean) process as "a continuous sequence of events that derives its unity or internal order (distinguishing it from other processes) from a final cause, which directs the sequence to some end state which itself may evolve."

Four generally subsidiary requirements for a theory of causation are: it must explain (1) whether causation involves *continuity*, (2) the relationship between causation and *time*, (3) the relationship between causation and *teleology*, and (4) the *status* of causes and effects as such, and the relationship between causes and effects.

1. *Continuity*. It has already been established that causation is a productive constituent of a *continuous* sequence of events, deriving its unity teleologically from one and the same final cause. It was established, furthermore, that between any event A, apparently causing another event B, there is an indefinite series of events affecting B, which are also involved in causing B. Describing A as *the* cause of B, abstracts from the aspect of continuity involved in causation.

   Peirce made it clear that causation must somehow involve a continuous transmission of forms. Given that such transmission cannot be explained on the basis of a substance ontology, and given that our most concrete experiences concern events rather than substances, it would appear that the explanation of causation requires an event ontology in the strictest sense of the word.

   Such ontology is provided by Peirce's categoreal scheme; within the context of this scheme, the urgent question of how forms be transmitted from causes into effects must be understood in terms of the question of how Thirdness may be transmitted. And since semeiosis provides the formal structure of genuine Thirdness, the solution to this problem must be thought of in terms of Peirce's conception of semeiosis.

   Peircean forms are general rules, which have the logical structure of a material implication: if *p*, then *q*. Thus, unlike Aristotelian formal causes, Peircean forms are *relational structures* rather than something embodied in a substance. It has been shown that the transmission of forms in causation is basically a semeiotic process in which the function of a sign is to mediate in the transmission of a form from the object to the interpretant. Thus, effects may function as signs that point to their causes, and 'causes' may function as signs of future effects, because the transmission of forms from causes to effects is a semeiotic process.
2. *Time*. According to Peirce, the most distinctive character of time consists in its asymmetry. While earlier members of an event-sequence may affect later members, the converse does not hold. Similarly, simultaneity is defined in terms of unaffectibility: two events are contemporaneous if and only if "each is absolutely unaffectible by the other" (EP I: 323, 1892). But since affectability is a causal concept, it would appear that Peirce held the view that time is derivative with

respect to causation. If this is correct, then Peircean events do not happen in time, but they are the condition of time, whatever time may be.

Whether this is a correct interpretation of Peirce's view or not, we accept it as a valid interpretation of time per se; for, if our hypothesis is correct that the present act of causation (that is, the present productive event) is the most determinate level of reality, and that temporal ordering is generated by relating events to a present productive event, both dimensions of time, the past and the future, are abstractions of an ongoing sequence of productive events. Thus causation, rather than time, is ontologically primary.

Our analysis of causation conceived as a productive event implies not only that causes and effects are abstractly separated elements of an ongoing causal process, but also that *causes precede their effects*. For were there no temporal distance between the two, the cause would co-exist with the effect, and the effect of that effect would co-exist with it, and so forth. In such a case, nothing would occur, neither would there be any time.

To be sure, it is often suggested that there are clear examples in which causes and effects are simultaneous, such as a ball pressing on a cushion, or a hand raising a pen. However, we think that Paul Weiss's suggestion is correct, that in those examples the simultaneity of cause and effect is only apparent, and that what we are inclined to call the cause, is nothing but an analytic component of the effect (Weiss 1947/1983, 4). Thus, the movement of the arm is not the cause of the movement of the pen, but (for example) the decision to raise the pen is the cause.

3. Teleology. An important part of the problem of teleology is the question of whether all natural processes can be adequately explained in terms of efficient causality. Here, Peirce forcefully defends the position that all natural processes are to some extent teleological, and can only be explained on the basis of final causes, which are general principles. The empirical fact that all (physical, biological, psychological and sociological) processes show a definite tendency toward a general end state whether it be described as chaotic or as involving a higher level complex order—cannot be explained without referring to certain general principles that somehow direct the process. As far as we know, Peirce's theory of teleology is the only theory that explains this fact. (see Chapter "Peircean Semiotic Indeterminacy and Its Relevance for Biosemiotics"). In Peirce's account, teleology becomes again a key-concept in the theory of causation. For there can be causation only inasmuch as there are boundaries set forth by certain general principles.
4. *The status of causes and effects as such, and the relationship between causes and effects*. It was established that, while *causation* involves a continuous sequence of events, *causality* involves a relationship between facts, which are abstractly separated elements of events. Hence, describing A as the cause of B abstracts from the aspect of continuity involved in causation; in other words, one changes his perspective from causation to causality, and thereby from events to facts. Thus, the status of causes is that they are both objective and perspectival; they are *facts* rather than events, because they are abstractly derived aspects from the

concrete, objective stream of events. As such they are not only determined by concrete reality, but also by the abstractive power of our thought.

It was established that the 'logic of events' is such that, though future events are not necessitated in their full particularity, certain more or less general features are determined in advance. Thus, given a certain event, it is necessary that a certain *type* of event will occur, while it is not necessary that this rather than some other event will occur. Given certain causal conditions (that is, a cluster of previous events and the relevant final cause), there is always only a limited range of *possible* events that can occur. While the common character of all these possible future events (effects) is necessary, the concrete, actual effect is not.

The necessity involved in causation is related to the transmission of Forms in the Peircean sense, which are final causes. They determine to what type or natural class a new event will belong. As such they help explain the relative stability of the world.

The Relevance of Peirce's Theory of Causation to Biosemiotics

Biosemioticians embarrassed by Peirce's teleological approach to causation prefer to skip the whole discussion and use only the excerpts where Peirce deals only with semeiotic terminology, apart from such philosophical musings.[11] Others prefer to learn Peirce's doctrine of signs second-handedly through commentators that have hidden Peirce's ideas about causation and teleology. Both options lead invariably to huge mistakes. For instance, one of the most famous and quoted definitions of the "necessary and sufficient condition for something to be a semiosis"—found in many articles, books and papers on biosemiotics—is:

> A interprets B as representing C.
> In this relational characterization of semiosis, A is the interpreter, B is some object, property, relation, event, or state of affairs, and C is the meaning that A assigns to B. (Posner et al. 1997, p. 4)

In Peirce's terminology of the triadic relation among sign, object and interpretant, when the above authors say that "A interprets B as representing C" we might understand that A is the place of the **interpretant**, B is "some object, property, relation, event, or state of affairs" that functions as the **sign** and C is the **object** being represented.

No problem up to this point. But we get into muddy waters when the authors define C as "the meaning that A assigns to B)". This is wrong in Peircean semeiotics because it does not take into account Peirce's semeiotic structure of causation. If by C we understand the **dynamic object**, that is, the object which the sign professes to represent because it conveys some information about it, then C cannot be the **mean-**

[11] About this, see "Teleology: A Peircean Critique of Ernst Mayr's theory of Teleology" in Hulswitt 2002, pp. 88–95.

ing of semeiosis in Peircean terms because meaning is for Peirce just another name for the interpretant of the sign:

> A 'sign,' I say, shall be understood as anything which represents itself to convey an influence from an Object, so that this may intelligently determine a 'meaning,' or 'interpretant'. (MS 318, 1907)

Putting the meaning of a sign under the realm of the interpretant is the kernel of Peirce's pragmatism. Indeed, his maxim states that the 'entire meaning and significance of any conception lies in its conceivably practical bearings' (EP II: 145), that is, the effects and consequences of accepting the truth of such sign under conceivable circumstances.

What is missing then in the above definition is, precisely, that interpretation is a teleological process of production of effects through the action of signs, or semeiosis. The dynamic object might very well be the efficient cause of semeiosis, but meaning can only be given by the immediate action of the sign on the interpreting mind. The meaning is the form, or immediate object, that the sign conveys as to determine at least an immediate interpretant. This form is the purpose that governs the sign action, i.e., the rule or habit of action that sign transmits from the dynamic object to the final interpretant.

The Symbol is a Living General

It remains to be explained how the sign is able to internalize information and develops itself as it represents the form of its dynamic object. If we go into the classification of signs, the symbol is the only type of sign able to transmit a rule from the object to the interpretant—and therefore embodying the Form of a conditional proposition, which is its purpose:

> A *Symbol* is a Representamen whose Representative character consists precisely in its being a rule that will determine its Interpretant (….) A Symbol is a law, or regularity of the indefinite future. (EP II: 274, 1903)

The functioning of a symbol in relation to its replicas is similar to the species in relation to the stock, or set of living organisms, belonging to a given species in a given period of time. In fact, any organism might be defined as a replica whose body obeys the general and evolving rules or forms prescribed by its species. The stock is the body of a species as the replicas are the bodies of a symbol, "(…) or every symbol is a living thing, in a very strict sense that is no mere figure of speech. The body of the symbol changes slowly, but its meaning inevitably grows, incorporates new elements and throws off old ones"(CP 6.138).

As much as a word grows and is continuously transformed by its day to day use in the community of its users, embodying new layers of meaning, a species shows evolution by the concrete experience of its specimens, or replicas, in relation to the environment where the experience takes place.

Conclusion

We have seen that in his critique of the principle of causality Peirce proposed to strictly distinguish the concept of *cause* from the concept of *force*.[12] He reserved the former term for the whole of human experience and of nature, including living processes, and he restricted the latter to the formal laws of physics. Whereas 'cause' pertains to irreversible processes, 'force' deals with reversible processes; whereas the former term deals with concrete evolving reality, the latter deals only with abstractions. We then discussed Peirce's conception of causation, according to which each act of causation involves a teleological, an efficient and a chance component, a triadic relationship that can be identified in any genuine biological event.

We have seen, furthermore, that Peirce's conception of efficient cause holds a middle way between the Aristotelian Conception and the Scientific Conception of cause. According to Peirce, the causal relata may be either events or facts, depending on the context of the discussion. While Peirce insisted that the causal relata are facts within an epistemological context (or context of *causality*), they appear to be events within an ontological context (or context of *causation*). Causation is conceived as a productive event in an ongoing process of growth analogous to the development of symbols, defined by Peirce as the only signs that can determine general interpretants of the nature of habits. Symbols are driven by general purposes, which are their final causes, but need replicas (which are their bodies) to feel the experiences of the world and continuously incorporate changes as they develop.

Thus, symbols are truly living signs capable of gathering information during the experience lived by their replicas, or stock, and of conveying it to their interpretants so as to produce general habits of conduct attuned with the causal patterns, or laws, that govern reality. The flow of causation that organisms perceive in reality provides the *form* of a genuine and continuous triadic relation, which is the predicate of every true proposition.

This form is the very breath of life. And any living species is analogous to a true proposition as much as it is attuned to the flow of causation that grounds the real and allows its evolution and permanence.

References

Emmet, D. (1992). *The passage of nature*. Hampshire: MacMillan.

Kruse, H. (1991). *Genuineness and Degeneracy in Peirce's Categories.* Transactions of the Charles S. Peirce Society. *A Quarterly Journal in American Philosophy,* 27: 267–298.

Peirce, C. S. (1998). *The Essential Writings.* New York: Prometheus Books.

Posner, R., et al. (1997). *Semiotik/semiotics: A handbook on the sign-theoretical foundations of nature and culture* (Vol. 1). Berlin: Walter de Gruyter.

Ransdell, J. M (1986). Peirce, Charles Sanders. In T. A. Sebeok (ed.), *Encyclopedic Dictionary of Semiotics* (pp. 673–695). Berlin: Mouton de Gruyter.

[12] Peirce himself was not always consistent in the application of this distinction, however.

Schaeter, D. L. (1996). *Searching for memory: The brain, the mind, and the past*. New York: Basic Books.

Uexküll, J. von (1934). A stroll through the worlds of animals and men. In C. H. Schiller (Ed.), *Instinctive behavior. The development of a modern concept*. New York: International Universities Press.

Weiss, P. (1947/1983). *Nature and man*. Lanham: University Press of America.

The Ineffable, the Individual, and the Intelligible: Peircean Reflections on the Innate Ingenuity of the Human Animal

Vincent Colapietro

> (…) just as there are three orders of Thirdness, so there are three kinds of synthetical consciousness. The undegenerate and really typical form has not been made so familiar to us as the others, which have been more completely studied by psychologists; I shall therefore mention that last. Synthetical consciousness degenerate in the first degree, corresponding to acidental Thirdness, is where there is an external compulsion upon us to think things together. Association by contiguity is an instance of this; but a still better instance is that in our first apprehension of our experiences, we cannot choose how we will arrange our ideas in reference to time and space, but are compelled to think certain things as nearer together than others. It would be putting the cart before the horse to say that we are compelled to think certain things together because they are together in time and space; the true way of stating it is that there is an exterior compulsion upon us to put them together in our construction of time and space, in our perspective. Synthetical consciousness, degenerate in the second degree, corresponding to intermediate thirds, is where we think different feelings to be alike or different, which, since feelings in themselves cannot be compared and therefore cannot be alike, so that to say they are alike is merely to say that the synthetical consciousness regards them so, comes to this, that we are internally compelled to synthesize them or to sunder them. This kind of synthesis appears in a secondary form in association by resemblance. But the highest kind of synthesis is what the mind is compelled to make neither by the inward attractions of the feelings or representations themselves, nor by a transcendental force of necessity, but in the interest of intelligibility that is, in the interest of the synthesizing "I think" itself; and this it does by introducing an idea not contained in the data, which gives connections which they would not otherwise have had. This kind of synthesis has not been sufficiently studied, and especially the intimate relationship of its different varieties has not been duly considered. The work of the poet or novelist is not so utterly different from that of the scientific man. The artist introduces a fiction; but it is not an arbitrary one; it exhibits affinities to which the mind accords a certain approval in pronouncing them

---

This chapter grew out of a presentation to a session of the Metaphysical Society of America (December 27, 2011), held in conjunction with the Easter Division of the American Philosophical Association. At this session, I received very helpful suggestions and criticisms for various members of the audience, but especially from Robert Innis, John Lysaker, and John Stuhr.

V. Colapietro (✉)
Pennsylvania State University, University Park, USA
e-mail: vxc5@psu.edu

V. Romanini, E. Fernández (eds.), *Peirce and Biosemiotics,* Biosemiotics 11,
DOI 10.1007/978-94-007-7732-3_7,

> beautiful, which if it is not exactly the same as saying that the synthesis is true, is something of the same general kind. The geometer draws a diagram, which if not exactly a fiction, is at least a creation, and by means of observation of that diagram he is able to synthesize and show relations between elements which before seemed to have no necessary connection. The realities compel us to put some things into very close relation and others less so, in a highly complicated, and in the [to?] sense itself unintelligible manner; but it is the genius of the mind, that takes up all these hints of sense, adds immensely to them, makes them precise, and shows them in intelligible form in the intuitions of space and time. Intuition is the regarding of the abstract in a concrete form, by the realistic hypostatization of relations; that is the one sole method of valuable thought. Very shallow is the prevalent notion that this is something to be avoided. You might as well say at once that reasoning is to be avoided because it has led to so much error; quite in the same philistine line of thought would that be; and so well in accord with the spirit of nominalism that I wonder some one does not put it forward. The true precept is not to abstain from hypostatization, but to do it intelligently…. (CP 1.383)

Introduction

Our embodied minds frame and act on hypotheses as spontaneously as our lungs inhale and exhale. The ceaseless rhythm involved in the inhalation and exhalation of breath might itself be taken as a metaphor for the framing and testing of hypotheses.[1] The give-and-take of this physiological function exhibits what the semiotic function of human conjecture *functionally* is—a dialogue with the world, truly a give-and-take. In being semiotic, however, this function does not cease to be somatic; indeed, it is at once a semiotic, somatic, situated, and historical function. So, even our most causal, hence our most unreflective movements, *embody* our hypotheses—for example, our hand assuredly reaching for the handle of a door is guided by a conjecture, no less than that same organ frantically trying to clutch a limb as we are tumbling down a steep incline. Explicitly formulated and critically assessed hypotheses are the exceptions to the rule, whereas thoroughly tacit, irreducibly somatic conjectures define the rule in this case. Our movements are in effect conjectures. The space in which they take place is ineluctably a space of signs (an expression deliberately used to recall the use of "the space of reasons" by Wilfrid Sellars and other philosophers).[2]

In the spontaneous exercise of our abductive disposition (our innate tendency to generate experientially prompted guesses), we exhibit a drive to render *intelligible* what we encounter in experience. As the cumulative result of the ceaseless exercise of this instinctual disposition, experience becomes *funded*.[3] Our funded experience tends to enhance (at least, in a broad, rough manner) our abductive skills. Thus, ex-

[1] Cf. R. W. Emerson (1982, pp. 225–238).

[2] See my Colapietro (2003, pp. 157–179).

[3] Though this is a word used more memorably by John Dewey than either Charles Peirce or William James, it is also used by Peirce and James. Moreover, it is used to make the point being stressed by Dewey in his deployment of this term.

perience is for us far from "a tale told by an idiot, full of sound and fury, signifying nothing." It is much rather (in effect) a tale told and inevitably *re*-told by more or less competent but ultimately unwitting agents.[4] Even the most competent agents are thrust, by the very exercise of their agency (hence, by the demonstration of their competency), into circumstances in which they are forced to confront the limits of understanding, indeed to taste the fruit, often extremely bitter fruit, of their own fallibility.[5]

We have proven ourselves able to make sense out of our world, at least our world as it is encountered in our experience.[6] This world is first and foremost an arena of

[4] "If any of us knew what we were doing, or where we are going," R. W. Emerson stress in "Experience," "then when we think we know best! We do not know to-day whether we are busy or idle" (Emerson 1982, p. 286). In "the Poet," he suggests: "All that we call sacred history attests that the birth of the poet is the principal event in chronology. Man, never so often deceived, still watches for the arrival of a brother who can hold him steady to a truth until he has made it his own" (265). The insights provided by the utterances of such a poet "will reconcile me to life and renovate nature, to see trifles animated by a tendency, and *to know what I am doing*" (emphasis added)—i.e., to allow me to come to know what I have been and am now doing. "Life [as a consequence] will no more be a noise." It will be rather a sound possessing significance, indeed a self-understood significance. Even so, it is imperative to recall (returning to "Experience"): "The results of life are uncalculated and uncalculable. The years teach much which the days never know. The persons who compose our company converse, and come and go, and design and execute many things, and somewhat comes of it, but an unlooked-for result" (300). Regarding this result, Emerson goes so far as to say (at least in this context) the "individual is always mistaken" (pp. 300–301). Without question, "something is done," even advances (though little) are made, "but the individual is always mistaken." For it "turns out somewhat new and very unlike what he promised himself" (301).

[5] In a later reformulation of his pragmatist position Peirce noted: The indubitable beliefs acknowledged by the critical commonsensist "refer to a somewhat primitive mode of life, and while … they never become dubitable in so far as our mode of life remains that of somewhat primitive man, yet as we develop *degrees of self-control* unknown to that man, occasions of action arise in relation to which the original manner of means, but in our highest activities" (5.511). The innate dispositions of the human animal (i.e., beliefs, if stretched to cover them, have no sufficient authority. In other words, we outgrow the applicability of instincts [including our instinctual beliefs or innate dispositions]—not altogether, by any human *instincts*) tend to be effective in the narrowly circumscribed situations for which these behavioral tendencies have proven advantageous. But, "man is so continually getting himself into novel situations that he needs, and is supplied with, a subsidiary faculty of *reasoning* for bringing instinct to bear upon situations to which it [instinct] does not apply directly" (6.497). This faculty or capacity is first and foremost that of conjecture, deduction and inducting being subservient to abduction.

[6] We are ineluctably haunted by the possibility that the qualitatively resonant world of our "immediate" (or direct) experience is somehow not identifiable with the real world (cf. Stanley Cavell on skepticism). The world as it is disclosed in our experience seems itself to point to a world beyond anything we can even imagine or conceive. For the pragmatist, however, any distinction between the world of our experience and *the* world needs to be seen as one *we* draw—and do so in the context of experience itself, for the sake of rendering the world of our experience more luminous, secure, fulfilling, inhabitable, and indeed intelligible. In *The Quest for Certainty* Dewey (1929/1990) emphatically asserts: "The world as we experience it is the real world. But it is not in its primary phases [or primordial character] a world that is known, a world that is understood, and is intellectually coherent and secure. Knowing consists in operations that give experienced objects a form in which the relations, upon which the outward course of events depends, are securely experienced. It marks a transitional redirection and rearrangement of the real. It is intermediate

action, one in which the success of no endeavor is guaranteed and even the most monumental achievements in human history are precarious.[7]

What renders the world intelligible to us is not the intuitive grasp of immutable forms (cf. David Weissman), or any other form of intuition,[8] but simply our fallible understanding of a rather narrow range of natural phenomena and, on the basis of this understanding, the flights of our imagination (see, e.g., Peirce on scientific imagination, *CP* 1.46-48). Though it is certainly possible to construct at least a plausible evolutionary account of how animal intelligence (hence, how human reason as a culturally modified form of animal ingenuity) secured a niche for itself in a world shot through with hazards and obstacles, and though Peirce himself is disposed on various occasions to reconstruct just such a narrative, he also took the conjunction between intelligence and intelligibility to be in a certain respect irreducible. The evolution of our intelligence is, in part, the story of the emergence of a set of capacities at once linked to animal life and (to some degree) exhibiting a distinctive autonomy (cf. [T. L. Short]). The roots of our intelligence—and indeed the soil in which these roots can thrive—need to be acknowledged, but no less so do the fruits and flowers appearing at a distance from those roots. The *emergence* of intelligence amounts to nothing less than the emergence of *intelligence*. The intelligibility of the world, at least for an animal such as us, cannot be gainsaid, however much nominalists attempt to strip reality of those very features allowing for our intelligence to obtain a foothold in the world. So, we are (if we accept Peirce's argument) moved to assert that nothing more is needed than the *acknowledgment* of animal intelligence and immanent intelligibility.[9] For example, it cannot be reduced to an instance of the will to power or (without qualification) the stochastic outcome of countless species proving their adaptability over the course of a millennial history. Rather he boldly claims: "Our physical science … seems to have sprung uncaused except by man's intelligence and nature's intelligibility" (*CP* 2.13). This is a truly remarkable and, without question, also deeply puzzling assertion. But, for the moment at least,

---

and instrumental; it comes between a relatively casual and accidental experience of existence and one relatively settled and defined. The knower is within the world of existence; his knowing, as experimental, marks an interaction of one existence with other existences" (*LW*4, 235-36). This knowing, precisely as experimental is in the fist instance conjectural. It is the offspring of hypotheses. No matter how secure it becomes in the tangled course of our intellectual history, the value of our knowing remains in large measure that of providing basis for ever more fruitful hypotheses.

[7] "No human acquisition is," José Ortega y Gasset (1957) notes in *Man and People*, "stable. Even what appears to us most completely won and consolidated can disappear in a few generations" (p. 25).

[8] The word *intuition* of course is ambiguous. I am using it (following the lead of Peirce in "Certain Faculties Claimed for Man") to designate an immediate cognition, i.e., a cognition not mediated by signs or concepts.

[9] This involves a paradox, since it leaves the relationship between intelligence and intelligibility unmediated and, thereby, unintelligible. But this paradox is generated by the necessity, on the one hand, to avoid reductionism (i.e., to acknowledge the irreducibility of the relationship between intelligence and intelligibility) and, on the other, to render this relationship intelligible in the accredited modes of human understanding (e.g., by offering an evolutionary account of the emergence and functions of animal intelligence).

let it stand. Physical science and, more generally, human understanding trace their origin to the encounters between an irrepressible intelligence and the intelligible relationships woven into the very fabric of the experiential world. Our experience is, as much as anything else, a series of such encounters and, in addition, the funded result of these fleeting affairs.

According to Peirce, we *act* in the interest of intelligibility and our endeavors in this regard are the source for the most significant syntheses wrought by human ingenuity. He does indeed offer a thoroughly naturalistic account of human intelligence as a distinctive kind of animal ingenuity, stressing (for example) that "one, at least, of the functions of intelligence is to adapt conduct to circumstances, so as to subserve desire" (*CP* 5.548).[10] In another place he goes so far as to disclose: "I hold … that man is so completely hemmed in by the bounds of possible practical experience, his mind is so restricted to being the instrument of his needs, that he cannot, in the least, *mean* anything that transcends those limits" (*CP* 5.536). But this does not preclude human agents from engaging in purely theoretical inquiry; it only rules out their ability to frame intelligible conceptions completely untethered to the inevitably indeterminate possibilities of our practical experience. That is, our practical experience is itself no narrowly or rigidly bounded affair, but is by its own inherent dynamism a self-transcending and self-transforming process. For example, our practical experience encompasses the passionate participation of scientific inquirers in the historically evolved and evolving practices with which these agents identify. In brief, theory is itself a form of practice and the participation of a theorist in a practice (e.g., the mathematician engaged in the task of constructing a proof) affords that heuristic agent invaluable opportunities for the acquisition of "practical experience" (the experience to be derived from participation in this or that form of practice, including the practice of theorizing).

"We naturally make," Peirce suggests, "all our distinctions too absolute" (*CP* 7.438). This can be seen in the case just noted (the distinction between theory and practice), but it is also evident in countless other cases. One instance especially pertinent to our concerns is the way we tend to separate cognition, emotion, and volition. In "The Guess at the Riddle,"[11] however, Peirce suggests, on the one side, "every phenomenon of our mental life [e.g., emotion, perceptions, and volitions] is more or less like a cognition" (*CP* 1.376) and, on the other, "every kind of consciousness enters into cognition" (*CP* 1.381). The modes of consciousness in which he is most interested are exemplifications of thirdness. Immediate consciousness (that is, affective consciousness)—that mode for which Peirce reserves the word feeling—in principle is indivisible (there are no categoreal distinctions to

[10] "It is," Peirce suggests, "plain that intelligence does not consist in feeling in a certain way, but in acting in a certain way. Only, we must acknowledge that there are inward actions—what might be called potential actions, that is, actions which do no take place [in the outward world], but which somehow influence the formation of habits" (*CP* 5.286; cf. 5.479). Cf. Chapter 5 of my Peirce's Approach to the Self.

[11] In particular, I am focusing on the section of this text devoted to "The Triad in Psychology," that is, how the categories of firstness, secondness, and thirdness appear in the context of the experimental study of consciousness and (more broadly) mind.

be drawn *within* this category of mental phenomenon), whereas polar and synthetic consciousness are categoreally divisible. There are two forms of polar consciousness (those roughly corresponding to what we can call sensing and willing), while there are three forms of synthetic consciousness.

In general, the highest mode of animal consciousness encompasses "the consciousness of a process," one ultimately identifiable with "the sense of learning" (and, for Peirce, this means above all else the sense of acquiring new habits) (*CP* 1.381). Some of the details of this account are not directly pertinent to our purpose, so I will be succinct in my summation of them. The lowest grade of synthetic consciousness[12] is an instance of accidental thirdness (*CP* 1.383). It involves external compulsion. It coercively determines us to arrange mental phenomena in certain relationships (or certain patterns). The next degree of such consciousness (the intermediate grade of intelligent synthesis) is our irrepressible tendency to conjoin and disjoin whatever presents itself to our minds with other phenomena. It involves inward compulsion.

"But the highest kind of synthesis is," Peirce stresses, "what the mind is compelled to make neither by the inward attractions of the feelings or representations themselves [as it does in the intermediate form of synthetic consciousness], nor by a transcendental force of necessity [as it does in the lowest form], but *in the interest of intelligibility* ..." (*CP* 1.383; emphasis added).[13] In the interest of intelligibility, then, we are *compelled* to bring together disparate ideas for the sake of intellectually powerful syntheses, ones more powerful than anything supplied by either external or inward compulsion, also by anything yet achieved in even the most powerful syntheses so far wrought by human ingenuity in its passionate devotion to an ever more expansive intelligibility (more briefly, to what our intelligence has so far established in the interest of intelligibility). In this context, it is especially instructive to recall that, in 1890 or thereabouts (the time when he penned "The Guess at the Riddle"), Peirce did not think that this manner of synthesis had been "sufficiently studied"; moreover, he took "the intimate relationship of its different varieties" not to have been "duly considered." He makes his meaning much clearer when he asserts: "The work of poet or novelist is not so utterly different from that of a scientific man." His reason for asserting this is that: "The artist introduces a fiction; but it is not an arbitrary one; it exhibits affinities to which the mind accords a certain approval in pronouncing them beautiful, while if it is not exactly the same as saying that the synthesis is true, is something of the same general kind" (*CP* 1.383). The

[12] Using an analogy from geometry, Peirce describes the lower modes of thirdness as *degenerate*. The mode of synthetic consciousness being identified here is thus indicated to be "[s]ynthetical consciousness degenerate in the first degree" (*CP* 1.383).

[13] By the expression "a transcendental force of necessity," all Peirce means here is an external force (i.e., a force outside of consciousness itself, thus one in a sense transcending consciousness), one capable of exerting such control over our minds or consciousness that, in given circumstances, we cannot think otherwise than we are compelled to do as a result of this force upon us. E.g., the object in the foreground of my vision is near, while those in the background are far from me, my experience of the relative positions of these visual objects being a paradigmatic instance of the kind of brute compulsion (external "necessity") intended by Peirce in this context.

scientist no less than the artist introduces a fiction, a figment of the imagination deliberately forged for the sake of rendering intelligible what otherwise would be enigmatic or inexplicable.

Reality as it is encountered in our experience calls for the exercise of imagination, in the interest of intelligibility. "The realities compel us," Peirce admits, "to put some things into very close relation [to each other] and others less so, in a highly complicated and … unintelligible manner …" The "genius of the mind" (arguably, simply the irrepressible ingenuity of animal intelligence) "takes up all these hints of sense, adds immensely to them, makes them precise, and shows them in intelligible forms in the intuitions of space and time" (ibid.). But it does not limit itself to supplying just these intelligible forms. In every sphere of human engagement, the imaginative work enjoined by the highest kind of intellectual synthesis is discernible. The growth of concrete reasonableness is, on the most basic level of human practice, nowhere more manifest than in the proliferation of intelligible forms, ones potentially providing clues for how more fully and finely to make sense out of the world of our experience. The growth of reason is one with the strengthening and proliferating of forms of imagination, hence, (to a large extent) the institution and elaboration of *iconic* signs having the greatest fecundity (think here of the diagrammatic signs of formal logic or innovative techniques in cartography).

On this picture, then, the dynamic conjunction of a world brutally asserting itself and an intelligence imaginatively articulating its understanding of this world (in no small measure, an intelligence doing so in response to these "assertions") provide us with a promising perspective from which to look at the question of intelligibility. Whether it makes sense to envision the relationship of the world to us in terms other than brute compulsion, however, needs to be considered. So, too, whether it is adequate to conceive of human intelligence as so interest-bound as to preclude the possibility of an objective understanding of the actual world deserves to be discussed.

In any event, a fruitful way to address this question is to consider the complementary aspects of what is, at bottom, a historical process—hence, an *ongoing* process. These are strictly correlative terms not simply complementary ones. Even so, what I am proposing involves considering the dynamic conjunction of sign-using animals and (to some degree) a sign-proliferating world.[14] Such animals and such a world are what I mean by the complementary facets of the historical process in which effective understanding is attained, augmented, and revised. We might focus on the most minimal conditions for the possibility of *making* the world intelligible (cf. Aristotle on *nous poiētikos*), but then we might spotlight the ideal (thus, maximal) conditions enabling human agents to render their world as fully luminous as possible. Whatever our focus, however, intelligibility (at least as I envision it) is not

[14] In "Religious Insight and the Cognitive Problem," John E. Smith writes: "Under what circumstances, in a world of whirling atoms, would systems of meaning, of intelligibility and purpose emerge unless there were some form of co-ordination between the types of order we encounter in every region of existence and the human mind with its capacity for symbolic representation?" (Smith 1971, p. 109). He adds: "Experience makes its appearance only when there occurs a [more or less] precisely co-ordinated interaction between a structured world and a sign-using animal capable of remembering what has happened and of anticipating what may come" (110).

a static feature of a completely determinate world, but the historical achievement of intricately interwoven forces. Put otherwise, intelligibility is a feature not of things made, but of things *in the making.*[15] (James). This is how I propose to address this question on this occasion. In doing so, I will draw heavily upon Peirce and, more generally, pragmatism, almost as heavily upon Emerson,[16] but my objective here is not to offer an explication of their views on this topic, but to sketch what I as a pragmatist am disposed to say about intelligibility. This is what I am disposed to say, in light of what they have written, but also in light of what has transpired in the second half of the twentieth century and the opening decades of the twenty-first.

Please allow me an aside at this juncture. There is an important but (outside of the circles of those in the know about knowledge[17]) a still largely unnoticed shift, a shift *from* a virtually exclusive preoccupation with knowledge *to* an emerging interest in understanding. Analytic epistemologists of the most rigorous stripe have realized of late that the attainment of understanding, not only that of knowledge, is requisite for an adequate account of our epistemic practices. We strive to understand (as least) as much as to know and it does not seem that understanding is reducible to knowledge.[18] Our form of life imposes upon us as much tasks of understanding, often ones intimately connected to the work of acknowledge (cf. Stanley Cavell), as the exigency to discover explanations of occurrences or simply the discovery of facts (including of course purely singular, contingent facts).

What is the context in which we take up our epistemic tasks, richly envisioned? What is the form of agency allowing us to take up and carry through these tasks? Let me begin with the question of agency. On the one side, then, there are sign-using animals. This category is far more inclusive than *Homo sapiens*, though I will for the most part focus here on the human animal. But, to suggest just how much more inclusive this category is in my judgment, let me confess to being sympathetic to such arguments as those encountered in, say, Eileen Crist's "The inner life of earthworms: Darwin's argument and its implications" (Crist 2002).[19] It is only appropriate that Darwin's deep fascination with these seemingly lowly creatures (see,

[15] In *A Pluralistic Universe*, William James (1977) asserts: "What really *exists* is not things made but things in the making" (117).

[16] In doing so, I am not implying that Emerson is a pragmatist, but rather that the pragmatists were (among other things) his progeny. See Stanley Cavell's (1998).

[17] Lest I be misunderstood, let me point out that I am speaking ironically: those in the know about knowledge—professional epistemologists—are as often as not guilty of knowingness (the presumption to be in an ideal position to determine—in some instance, truly to dictate—what knowledge is or even what it must be.

[18] In *Achieving Knowledge: A Virtue-Theoretic Account of Epistemic Normativity*, John Greco (2010) notes: "One benefit of the distinction [between knowledge and understanding] is that it points to a richer plurality of epistemic goods. Truth and knowledge have epistemic value, but so do understanding, wisdom and other intellectual goods. It should not be unexpected that the values manifested in this plurality are distinctive, and therefore require distinctive treatment in a complete epistemology" (pp. 7–8).

[19] In "What Does the Maker Mind Make?" (an essay on Aristotle's doctrine of *nous poiētikos*), L. A. Kosman (1992) asserts, "animals have a rudimentary form of nous in the general capacity of discrimination which is aesthesis" (p. 357).

e.g., *Darwin's Worms: On Life Stories and Death Stories* [1999] by Adam Phillips) should be taken up today with his own sharp eye for the overlooked detail and an equally sharp sense of the radical implications of his own theoretical discoveries, but it is, alas, even more appropriate on this occasion to limit our attention mostly to the human animal.

The presence of an organism in an environment is (among other things) a set of demands (a more or less integrated *set* of demands, also more or less imperative and insistent demands), hence its life is a drama of frustrations and fulfillments (cf. *LW* 1,232 Dewey's *Experience & Nature* for a suggestive account of the *dramatic* condition of human life). It might of course be a very short drama, one in which the frustrations are so massive and unavoidable as to spell death upon entrance into the world. By its very structure or constitution, the organism demands nourishment and much else. What William James proclaimed about us is in fact true of all organisms: we are in the game. Let us recall the passage in which James so forcefully drives this point home:

> I, for my part, cannot escape the consideration, forced upon me at every turn, that the knower is not simply a mirror floating with no foot-hold anywhere, and passively reflecting an order that he comes upon and finds simply existing. The knower is an actor, and coefficient of the truth *on one side* [emphasis added], whilst on the other he registers the truth which he helps to create. Mental *interests* [emphasis added], hypotheses, postulates, so far as they are bases for human action—action which to a great extent transforms the world—help to make the truth which they declare. In other words, there belongs to mind, from its birth upward, a spontaneity, a vote. It is in the game, and not a mere looker-one …" ("Spencer's Definition of Mind" [1878]; *WJ: Essential Writings*, ed. Wilshire, 23–24) [*Essays in Philosophy* (Cambridge, MA: Harvard University Press, 1978), 21]

What most needs to be stressed here is that *we* as organisms make demands upon the world. We do so incessantly, (for the most part) unreflectively, and (alas) fatefully. In our case, the range of these demands is far from fixed, once and for all, by our biological or cultural inheritances; it is rather revisable and expansive. Arguably, nothing is more indicative of our identity than the most imperative demands we make upon the circumambient world and (from very early in our development) our alterable selves (selves caught up in process of becoming—thus being—otherwise).

So, *on the one side*, we have sign-using animals which make demands upon the world, not the least of which is that of making sense. The sensory experience of living organisms is, at least in its most primordial forms (cf. Innis), not bare noting, but a punctuated process of *making* sense. Perception is attuned to salience, while our sense of salience is itself refined and even altered by the disclosures of our perception. Our sense of salience is, moreover, indicative of the interests, concerns, and hypotheses we *bring* to our encounters with the world (in a word, to our experience).[20] These are in effect demands, if only of a very gentle, tentative character (though not infrequently also of an insistent, even obsessive nature). Life

[20] In his *Philosophical Investigations*, Wittgenstein (1953) notes: "Concepts lead us to make investigations; [they] are expressions of our interests, and direct our interests" (#570). They are shaped in no small measure in accord with out interests; but, in turn, our interests themselves are directed and even transformed by our reliance upon concepts.

itself is a set of demands upon the world regarding reparation, nourishment, motility, and much else. *On the other side*, there are the claims (or demands) the world itself makes on us. I am not unmindful of the anthropomorphism in my own claim regarding the world making claims on us. Indeed, this is a point on which I will comment eventually.

In the first instance and (I would argue) also the final analysis, everything turns upon *describing* these complementary facets of an ongoing process in a phenomenologically faithful and pragmatically suggestive manner. In part, this requires us to go back and guard against possible misunderstandings generated by seemingly innocent formulations (not least of all, the rhetorical device of juxtaposing sign-using animals, *on the one side*, and sign-proliferating reality, *on the other*). The fact of the matter is that such animals and such reality are, always in some manner and measure, *on the same side*. Even antagonists are in some respects allied (for two football teams to compete, they have to show up at the same field). That is, one can easily overdo the degree to which organism and environment are at odds with each other. Of course, there are countless and dramatic episodes in which the very survival of an organism is immediately threatened by its environment. But for untold—indeed innumerable—species and their representatives there are occasions when dangers are eluded, mates attracted, offspring born, nourishment found, and myriad other needs and desires fulfilled.[21] Whether or not we agree with how he uses this point, Richard Rorty is certainly right in insisting that language does not mediate between humans and their world in such a way as to bring together what was, in the first instance, disparate or disjoined. There is no need for language or anything else to join the organism and the environment, the historical actor and the unfolding drama, together, since they were never separate. Indeed, they are inseparable. Our life is one with an entanglement with the world, the world as anything to which we might meaningfully refer is one with our form of life.

Regarding how to describe the relationship between sign-using animals and sign-generating situations but also much else, consequently, everything turns on exerting the utmost care at the very outset of what cannot avoid being *repeated* attempts to execute this delicate task. We cannot help but fail, to leave something salient out of account or out of focus. Hence, we must return to the task, time and again, in the hope of (to invoke the words of Samuel Beckett) failing better. Our intellectual fingers are too coarse and clumsy at first simply to hold the elusive threads,[22] too unskilled and unpracticed then to weave them into a coherent pattern (cf. Emerson). Time and again, we need to try to pick up anew the threads so frustrating in their capacity for eluding the grasp of thumb and index finger. As Thomas

[21] In *Philosophy in a New Key*, Susanne K. Langer (1966) asserts: "The first thing we instinctively try to conceive is simply the experience of being alive. Life is a network of needs and fulfillments and further needs, with temporary frustrations here and there. If its basic needs are long unsatisfied, it ends [death ensues]. Our first consciousness is the sense of need, i.e., desire. Therefore our most elementary conceptions are of objects for desire" (130). Cf. Dewey (1925/1990, p. 194 ff.)

[22] "I take this evanescence and lubricity of objects," Emerson suggests, "which lets them slip through our fingers then when we clutch hardest, to be the most unhandsome part of our condition" (Emerson 1982, p. 288). Cf. Cavell.

Aquinas notes, however, a little error in the beginning can amount to a great one in the end (*De ente et essentia*). Hence, the most painstaking care must be taken with our inaugural descriptions.[23]

Part of the difficulty here (a critical part of the specific difficulty) concerns the constitutive role of human interests, while another part concerns the most apt description of cosmic demands (the demands the world itself makes on us, beginning of course with the use of demand and analogous terms in reference to the world). In divesting the human organism of its animating interests (an endeavor typically launched in the name of "objectivity"), we effectively divest that organism of nothing less than mind.[24] In situating the human organism in the causal nexus in which it is unquestionably caught (especially if we adopt as wholly adequate a Humean conception of causal relations), we run the risk of failing to see the extent to which the biosphere is truly a semiosphere (see, e.g., Jesper Hoffmeyer). Both tendencies are as widespread as they are misleading, so it is imperative to treat them critically. Let us, accordingly, take up each part in turn, beginning with the constitutive role of human interests and then turning to the demands exerted on us by the world itself. As an expression to refer to this role, I will invoke a famous expression used by William James.

"The Trail of the Human Serpent"

Upon recognizing the pervasive role of human interests in even the most abstract forms of theoretical inquiry, say, mathematics (the demands for consistency, economy, elegance, and a wide array of other cognitive interests, but *human* interests nonetheless), we might allow this recognition to eclipse what possibly are equally weighty considerations, e.g., the frustration of an endeavor to realize or enact those

[23] In *Modes of Thought*, Whitehead (1938/1966) identifies the activity of assemblage. Specifically, he urges that the task of systematic articulation be preceded by that of intellectual assemblage (the identification of the themes or ideas most worthy of being integrated into a systematic articulation of what strives to cover the entire range of human experience). See Colapietro (2010).

[24] "Interests which we bring with us, and simply posit or take our stand upon, are," James notes, "the very floor out of which our mental dough is kneaded. The organism of thought, from the vague dawn of discomfort or ease in the polyp to the intellectual joy of Laplace among his formulas, is teleological through and through. Not a cognition occurs but feeling is there to comment on it, to stamp it as of greater or less worth" (*Essays in Philosophy*, 18). In reference to the human organism, as an engaged participant rather than aloof spectator, James emphasizes: "its judgments of the *should-be*, its ideals, cannot be peeled off from the body of *cogitandum* as if they were excrescences, or meant, at most, survival" (21). "Every actual existing consciousness," James (1983) proposes in his *Principles of Psychology*, "seems to itself at any rate to be a *fighter for ends*, of which many, but for its presence, would not be ends at all. Its powers of cognition are mainly subservient to these ends, discerning which facts further them and which do not" (144). The success of the ends for which an organism fights, and by which he practically defines its identity, are far from guaranteed. This makes of the world not only (as is implicit in his characterization of consciousness as a fighter for ends) an arena of action but also (as he states explicitly in *The Varieties of Religious Experience*) "a theatre for heroism" (290).

interests, the power of factors other than our own agency to thwart the pursuit of our purposes. The realization of our purposes, the fulfillment of our interests, depends on more than those purposes and interests, effectively allied with our intelligence and agency. It depends upon luck and, in addition, the assistance of forces exceeding our comprehension even more than our control. This is, in a sense, a fact *about us*; it points to an insurmountable dimension of human finitude. But the experience of our finitude in effect is an acknowledgment of others—of *other* beings, forces, and factors—at least as critical to the realization of our interests as the exercise of our own agency or the reach of our own intelligence. "The trail of the human serpent is," as James so memorably wrote, "… over everything" (*Pragmatism*, 37). While this trail is discernible or discoverable (without question) only along paths defined by the movements of that serpent (movements expressive of hunger, and aggression, and abandon, also caution and cunning and interests far too alien from our own for us to be remotely in a position to identify, even to suspect, them), though in a world in which it haphazardly evolved and *to* which it must be intelligently responsive.

There is a passage from R. W. Emerson's "Experience" which bears upon the realization of our presence in the world, our irreducible presence in a seemingly enveloping cosmos having a tenure and nature *quite apart from* human comprehension. He goes so far as to identify the discovery of self with the fall of humankind.

> It is very unhappy, but too late to be helped, the discovery we have made that we exist. The discovery is called the Fall of Man. Ever afterwards we suspect our instruments. We have learned that we do not see directly [and, as I would rather say, that we do not see immediately], but [see] mediately, and we have no means of correcting these colored and distorting lenses which we are, or of computing the amount of their errors. Perhaps these subject-lenses have a creative power

What Emerson immediately adds is no less worthy of being recalled on this occasion than the sentences just cited: "Once we lived in what we saw; now, the rapaciousness of this new power, which threatens to absorb all things, engages us." The consequence is unquestionably momentous and potentially debilitating, even far more destructive than most of us are likely to appreciate: "Nature, arts, persons, letters, religions, objects, successively tumble in, and God is but one of its ideas [i.e., but one of our ideas]. Nature and literature are subjective phenomena; every evil and every good thing is a shadow which we cast" (Emerson 1982, pp. 304–305).

But does the trail of the serpent simply and completely annihilate that over which the traces of this trail are presumable discoverable? Can it be the case that self-correction is so thoroughly bound by our interests and instruments that it amounts to no correction at all? Is the metaphor of wearing lenses so inextricably part of our very being that to take them off is to negate ourselves (our "subject-lenses") an apt one? Is this not rather an instance of a picture holding us captive, a desire for freedom from that which alone makes freedom possible? Is there not wisdom in George Santayana's emphatic assertion—"On each animal species, on each man and nation, nature imposes a special way of thinking, and they will be foolish to quarrel with their endowment; they will not attain truth, or anything else [least of all happiness], by eluding it" (*The Realm of Essence*; *The Philosophy of Santayana*, ed. Edman, p. 520)?

My quarrel is actually not with Emerson, since there are texts in his writings on the other side of this issue (ones just as eloquently making the complementary point to this passage regarding the Fall of Humankind). Indeed, he is as much interested in enabling us to recover our capacity *to live in our perceptions* (to live in what we are able to see and perceive in other modalities), *without* the distracting or deflective consciousness of self,—he is as much interested in this—as he is in underscoring just how fateful was the discovery that "we exist." The discovery of subjectivity can threaten the objectivity of the world and, indeed, it *must* trouble our seemingly instinctual or natural confidence in our inherited ability to frame reliable judgments. Skepticism is the vestibule of the temple, nothing more (but also nothing *less*) than that part of the building by which we must enter into a very different space. (Emerson; cf. J. W. Miller). "People forget," Emerson reminds us, "that it is the eye which makes the horizon" (Emerson 1982, p. 305). Just as truly, the horizon defines the field in which the eye operates.

Accordingly, the pervasive, constitutive, and ineliminable role of human interests needs to be accorded its due. What counts as reality—indeed, what counts as the cosmos or world—for us does so in large measure by virtue of the interests animating our endeavors to make sense out of the disclosures of our experience. But, *given* anything approximating an adequate phenomenology of interest, what attracts, absorbs, and holds our interest—even more basically, what engenders, facilitates, and multiplies our interests and thus our susceptibilities and sensitivities, our receptivity and attunements—also need to be given *its* due. The capacity of the biologist simply to see the fossil (i.e., the pattern inscribed in stone) as a sign is a truly remarkable exercise of theoretical imagination. But, so too, is the foothold for intelligence or imagination provided by the fossil itself. Originally, we might not be in a very different position from Hamlet (save for the ironic and indeed malicious playfulness animating his suggestions) when he says to Polonius: "Do you see yonder cloud that's almost in the shape of a camel?" To this Polonius responds: "By th' mass, and 'tis like a camel indeed." Then Hamlet immediately shifts ground: "Methinks it is like a weasel." And here again Polonius is prone to give immediate consent to this alternative reading of the cloud mass: "It is backed like a weasel." And once again the young prince offers another interpretation: "Or like a whale," with his elder reply, "Very like a whale." (*Hamlet*, Act 3, scene 2). The shape of the cloud allows us to interpret it in any number of ways and the limits of interpretation are not tightly constrained (if constrained at all) by the features of the object being interpreted. In fact, however, there are constraints (however unperceived, perhaps even to some extent imperceptible) on our imagination and, hence, on the actual range of imaginable readings (or interpretations). Without question, the shape of the cloud is variously interpretable, but it would seem that certain readings are simply silly or misplaced or unlicensed. The significance of the things and events having a more direct and inescapable bearing on the flow of our activities, however, is arguably at a far remove from the "significance" to which we can attach to the shape of a cloud. In this flow, our interests operate in narrower constraints, with more immediate accountability (not least of all, accountability to the actual contours of the practical situations in which human actors are destined to pursue their pursues and fulfill their interests).

While interests are extremely varied, they are functionally one in principle. All interests, Dewey stresses, "mark an identification in action, and hence in desire, effort, and thought, of self with objects; with, namely, the objects in which the activity terminates (ends) and with the objects by which it is carried forward to its ends (means)" (*Interest & Effort in Education*, 90). Indeed, the "root idea of the term [interest] seems to be that of being engaged, engrossed, or entirely taken up with some activity because of its recognized [at least because of its *felt*] worth." This becomes evident when we recall that: "The etymology of the term *inter-esse*, 'to be between,' points in the same direction. Interest marks the annihilation of the distance between the person and the materials and results of his action; it is the sign of their organic union" (17), their dynamic conjunction. In *Democracy and Education* Dewey goes so far as to claim: "self and interest are two names for the same fact: the kind and amount of interest actively taken in a thing reveals and measures the quality of selfhood which exists" (Dewey 1912–1914/1985, pp. 361–362).[25]

Interests by their very nature pay homage to the objects by which they are elicited, sustained, and deepened, even if these "objects" turn out upon a sufficiently developed and detailed analysis to be quite far from what we commonly understand by objects. The significance and, not infrequently, the very identity of objects and events tend to become altered in the course of our experience with these objects and occurrences. The history of our encounters and engagements is, accordingly, one of revision, often radical revision, for we come to understand in novel ways even the most familiar phenomena.

To return to James's metaphor, the trail of the human serpent is discoverable at every turn (its various turns are manifest to us by the enduring, perceptible traces of its improvised movements). But the traces of this trail are themselves (in effect) inscriptions on or beneath the surface of the Earth. They unwittingly tell a tale and they do so in a medium with which the serpent itself is presumably unaware, at least at the level of description and interpretation offered by sign-using animals such as ourselves. The tale is not only that of an earth-bound animal, but also a life-sustaining earth. The traces of the serpent are inscriptions on the surface of the earth and, as such, they disclose not only the path of the serpent but also properties of the environment through which this slithering animal has devised its ingenious routes. The trail of the serpent might indeed be over everything, but that which upon the traces of this trail are legible deserves consideration in its own right.

[25] "Bear in mind that interest means the active or moving *identity* of the self with a certain object" (*MW* 9, 362). If we bear this in mind, then there is no need *either* to assume a need to propel the self into motion (as though a living being were an inert mass until pushed or pulled by an utterly external force) or to take the objects with which the self, in its very identity, is bound up to be fixed.

Our Primordial Relationship to the Natural World

We exist. We do so as interested parties and, hence, as implicated agents, i.e., as improvisational actors implicated in more or less identifiable situations defined in large measure by our animating purposes and pressing interests. This makes of our existence (as Dewey suggests) an inter-esse. The discovery of our own existence however does not decisively or inescapably mean the forfeiture of the world; rather it makes of our relationship to the world as much (if not more) a *task* than an inheritance. In particular, it makes this relationship the ongoing *work* of acknowledgment as much as an inherited set of affordances (J. J. Gibson).

To stand on the ground of interest is to occupy the place where self and world (organism and environment) are always already conjoined in a more intimate, dynamic, and fateful manner than we can begin to imagine. Mediation accordingly does not mean conjoining what philosophers and other theorists have rent asunder; rather it involves instituting the means by which even more intimate, dynamic, and transformative conjunctions, attuned to those already operative, can be realized. Closely connected to this point, however, is the task of describing the *very form of our relationship to the world*.[26] It is to this task that I now turn.

As we have already noted, there is, at least on the part of many theorists (especially those committed to one or another form of naturalism) a tendency to conceive the relationship of the world to us in exclusively causal terms and, moreover, to conceive these causal terms in a narrowly Humean manner. (One sometimes feel—at least I occasionally do—the need for a recovery of formal causality, but this is not the place to consider this need). There is at least as prevalent a tendency to conceive the relationship of the self to the world as that of a knowing subject to knowable objects (at least, objects about which the knower naively presumes the capacity to know[27]).

But a scheme of categories such as that devised and defended by Peirce allows us to see a wider range of experiential possibilities than either our inherited theories or present preoccupations allow. Peirce's categories are themselves intimations of intelligibility beyond anything we have yet noted or even suspect. That is, they are by design instruments to be used for goading and guiding inquiry, especially in new directions. It would be hard to imagine more appropriate tools for reformulating our understanding of our relationship to the world than Peirce's categories of firstness, secondness, and thirdness, especially in their recursive function and immanent drive to bring into sharpest focus intricate meshworks of complex relationships. Though I cannot here develop this point in much detail (let alone prove it!), I feel compelled

[26] In *Adventures of Ideas*, Whitehead (1933/1967) suggests that the subject-object relationship is not first and foremost the knower-known relationship. The primordial relationship of subjectivity to the world in which it is enmeshed is affective, not epistemic—a relationship of "concern" in the Quaker sense, not knowledge in its typically abstract meaning (pp. 175–176).

[27] The reason for this parenthetical remark is to avoid begging the question of skepticism. It may be the case that the knower is animated by a desire forever frustrated by some invincible defect in it or some insurmountable elusiveness in the objects on which it sets its sight.

to touch upon it. Peirce's categories are, in a sense, suggestions made by phenomena themselves: they are not only ubiquitous features of virtually all phenomena but also hints or clues regarding how any object of investigation, indeed any phenomena whatsoever, might be interrogated. The phenomenon of intelligibility itself invites a systematic deployment of the Peircean categories, a deployment promising to illuminate what is at the center of this paper—the relationship of a remarkably ingenious animal to a staggeringly complex world.

The Intimations of the Peircean Categories

As I have just indicated, then, the main function of the Peircean categories,[28] much like Aristotle's categoreal framework, is heuristic.[29] They were designed and deployed to guide and goad inquiry, to open new fields of investigation (*Essential Peirce*, volume 2, 413) but also to make significant advances in already established fields. In their most vital form,[30] they are not answers, but prompts to questions (as is—or should be—obvious in the case of Aristotle's categories, for the proper form of the Aristotelian categories is, arguably, their interrogative form[31]: What kind of being is this? What is its quantity? its quality? Where is it located? When did it come into being and when did it cease to exist, if temporal duration is relevant in a given case? What is its position? state? action? affection?). In one place Peirce

[28] Our understanding of the categories, Peirce insists, "must grow up in the mind, under the hot sunshine of hard thought, daily, bright, well-focussed, and well-aimed thought; and you must have patience, for long time is required to ripen the fruit. They are no inventions of mine [cf. Peirce's letter to James, *CP* 8.264]. Were they so, that would be sufficient to condemn them. Confused notions of these [distinct] elements [or aspects] appear in the first infancy of philosophy, and they have never entirely been forgotten. Their fundamental importance is noticed in the beginning of Aristotle's *De Caelo* ..." (*CP* 1, 521). In a letter to Victoria Lady Welby (12 October 1904), Peirce confessed that, "after only three or four years' study," he was led to the categories of firstness, secondness, and thirdness. "This sort of notion is," he stressed, "as distasteful to me as to anybody; and for years, I endeavored to pooh-pooh and refute it; but it long ago conquered me completely. Disagreeable as it is to attribute such meaning to numbers, and to a triad above all, it is as true as it is disagreeable" (*CP* 8.328).

[29] See Colapietro (2001, 2008).

[30] There are often dramatically what William James calls life answering to life, e.g., a smile returned or an embrace accepted and answered by an embrace. The sense of a threatening or a beneficent being is one of our most rudimentary yet crucial abductive propensities.

[31] In *Aristotle: The Growth and Structure of His Thought*, Lloyd (1968) notes: Aristotle "evidently arrived at his classification [of reality] mainly from a consideration of linguistic facts, that is of what kinds of things may be said, and of *what kinds of questions may be asked*, about anything, and this is natural enough, for he assumes that the logical distinctions that these data suggest will reflect and reveal the real distinctions between [or among] the real things in which he is interested" (114; emphasis added). I would make the even stronger claim that Aristotle derived his categories, first and foremost, from the array of questions any responsible inquirer, regardless of the specific field being explored, must ask. The categories are instruments of inquiry and, hence, means of interrogation.

suggests their value is simply in eliciting potentially illuminating suggestions: "This is all the categories pretend to do. They suggest a way of thinking"[32] (*Collected Papers*, 1.351), thinking being in the first instance *query* (to Justus Buchler's word for probing).[33] In other words, they simply intimate directions in which we might look, hence paths down which we might tread. They are not so much a method of inquiry, not even the sketch of an outline of such a method, as the categories are a meta-method—a method designed to facilitate the task of devising specific methods for diverse fields of experimental investigation. Viewed in this light, they are obviously connected to what Peirce identified as his quest of quests.

Their significance and value are methodological rather than epistemological. They serve not to justify knowledge allegedly already in our possession, but rather to advance inquiry so that we might be in a better position to obtain knowledge manifestly *not* in our possession. They are useful as means not for overthrowing skepticism, but for making discoveries. In other words, they have less to do with security than with fecundity (or uberty). The question of securing our beliefs is not unimportant, but for an inquirer that of elaborating a set of ideas critical for generating hypotheses is far more important. Our instinctual tendency to make guesses has over the course of centuries been tempered and tutored by the experiential consequences flowing from our often irrepressible impulse *to act* on those guesses. The vast, vague background against which our guesses make sense, at least provisional and contextual sense, provides us with a felt sense of an intelligible world in which our cognitive risk-taking itself makes sense. In a deep, abiding, yet tacit form, this

[32] Peirce goes on to claim: The very "possibility of science depends upon the fact that human thought necessarily partakes of whatever character is diffused through the whole universe, and that its natural modes have some tendency to be the modes of action of the universe" (CP 1.351). In *Psychology: the Briefer Course*, James makes an analogous point when he writes: "Mind and world ... have evolved together, and in consequence they are something of a mutual fit" (19).

[33] While inquiry is a species of query, query is hardly exhausted by the forms of probing properly identified as inquiry or investigation. Moral and artistic query ought not to be assimilated too quickly or too completely to scientific query. In the chapter in *Nature and Judgment* (Buchler 1966) devoted to explaining query bu noting: "The tradition term 'inquiry' has come to be used in a very broad sense. ...This breadth of usage for 'inquiry' is booth good and bad. It is good because it suggests that an important form of conduct is not limited to the profession of a discipline. But it is bad because it forces the mold of assertive judgment over the other modes of judgment [these being, according to Buchler, active and exhibitive]. To speak [for example] of art as inquiry is misleading and awkward. The term is applicable to science and to one function of philosophy; but art is contrivance [i.e., a mode of articulation primarily involving exhibitive judgment]. Inquiry may enter into art, contrivance into science. Contrivance in natural science takes the form of physical experimentation, and in mathematics it takes the form of symbol legislation and symbol organization. ... In philosophy, contrivance is an end in itself; it takes the form of arranging categories into an order of judgments which compels as an order and not only as a means of assertive [i.e., the exhibitive function of philosophical discourse is as important as the assertive dimension]. Like art, philosophy contrives to exhibit traits; like science, it aims to affirm truth. Science, art, and philosophy are equally modes of invention. But 'query' is a fuller term than 'invention'. 'Invention' primarily suggests the emergence of a product: 'query' the process of advance, the nature of this process and the product as a *relative* termination. 'Query' bears the sense of activity persisting beyond a given product. ... The freedom and sacredness of query, not merely of inquiry, is what must be meant in the account of the struggle between reason and unreason" (58–59; emphasis added).

felt sense serves as a practical orientation to the experiential world: it in effect expresses an instinctual confidence (perhaps an animal faith) in our epistemic capacity to discover the significance of objects and events. Our dispositions, including our instinctual or innate ones, constitute, to some extent, an *attunement* between our world and ourselves. This attunement is one with our sense of the intelligibility of the cosmos in which we live and move and have our being.

Peirce's categories however provide us with resources for not only distinguishing different levels of intelligibility but also identifying different factors contributing to the availability of the world for our understanding. In some manner and measure, the world avails itself: it invites and supports our epistemic endeavors and conjectural ventures. Specifically, what Peirce's categories intimate in the context of our exploration of intelligibility is, I propose, a threefold suggestion (one indicated in the title of my paper). In light of our experience, we cannot make sense out of intelligibility except in reference to what might appear to undermine its very possibility—the ineffability of qualitative immediacy and the *hecceity* of irreducible individuality.[34] That is, intelligibility would be, *for us*, unintelligible were it taken to stand on its own, apart from all else (but especially apart from the invincible elusiveness of immediate qualities[35] and the brute facticity of individual existents). The firstness of intelligibility cannot be anything discursively articulated; it must be immediately felt. So, too, the secondness of intelligibility cannot be anything conceptually expressible; it must be directly experienced. And it is experienced as either the shock of disruption (the disruption of our habits identifiable as doubt) or that of recognition.

While thirdness is irreducible, it is also indissociable from firstness and secondness. Hence, while intelligibility (a paradigm of thirdness) is irreducible, it is in our experience inextricably intertwined with ineffability and individuality. From a Peircean perspective, the consideration of intelligibility cannot be severed from that of either ineffability or individuality. The question of intelligibility is, in other words, inseparable from that of the limits of—and constraints upon—intelligibility, those limits and constraints being discoverable only in reference to the qualitatively immediate and the brutally oppositional features of our direct encounter with what is, nonetheless, an intelligible world.

Our relationship to the world is one in which the immediacy of intelligibility must be given its due, but no less so we must do justice to both must we do justice to the brutal oppositions and unbounded intimations of an expansive intelligibility so integral to make sense out of what is happening, who we are, and what the world itself is. In other

[34] "Indeterminacy, then, or pure firstness, and *hæcceity*, or pure secondness, are facts not calling for and not capable of explanation. Indeterminacy affords us nothing to ask a question about; *hæcceity* is the *ultima ratio*, the brutal fact that will not be questioned. But every fact of a general or orderly nature [and intelligibility is such a fact] calls for an explanation" (*CP* 1.405).

[35] The idea of the absolutely first "precedes all synthesis and all differentiation; it has no unity and no parts. It cannot be articulately thought; assert it, and it has already lost its characteristic innocence; for assertion always implies a denial of something else. Stop to think of it, and it has flown!" (*CP* 1.357). Along these same lines, Peirce stresses, "the conception of the absolute first eludes every attempt to grasp it" (*CP* 1.362). While the "idea of second must be reckoned as an easy one to comprehend [at least, to designate or point out], "[t]hat of first is so tender that you cannot touch it without spoiling it" (*CP* 1.358), i.e., with destroying it.

words, the firstness, secondness, and thirdness of intelligibility are clues about how we might begin to elaborate an account of intelligibility—hence an account inclusive of ourselves as sign-using animals and the world as a sign-generating matrix. Things and events in their qualitative immediacy, their brute facticity, and their inherent intelligibility reveal both the limits and the bases of intelligibility in its distinctively human form (thus, in their orienting implications for the ongoing efforts of the human animal to cultivate habits more fully and finely attuned to the dispositional properties of experiential reality). If Peirce's categories intimate directions in which we might look in our efforts to make sense of things and events, they are themselves attempts to articulate the intimations made by phenomena themselves. If Peirce's categories have ontological and cosmological deployments (and I am convinced that they do), these particular deployments are derived from the phenomenological of the Peircean categories: in identifying the ubiquitous features of phenomena, the categories serve as systematic clues for discovering the more or less disguised form of those features in specific heuristic contexts. They are intimations of the bases and limits of intelligibility because they are, in themselves (in their firstness), attempts to register or record the intimations of phenomena themselves. Our primordial relationship to the world calls for a categoreally nuanced form of articulation. But the categories best designed to carry out this task are those forms which can trace their roots to the world itself, as this world makes itself available to us in its myriad manifestations.

Conclusion

Reality bears some relationship to us, not only from our side but also from its side. We are, without question, in *this* together—and *this* calls time and time again for re-identification and re-description.[36] It is that *to* which we can emphatically and (on countless occasions) successfully *point* (Dewey's denotative method[37]), but that *about* which we can never say anything as penetrating or reliable as we so desperately de-

[36] In "Religious Insight and the Cognitive Problem," Smith stresses: "Whatever may be the shortcomings of modern idealism, it has one distinctive insight indispensable for understanding man and the world. From that standpoint, the real is not adequately represented by a 'given' external world on one side and an internal knowing subject on the other. The fact of man and the fact of the world do not exhaust what there is; in addition there is the peculiar relation of intelligibility holding between the two and this relation is just as real as its terms. The total reality we find must be represented as man-in-the-universe-interpreting the universe" (110-11; cf. William James (1902/1985) in *The Varieties of Religious Experience* on the "full fact," 393). Though Smith almost certainly did not have John William Miller in mind when he made this point, Miller was one of a host of idealists who tried to stake a position (to use W. M. Urban's expression, "beyond realism and idealism"). (See Smith's article in *Review of Metaphysics* on Urban). Moreover, he was emphatic in refusing to sunder what in our experience is integrally related. Self and world cannot be torn apart; they are equi-primordial. In *The Midworld of Symbols*, Miller (1982) affirms: "What I am unwilling to say is, 'There is the world, and here are the signs." That seems to me impossible…" (75).

[37] Dewey (1925/1988) *Experience and Nature* (*Later Works*, volume 1, see especially 371–372, 386–387, and 389–391).

sire to say. Even the most sophisticated instances of our most reliable utterances are remote cousins of the sounds of (say) a wolf howling at the moon. It however does matter that the contrast between our experiments in articulation and the sounds of this animal is one we can make. Affirming continuity here does not preclude discerning differences, indeed, multiple and profound differences. But the very acknowledgment of continuity implies that our most nuanced, adequate, and defensible utterances are, in some respects, not more than an animal form of braying at an object far transcending its understanding (cf. Santayana), also that the seemingly irrational cries of an animal exhibiting no evidence of reflexive symbolization express a pathos and arguably understanding far outstripping the consciousness of that animal.

We are enmeshed in causal nexus, but we come to this realization as a result not so much because things act on us as but because we act—at least try—to act on them, our lives being a kaleidoscopic jumble of dramatic situations in which the ineluctable exercise of our embodied agency issues in as often as not unintended consequences as anticipated fulfillments and feared frustrations. Things (objects and events) claim our attention and they do so primarily by the intrusive insistence so characteristic of indexical signs (the unexpected crash outside the room, the unwelcomed knock on the door, the desired indication of one color or another on a pregnancy test). The causal relationships in which we are so fatefully caught up, ones critically related to our being as agents, however, need to be seen as part of a largely unknown set of more complex relationships.

We make demands on the world but in turn the world (quite apart from the intentional claims of other rational agents) makes claims on us. The terms in which these mutually conflicting, coinciding, and evolving demands are framed are ones needing to be renegotiated, time and again, at least by us. How are we to understand our capacity to understand the world, including our own efforts to understand? My ultimate suggestion is at once a modest and bold proposal: we are to do so as an unavoidable task, that of taking up anew the task of renegotiating the terms in which I assert my demands on the world and, in turn, those in which the world makes is claims on us (for I am always a part of some we and thus the emphasis should fall on the social—not the isolated—self). More specifically, I am claiming that the language of claims (despite—perhaps *because*—of its anthropomorphism) is appropriate here.

The personification of Nature was not simply foolish or mistaken; it was (from all appearances) inevitable and, beyond this, critical for the development of an intelligence struggling to find a foothold in the world by which that intelligence could begin an ascent to levels affording more synoptic vistas than those at which human imagination transformed into disciplined forms of human intelligence began its millennial trek (see, e.g., Langer's *Mind: An Essay on Feeling*).[38] We are *in effect* addressed by beings incapable of intentionally uttering signs. What are *in effect* the signs of quasi-utterers are in more respects than I can count different from what are *by intent* signs put forth by intelligent agents. But they are nonetheless signs. It is

[38] Cf. William Ernest Hocking (1959) in *Types of Philosophy* [Third Edition] on spiritualism (Chapter "The Continuity of Life: On Peirce's Objective Idealism, "The Enduring Pre-Philosophy: Spiritualism"; also *Before Philosophy: The Intellectual Adventure of Ancient Man* by Henri Franfort et al.

not completely misleading or deluded to take ourselves to be addressed by beings deprived of the capacity to utter signs in the manner of a dog evincing a territorial growl or a cat eliciting the stroke of a human hand. Peirce: "Some [signs]," Peirce observes, "address themselves to us [e.g., the words of another human being in a face-to-conversation], so that we full apprehend them [as signs directly addressed to us]. But it is a paralyzed reason that does not acknowledge others that are not directly addressed to us, and that does not suppose [or imagine] still others of which we know nothing definitely" (*New Elements of Mathematics*, volume IV, edited by Carolyn Eisele, p. 299).

An emancipate reason however takes ever greater pains to attune itself to the myriad forms of semiotic phenomena flowing from various sources, not only purposely agents. For such a reason, intelligibility traces its roots to the possibility of generating and interpreting signs, a process always exhibiting, however distantly, the form of a dialogue (Ransdell 1976), a give-and-take in which the very possibility of exchange is itself rooted in primordial forms of ontological togetherness. Animal intelligence in its most basal form is the capacity to interpret signs (cf. Hoffmeyer), while *rational* intelligence is the ability to do so in an ever more conscious, critical, and hence autonomous manner (cf. Colapietro 1989). Intelligibility points to signs as what affords—often in an insistent and (in effect) helpful manner—possibilities of interpretation. As Aristotle long before Kant realized, however, the bare possibility needs to be realized as such. That is, the possibility needs to be transformed by intelligence itself into a potentiality. Intelligibility as an abstract possibility needs to be rendered into a grounded possibility. The work of intelligence itself is required for this to take place. What the making intellect makes—or, in other terms, what creative intelligence institutes—is the world as actually intelligible (the intelligible world in actuality). The world itself, in myriad and arguably mysterious ways, affords a host of its offspring just this opportunity. Indeed, it lends itself, sometimes in a more generous and cooperative manner than most of us today are disposed to acknowledge, to being investigated and better known, interpreted and better understood.

Human intelligence as an evolving set of animating interests would never have evolved in a world completely indifferent, let alone one relentlessly hostile, to its appearance and presence. This does not make of the cosmos a home for us. It does however make for the possibility of making the enveloping universe an even more commodious habitat for animal intelligence, even if this is a habitat into which intelligent animals have been thrown, willy-nilly. This world neither envelops us as the arms of a loving parent nor assaults us as an unqualified antagonist. It emphatically—sometimes fatally—objects to our folly and misunderstanding, but it also underwrites and sustains our efforts to understand, our attempts to know. A patchwork of approximations is not the radiant gown of an absolute monarch who is master of all he surveys, but it might be sufficient to the day—or, more likely, the evening when comfort or warmth or protection is especially needed.

Not all forms of human intelligence conform to the dominant pattern (or differential perspective) of the theoretical inquiry. Being is more than being known, or even more than anything to be known.[39] It claims *our* reverence and awe as much

[39] In *Themes in American Philosophy* (Smith 1970), in particular in "Charles S. Peirce: Community and Reality," Smith charges (and, in my judgment ,justly charges that: "Peirce, for all of his under-

as elicits our wonder and curiosity (cf. Dewey's *A Common Faith*; also John E. Smith's *Experience and God*), inspires our solicitude and secures our witness as much as it provokes our anxiety and frustrates so many of our purposes.[40] At the level of abstract definition, it is not mistaken to define reality in terms of its independence of how we happen to conceive it. At the level of pragmatic clarification, however, it is perhaps not amiss to envision reality in explicit reference to human purpose, more precisely, in such reference to the wide array of constitutive purposes bound up with our *various* practices. For the purposes of inquiry, then, it is instructive to conceive of reality in reference to the ongoing work of experimental inquirers: reality is clarified by taking note of how it exerts the power to challenge our beliefs and thereby to engender doubts. In a sense, then, it is defined or (more accurately) clarified in reference to thought, but not your thought or mine, not that of any other finite individual or even limited community.

Much eludes us, but not so completely so that we are utterly at a loss regarding its presence or even absence. Much by its very nature invincibly resists our efforts to do anything more than fall back on feeling or to try moving forward in an inquiry or exchange with someone else by some *act* of indication (some instance of pointing[41]). The ineffable immediacies of unique qualities are no less integral to our sense of reality than the irreducible otherness of singular events and objects. But some facets of reality court our attention and, having won this attention, sustain it indefinitely. Moreover, in some instances, they not only begrudgingly allow themselves to be investigated but also graciously assist us in our clumsy efforts to attain a deeper understanding. If the intelligible is only one facet (or dimension) of reality, this is far from a defect or disfigurement of reality. Only the inordinate demands of an untempered rationalism (principally untempered by experience) would make this seem to be defective or unbecoming. The more reasonable demands of an experi-

standing of the need for an ontological theory, still belonged to that modern tradition in philosophy according to which the key to *being* is found through *being known*" (105). Later in this essay he details this charge by noting: "Peirce seems to have underestimated the differential character of the controlled, theoretical inquiry that is to issue in the real truth about things. The question may fairly be raised as to whether the knowing relation is the only [or even simply the primary] relation in which we stand to the world and to the things in it. Ethics, aesthetics, and religion point to dimensions of things that are excluded from the highly precise, and therefore abstract, considerations that are alone relevant for scientific inquiry. This is not to say that Peirce neglected these other dimensions n his thought; it is rather to say that they cannot easily be included in a theory of reality in which the ultimate truth comes from scientific inquiry alone. The reality of things is not exhausted in their being material for knowledge. This is the great error of much modern philosophy; Peirce was not free from it" (108). Cf. also Smith's (1981) "Philosophical Interpretation and the Religious Dimension of Experience" in *Logos* 2, pp. 5–20; and John Dewey, "Epistemological Realism: The Alleged Ubiquity of the Knowledge Relation," *The Journal of Philosophy*, volume 8, number 20 (September 1911), reprinted in *Middle Works of John Dewey*, volume 6, 111–122.

[40] In "Ego and Reality," one of the essays included in *Papers on Psycho-Analysis*, Hans Loewald offer extremely important insights into how certain conceptions of reality reflect distinctive stages of emotional maturity. In *Open Minded: Working Out the Logic of the Soul*, Jonathan Lear (1998) helpfully draws upon these insights (see pp. 123–147).

[41] This in no way contradicts what Ludwig Wittgenstein (2009) argues in *Philosophical Investigations* regarding ostensive definition.

entially tutored and tempered intelligence would, in contrast, recognize in such a description (or characterization) of reality precisely what ineffable feeling, brute experience, and semiotic ingenuity in *their* dynamic conjunction would dispose us to suspect. Felt qualities despite their ineffability, irreducible individuals *because* of their irreducibility, and immanent intelligibility both despite and because of its deceitful seductions and ecstatic consummations each need to be given their due. Peirce's categories suggest nothing less. Of far greater moment, our experience—not least of all our experience as mindful participants in an array of shared practices, practices standing in complex relationship to one another—also suggests nothing less. For a pragmatist such as myself, this matters greatly. The texts of Peirce allow us to read more critically but also more imaginatively the textures of our own experience. They allows us to see experience as a dramatic engagement of a sign-using animal with a sign-generating world, thus to re-imagine human rationality as a species of animal ingenuity and the world in which such ingenuity was emerged and evolved as one affording open-ended possibilities for the development of self-corrective practices. These texts invite us to accept our fallibilism and finitude without thereby abandoning the quest for ever wider knowledge and deeper understanding, unabashed by the demand of our rationality to provide itself with nothing less than an account of the cosmos, but sensitive to the demands of reality in carrying out this task in a responsibly manner. We are responsible to the extent that we are *responsive* to the demands of the world but also to the extent that we are willing to take up the task of acknowledging the demands we are making of the world and especially others and ourselves. The task of such acknowledgment carries with it the obligation to consider whether the demands we are making are wise or foolish, reasonable or not (perhaps even pathological or obsessional). It is not enough to say, "This is simply what I do": it is necessary, time and again, to consider, in light of our specific form of animal life and the normative constraints imposed by various aspects of our defining inheritances, whether the demands we are making but, no less, whether our interpretation of the demands made on us by the world are defensible. There is no place to stand in trying to adjudicate these matters than the Earth in whose being our lives are implicated. The human animal possesses an innate disposition to frame guesses, one inextricably intertwined with another instinctual tendency: to make sense out of the objects and events with which the life of such an animal is so dramatically, so fatefully, linked.

References

Buchler, J. (1966). *Nature and judgment*. NY: Columbia University Press.

Cavell, S. (1998). What's the use of calling Emerson a pragmatist? In M. Dickstein (Ed.), *The revival of pragmatism: New essays on social thought, law, and culture*. Durham: Duke University Press.

Colapietro, V. (1989). *Peirce's approach to the self.* Albany: SUNY Press.

Colapietro, V. (2001). A lantern for the feet of inquirers. *Semiotica, 136*(1/4), 201–216.

Colapietro, V. (2003). The space of signs. In D. Jacquette (Ed.), *Philosophy, psychology, and psychologism* (pp. 157–179). Dordrecht: Kluwer Academic Publishing.

Colapietro, V. (2008). A gloss on Peirce's categories. In S. Petrilli (Ed.), *Approaches to communication*. Madison: Atwood Publishing.

Colapietro, V. (2010). Toward a metaphysics of expression. In R. Faber, B. Henning, & C. Combs (Eds.), *Beyond metaphysics? Explorations in Alfred North Whitehead's later thought*. NY: Rodopi.

Crist, E. (2002). The inner life of earthworms. In M. Bekoff, C. Allen, & G. M. Burghardt (Eds.), *The cognitive animal: Empirical and theoretical perspectives on animal cognition*. Cambridge: Bradforfd Book [MIT Press].

Dewey, J. (1912–1914/1985). Essays on philosophy and psychology 1912–1914, Vol. 7 of *the middle works of John Dewey*. Carbonale, I: Southern Illinois Press. Cited as MW 7.

Dewey, J. (1925/1988). Experience and nature. *The later works of John Dewey*. Carbonale, I: Southern Illinois Press. Vol. 1. Cited as LW 1.

Dewey, J. (1929/1990). The quest for certainty. *The later works of John Dewey*. Carbonale, I: Southern Illinois Press. Vol. 4. Cited as LW 4.

Emerson, R. W. (1982). *Ralph Waldo Emerson: Selected essays*. In L. Ziff (Ed.). NY: Penguin Books.

Greco, J. (2010). *Achieving Knowledge*. Cambridge: Cambridge University Press.

Hocking, W. E. (1959). *Types of philosophy* (3rd ed.). NY: Charles Scribner's Sons.

James, W. (1890/1983). *Principles of psychology*. Cambridge: Harvard University Press.

James, W. (1902/1985). *Varieties of religious experience*. Cambridge: Cambridge University Press.

James, W. (1977). *A pluralistic universe*. Cambridge: Harvard University Press.

Kosman, L. A. (1992). What does the maker mind make? In M. Nussbaum & A. O. Rorty (Eds.), *Essays on Aristotle's De Anima*. Oxford: Oxford University Press.

Langer, S. K. (1966). *Philosophy in a new key* (3rd ed.). Cambridge: Harvard University Press.

Lear, J. (1998). *Open minded: Working out the logic of the soul*. Cambridge: Harvard University Press.

Lloyd, G. E. R. (1968). *Aristotle: The growth and structure of his thought*. Cambridge: Cambridge University Press.

Miller, J. W. (1982). *The midworld of symbols*. NY: W. W. Norton & Co.

Ortega, J.y G. (1957). *Man and people*. NY: W. W. Norton & Co.

Ransdell, J. M. (1976). Another interpretation of Peirce's semiotic. *Transactions of the Charles S. Peirce Society, 12*(2), 97–110.

Smith, J. E. (1970). *Themes in American philosophy*. NY: Harper & Row.

Smith, J. E. (1971). Religious insight and the cognitive problem. *Religious Studies, 7*(2), 97–112.

Whitehead, A. N. (1938/1966). *Modes of thought*. NY: Free Press.

Whitehead, A. N. (1933/1967). *Adventures of ideas*. NY: Free Press.

Wittgenstein, L. (2009). *Philosophical investigations*. Malden, MA: Wiley-Blackwell.

Instinct and Abduction in the Peircean Informational Perspective: Contributions to Biosemiotics

Lauro Frederico Barbosa da Silveira and Maria Eunice Quilici Gonzalez

Proposals for hypotheses inundate us in an overwhelming flood, while the process of verification to which each one must be subjected before it can count as at all an item, even of likely knowledge, is so very costly in time, energy, and money—and consequently in ideas which might have been had for that time, energy, and money, that Economy would override every other consideration even if there were any other serious considerations. In fact there are no others. For abduction commits us to nothing. It merely causes a hypothesis to be set down upon our docket of cases to be tried. I shall be asked, Do you really mean to say that we ought not to adopt any opinion whatever as an opinion until it has sustained the ordeal of furnishing a prediction that has been verified? In order to answer that question, it will be requisite to inquire how an abduction can be justified, here understanding by abduction any mode or degree of acceptance of a proposition as a truth, because a fact or facts have been ascertained whose occurrence would necessarily or probably result in case that proposition were true. The abduction so defined amounts, you will remark, to observing a fact and then professing to say what idea it was that gave rise to that fact. One would think a man must be privy to the counsels of the Most High so to presume. The only justification possible, other than some such positive fact which would put quite another color upon the matter, is the justification of desperation. That is to say, that if he is not to say such things, he will be quite unable to know anything of positive fact. In a general way, this justification certainly holds. If man had not had the gift, which every other animal has, of a mind adapted to his requirements, he not only could not have acquired any knowledge, but he could not have maintained his existence for a single generation. But he is provided with certain instincts, that is, with certain natural beliefs that are true. They relate in part to forces, in part to the action of minds. The manner in which he comes to have this knowledge seems to me tolerably clear. Certain uniformities, that is to say certain general ideas of action, prevail throughout the universe, and the reasoning mind is [it]self a product of this universe. These same laws are thus, by logical necessity, incorporated in his own being. For example, what we call straight lines are nothing but one out of an innumerable multitude of families of nonsingular lines such that through any two points there is one and one only. The particular family of lines called straight has no geometrical properties that distinguish it from any other of the innumerable families of lines of which there is one and one only through any two points. It

L. F. B. da Silveira (✉) · M. E. Q. Gonzalez
São Paulo State University, Marilia, Brazil
e-mail: lfbsilv@gmail.com

M. E. Q. Gonzalez
e-mail: gonzalezquilici@gmail.com

V. Romanini, E. Fernández (eds.), *Peirce and Biosemiotics,* Biosemiotics 11,
DOI 10.1007/978-94-007-7732-3_8,

is a law of ***dynamics*** that every dynamical relation between two points, no third point being concerned, except by combinations of such pairs, is altogether similar, except in quantity, to every such dynamical relation between any other two points on the same ray, or straight line. It is a consequence of this that a ray or straight line is the shortest distance between two points; whence, light appears to move along such lines; and that being the case, we recognize them by the eye, and call them straight. Thus, the faculty of sight naturally causes us to assign great prominence to such lines; and thus when we come to form a hypothesis about the motion of a particle left uninfluenced by any other, it becomes ***natural*** for us to suppose that it moves in a straight line. The reason this turns out true is, therefore, that this first law of motion is a corollary from a more general law which, governing all dynamics, governs light, and causes the idea of straightness to be a predominant one in our minds. In this way, general considerations concerning the universe, strictly philosophical considerations, all but demonstrate that if the universe conforms, with any approach to accuracy, to certain highly pervasive laws, and if man's mind has been developed under the influence of those laws, it is to be expected that he should have a ***natural light,*** or ***light of nature,*** or ***instinctive insight,*** or genius, tending to make him guess those laws aright, or nearly aright. This conclusion is confirmed when we find that every species of animal is endowed with a similar genius. For they not only one and all have some correct notions of force, that is to say, some correct notions, though excessively narrow, of phenomena which we, with our broader conceptions, should call phenomena of force, and some similarly correct notions about the minds of their own kind and of other kinds, which are the two suficiente cotyledons of all our science, but they all have, furthermore, wonderful endowments of genius in other directions. Look at the little birds, of which all species are so nearly identical in their physique, and yet what various forms of genius do they not display in modelling their nests? This would be impossible unless the ideas that are naturally predominant in their minds were true. It would be too contrary to analogy to suppose that similar gifts were wanting to man. Nor does the proof stop here. The history of science, especially the early history of modern science, on which I had the honor of giving some lectures in this hall some years ago, completes the proof by showing how few were the guesses that men of surpassing genius had to make before they rightly guessed the laws of nature…. (CP 5.602–604)

Introduction

The interrelations among *abduction*, *instinct*, and *information* do not seem to have received any specific treatment in Peirce's writings. Nevertheless, his work on them either individually or in pairs with abduction evolved significantly over the years. Since at least 1886, instinct acquired a growing role in that which Peirce considered useful to life, with convergence between humans and other animals while acknowledging different levels of complexity. In the abovementioned 1913 text[1], reasoning itself is viewed as being instinct, with the commonality between diverse animal species, including humans, serving as inductive proof of the ability of semiotics to provide a foundation for elucidation of the entire spectrum of conduct involving goals and the maintenance or furthering of life.

Based upon Peirce's views on the nature of abduction, in this chapter we present a detailed justification for the hypothesis that instinct, grounded on information, may express a manifestation of abduction in the unfolding of the semiotic process

[1] EP2:463–474.

that characterizes life. In its myriad of interconnected levels of communication, life will be depicted as a complex self-organizing informational process that causes dynamic webs of self-referential signs to be generated (and to vanish). An explanation of the role played by the Peircean concept of information in the understanding of how abduction and instinct may come together in the life process constitutes the main theme of the present chapter.

Within Peirce's published texts can be found the conceptualization of information, present since 1865 and carefully worked upon up until texts dated 1909. With respect to instinct, texts can be found dated from at least 1886 until 1913, the year immediately prior to his death. Our task here consists of examining some of these texts in search of the principal characteristics that touch upon abduction, instinct, and information, even though no explicit convergences amongst these themes were treated by Peirce.

It must also be considered that, in parallel with instinct, Peirce insists that abductive reasoning may operate contiguously with perceptual judgment, since both instinct and this kind of inference do not find their origin at the conscious level of the mind[2]. Therefore, after investigating the role played by instinct in mental life, we shall inquire into the connections established by Peirce concerning abduction and perceptual judgment, selecting those aspects that can contribute to our present research on the interrelations between instinct, abduction, and information in the domain of life.

The Peircean Concept of Abductive Reasoning and its Anchoring on Instinct and Perceptual Judgment

Abduction, according to its most common description formulated by Peirce, is a way of reasoning by which one infers to the best explanation of a problem or a surprising event. This type of reasoning initiates with the perception of an anomaly or a novelty, which gives rise to the search for hypotheses that, if true, would explain the perceived anomaly. Once reasonable hypotheses are created, they are submitted to inductive testing. The plausibility of the new hypotheses is confirmed (or not) by means of deductive inference to indicate their predictive and explanatory power in elucidation of the initially perceived anomaly.

Peirce argues that abduction starts whenever a surprising or anomalous fact is observed that interrupts a chain of well-established habits, introducing a real doubt in the mind. This process, which may occur not only in science, but also in almost all aspects of life, does not end until a new habit or hypothesis can be created that causes the initial doubt or feeling of surprise to vanish. As stressed by Gonzalez et al. (2007)[3], in the context of life this characterization of abduction presupposes an understanding of Peirce's ontology, according to which the universe is in a non-

[2] Cf.EP2:217–218.

[3] Gonzalez et al. 2007.

mechanical process of continual expansion, acquiring and modifying habits. Based on the law of habit formation, which describes the tendency for repetition of already existent events, the cosmos is conceived as a self-organizing semiotic process structured by a continuous flow of information available to all existent beings.

Given the above scenario, and following in Peirce's tracks, we shall provide evidence for the hypothesis that abduction plays an organizational role not only in the evolution of embodied *individual* agents rooted in the world, but also in *collective* structures constitutive of life in general. With this broad hypothesis in mind, our first problem is to understand the type of agents or structures that are allowed to participate in this evolutionary process. Are only humans and similar animal species capable of developing abductive reasoning, or can it also be found in organs, cells, colonies of bacteria, and so on? According to Peirce, wherever the possibility of habit formation and change exists, minds will begin their continuous evolutionary path.

Nourished by information and chance, as well as by failure and learning processes, living beings develop chains of stable webs of habits that direct and constrain their actions. Given the complex dynamics of the living world, organisms sometimes have to abandon or radically transform their habits in order to survive. In this circumstance, whenever the emergence of chance and novelty results in the disruption of stable existent habits, abductive reasoning starts and, Peirce argues, will not be interrupted until temporary stability can be restored.

In order to understand the possible operations of abduction among different existent entities, in this section we are going to present the Peircean concept of abductive reasoning, emphasizing those passages where he mentions instinct as an anchor for this type of reasoning.

In 1910, in a text entitled On The Foundation Of Ampliative Reasoning,[4] Peirce provides clarifications of three types of reasoning with which he had worked for many years, and attempts to better define *induction* and the expanding reasoning that he often termed *abduction.* Taking, in the human context, the case of an insurance company, it could be said that the statistical establishment of mortality tables, being explanatory, belongs to the realm of mathematical thinking, and is therefore deductive in nature. Nevertheless, this is different to the reasoning behind the decision to assume a risk; the latter consists in using reasoning by similarity with previous situations, and is considered by Peirce "*[...] the very Best kind of reasoning in the world,—better than the demonstrative reasoning of mathematics, because it assures us of what we have not previously known even in dismembered fragments; and although this assurance is not absolute and may be erroneous, yet perseverance in the same kind of reasoning will gradually correct every error, and will enable us to approach indefinitely near to certainty*"[5].

Peirce compares abductive reasoning with that which motivates reason to create its first hypotheses about a given phenomenon, and which being instinctive in its origin was recognized by Galileo as being derived from the *Il lume naturale*, common to humans and other animals. The importance of abductive reasoning is due to

[4] MS 660. NEM III.1:214.

[5] MS 660. NEM III.1: 214.

that which it creates as a basis for its own induction, which, being self-corrective, should greatly assist humans, whose instincts may be atrophied due to an addiction to rationalizing about everything.

The following year, in 1911, Peirce dedicated a text called Probability and Induction[6] to the three original forms of reasoning: *deduction*, *adduction* (or qualitative induction), and *retroduction* (or abduction, as corresponding to the raising of hypotheses). Deduction could be defined as the inferential process according to which true conclusions can be generated, whether always or in a certain proportion, from certain true premises. Through induction it can be inferred that a certain state of things is true, at least approximately. In contrast, retroduction (or abduction) is defined as a class of reasoning by which "*one only infers that a certain state of things may be true and that the indications of its being so are sufficient to warrant further examination.*"

It is important to stress here that, according to Peirce, retroduction (or abduction) is the only guide available to us for attainment of the truth, and that its instinctive character is shared with many other living beings. In proceeding according to his instinct, man recognizes that he is acting in accordance with good sense. As in other animals, our instincts are formed under the influence of the same laws of nature to which we are led to conform. Furthermore, instincts are not infallible, and there can be an illusion of apparently following our instincts, when in fact one is motivated by preconceptions. Hence, there is a need to use the inductive test to determine, through experience, if the conclusions obtained are reflected in practice.

The difference between the postulation of hypotheses and their qualitative confirmation is that the guarantee offered by induction for their conclusions is that the latter are provisionally accepted but could be shown to be false, should the confirmation process be continued for a prolonged period. In contrast, retroduction is sustained by the compulsion to accept its conjectures, which should therefore be submitted to the inductive test.

Finally, in 1913, Peirce wrote his famous text "An Essay Toward Improving Our Reasoning in Security and in Uberty."[7] Starting from the conception of the power of reasoning as the ability to produce inferences that tend towards the truth when the premises (or virtual assertions) are true, he investigates the processes that occur in the mind during reasoning and the production of science.

Reiterating his long-held conviction that what happens in the mind can only be recognized by its external manifestations (this being an application of the maxim of pragmatism to mental phenomena), Peirce argues that this can provide a means of explaining the logical structure offered by the mental phenomenon, and consequently also the 'security' offered by the reasoning. Although it may be important to recognize the security of reasoning, it is not possible to ignore the fact that an entity that is alive is in a process of development, and that security is based on the collateral experience provided by the perception of previous states.

[6] **L**231. NEM III.1: 203–204.

[7] MS 682. EP2:463–474.

It should also be emphasized that, in maturity, Peirce considers that it is by means of instinct that a mental fact is produced that tends towards the truth. In the passage below, he explains his understanding of instinct:

> The word 'instinct' itself is but a generalization of abstractions, one of the brood of language or of thought: there is no great difference between the two, as we shall see. When an animal responds to a stimulus in much the same way as almost any other individual of the same species or division of that species (such as a sex, for example), and does so not, so to speak, mechanically, but voluntarily, and when the response is of such a kind as generally to have a beneficial effect upon the same animal or its progeny, which effect, however, the animals that act so can hardly be supposed to have divined or, at any rate not to have ascertained by reasoning from any other facts within their knowledge, then we call the action 'instinctive', while the general habit of behavior, regarded as appertaining to the animal's consciousness we call 'an instinct'. It will thus be seen that one 'instinct' of an animal may be a special determination of a more general 'instinct' of the same animal; so that the former can be said in one sense to be a 'part' of the latter, while in another sense the latter may be called a 'part' of the former….. (EP2 p. 473)[8]

In addition to considering the non-mechanical aspect of instincts, Peirce reminds us that they are fallible and can betray us. Equally, instincts are subject to learning. Other animals are witnesses to this, in terms of both the fallibility of instincts and the capacity to acquire them by experience. We can see that the desire to please leads animals to learn new behavior and acquire new habits, as well as occasionally be deceived in their inferences. Both these aspects are manifested to a greater extent in humans, in whom the domain of instincts is broader, and potentialized by the domain of language, which assists us in checking the 'security' offered by reasoning.

Peirce asserts that in order to better evaluate the roles played by security and uberty in processes of knowledge acquisition, we should understand how we genuinely reason, and in what way both we and other animals are able to learn. Focusing on human knowledge, Peirce seeks to define that which is essential for security in scientific knowledge. Differentiating observation from mere sensation, he argues that observation must be attentive, since simply noting something is not sufficient, with the presence of a thought founded on information being necessary to evaluate and give form to that which is attentively observed. There still remains uberty, by means of which, he states, judgment is sustained by the reasoning instinct. If security and uberty are assured, the work of science could be considered the genuine operation of reasoning.

The end of the text reserves space for an alert to the reader:

> Please understand, Reader, that I do not in the least mean to represent that it is at all in this way that we actually do come to this conclusion; for our natural rational instinct (for Reason is a sort of instinct) makes us pretty well aware of it all in advance of any such reflections. I only sketch in the vaguest outline how we might come to the truth of the matter, even though we were in the most hardened state of necessity to the just authority of instinct… But what can that phrase mean? (EP2 p. 472–473)

Instinct and reason, then, seem to share a common origin: the belonging to a single 'nature' that seeks security in the information available in the environment, in order to initiate and maintain 'alive' the process of habit adjustment. Humans and other animals, equally possessed of instincts, both proceed rationally (or, better still,

[8] Peirce Edition Project (1998).

reasonably)! Instinct and reasoning are distinguished, but not separated; the same can also be said of security and uberty in our understanding of reality founded on perceptual judgment, where the gates of information open to enable the action of abductive reasoning.

Something very similar could be considered with respect to abductive reasoning and perceptual judgment. Peirce often insists that abductive reasoning can operate in continuity with perceptual judgments. During the course of a series of lectures at Harvard in 1903[9], he provides an ample discussion of the relations between abductive argument and perceptual judgment.

Considering the division of Philosophy into phenomenology, normative sciences (composed of aesthetics, ethics, and logic), and metaphysics, from a predominantly phenomenological perspective Peirce recognizes the common natures of abduction and perceptual judgment. He explains, nevertheless, that they are distinguished one from the other by the logical criticism to which abduction, but not perceptual judgment, may be submitted. Not being based at the conscious level, perceptual judgment, although fallible, is not subject to doubt, and still less to logical control.

In the phenomenological universe, perceptual judgment is a necessary condition for the construction of all knowledge. It is to this domain that Peirce applies the Aristotelian maxim that nothing may be found in the intellect that has not previously existed in sense. Peirce notes that by 'intellect' he does not only mean human intellect, but understands "*the meaning of any representation in any kind of cognition, virtual, symbolic, or whatever it may be.*" (EP2:226–227) In turn, by the term 'in sense', he intends to signify "*in a perceptual judgment, the starting-point of our first premise of all critical and controlled thinking.*" (EP2:227)

Peirce argues that perceptual judgment, understood as a first response that is sensitive to experience, can even be found in mathematics at its fundamental level. Although mathematical entities may not directly depend on being able to correspond to a certain phenomenon that is merely feasible, they can be effectively constructed in the form of observable diagrams, and can consequently be perceived by he who constructs them. Peirce therefore affirms that:

> All necessary reasoning without exception is diagrammatic. That is, we construct an icon of our hypothetical state of things and proceed to observe it. This observation leads us to suspect that something is true, which we may or may not be able to formulate with precision, and we proceed to inquire whether it is true or not. (EP2:212)

Here, our interest is to stress the dynamic partnership established between abduction, in its strict sense as a type of reasoning, and perceptual judgment: abduction never denies its being grounded in perceptual judgment, but given its common nature, maintains with it a process of continuity. On the other hand, since perceptual judgment is not sustained at the conscious level, preceding the discrete character of spatial-temporal experience (that applies to abduction), it does not involve, Peirce notes, a regression to the infinite, which would falsify reasoning (EP2: 227, 236–237).

In synthesis, perceptual judgment being a first sensitive response to experience, eminently fallible but undeniable, it could be considered, according to our understanding,

[9] EP 2:133–241.

the first access of a mind to information (understood, as we shall explain below, in the semiotic level of transmission of forms), which extends far beyond the human universe. Its role in abduction is fundamental, given that it could not be initiated without the ubiquitous information provided by perceptual judgment.

The following section presents the Peircean concept of information and its relation with signs, focusing on Peirce's characterization of sign as a vehicle for the communication of form, which allows abduction (differently from induction and deduction) to take place at a very fundamental level in the evolutionary flow of life.

The Peircean Concept of Information and its Relation with Sign: An Inquiry on the Unfolding Layers of Abductive Reasoning

In the previous section, we presented evidence for the assumption that abduction operates, according to Peirce, in continuity with perceptual judgment, thus dealing with ubiquitous entities belonging to different ontological categories. A question that so far remains unexplained is: What gives unity to abductive reasoning? Our hypothesis is that information, in the Peircean semiotics sense of form carriers, may play this unifying role in abductive reasoning, as characterized by Peirce in maturity. Starting at the level of perceptual judgment, where ubiquitous non-discrete information allows collateral experience to express itself as a first sensitive response to experience, the feeling of surprise may give place to abductive reasoning:

> The surprising fact, C, is observed,
> But, if A were true, C could be a matter of course;
> Hence, there is reason to suspect that A is true. (EP2: 231)

Abduction, after being initiated by the feeling of surprise, grounded on perceptual judgment, allows information from different sources to be expressed in the search for a new habit (or hypothesis); this, according to Peirce, if true, could help with the establishment of a new web of habits or hypotheses that would explain/accommodate the initially surprising fact. However, as he insists, "*A could not be abductively conjectured, until its entire contents is already present in the premiss, if A were true, C could be a matter of course.*" (EP2: 231)

We believe that the Peircean concept of information may help us to understand how the 'entire contents' may be revealed in the premise of an abductive reasoning. In order to explain this supposition, we propose a reading of the Peircean concept of information that highlights its two main descriptions, logical and phenomenological:[10]

[10] Peirce was probably one of the first among Western philosophers to develop a very detailed analysis of the ontological, epistemological, and logical nature of information. For the purposes of the present investigation, only two of his perspectives on information are presented.

1. **The formal logic perspective:** Information is characterized in terms of the depth and extension (CP 2.419) of a term.
2. **The ontological perspective:** Information is described in terms of its relation to signs, understood as vehicles of forms.

With respect to (1), Peirce explains that by the ***informed depth*** of a term he means: "*... all the real characters (in contradistinction to mere names) which can be predicated of it (with logical truth, on the whole) in a supposed state of information.*" (CP 2.408) By extension, he means all those elements to which a predicate can be attributed: "*By the **informed breadth** of a term, I shall mean all the real things of which it is predicable, with logical truth on the whole in a supposed state of information.*" (CP 2.407)

Different to established tradition, where growth in the extent of a concept was considered inversely related to growth in the understanding of that concept, Peirce analyzed case by case the correlation obtained between these two parameters. There could be cases when the inverse relationship was maintained, for example when specifying the subclass *rational animal* within the broader classification of animals. This attribution would deepen understanding of the concept of animal, restricting the extension of the concept thus formed to only a subset of those beings to which the concept of animal could be attributed. In such cases, there would be no variation in the information provided by the concept. However, cases could occur in which the increase of a concept, whether in terms of extension or of understanding, would not *ipso facto* imply diminution of the other quantity with which it forms a pair, so that there would be variation in the value of the information provided by the concept.

For example, in the case of attributing the predicate 'social' to every rational animal, the understanding of the concept of rational animal is deepened, although the extent of the concept remains unaltered, since it continues to be distributively attributable to the totality of the class to which the concept of rational animal is applied. In such a case, the product of extension and depth increases, and there is therefore an increase in the information conveyed.[11]

The attribution of information to the interrelation between the quantities of concepts is a constant feature of the realism adopted by Peirce. It submits the concepts (and their attributions) to the effective conditions of knowledge, configuring that which he terms *state of information*, understood as the knowledge permitted by the concepts attributable to the subjects that they predicate. Subsequently, Pierce affirms in the Harvard Lectures XI (1865) that "*every true symbol is applicable to some real thing. Hence every symbol whether true or not asserts itself to be applicable to some real thing. This is the denotation of the symbol*".[12]

Peirce adds to his affirmation, stating that '*every genuine symbol relates or purports to relate to some form embodied in its object. This is its connotation.*'

This reflection enables him to reach the following conclusion:

[11] Cf. W1: 275–279.

[12] MS 107. W1: 286–287.

> Thus, no matter how general a symbol may be, it must have some connotation limiting its denotation; it must refer to some determinate form; but it must also connote reality in order to denote at all; but all that has any determinate form has reality and thus this reality is a part of the connotation which does not limit the extension of the symbol. And so every symbol has information.

In Speculative Grammar (1867),[13] Peirce develops the concepts of extension, depth, and information attributable, or quantities to symbols, or more generally to terms. In the relations of the symbols with their objects, which are general classes of phenomena, the informed extension of the symbol is considered; the informed depth of the symbol will be considered in the reference of the symbol to its origin, as well as to the common characteristics of the objects represented in the symbol. From that is derived information pertaining to the symbol as the sum—later corrected to the product or area—of the propositions in which the symbol is subject or predicate.[14]

In synthesis, in the logical perspective (1), which is only briefly outlined here, it is important to stress that, according to Peirce, with every increase of information there will be increase of the depth or extension of a symbol. In the absence of information, there will be no extension or depth. The symbol that does not present depth or extension (or both) would bring no information, since the product of both the quantities of the concept or the term would be null.

In the context of the strictly formal logical analysis of information, abductive reasoning is characterized as a rational procedure that leads to growth of information arising from the increase of depth of the concept. Faced with an anomalous fact, hypotheses are elaborated considering that—should this fact be the occurrence, for example, of a determined class of phenomena not exempt from collateral experience (such as those belonging to a known theory)—it would be possible to attribute to it the predicates characterizing that class of phenomena. Such an attribution would render intelligible the initially unexplained fact, hence dissipating the feeling of surprise that initiated the abduction.

While perspective (1) of Peircean information, outlined above, may be fundamental for an understanding of the unifying role of information in abductive reasoning, of especial interest to us here is the second, phenomenological, concept of information, and its relation with the sign, as a means of communication of forms. As we shall see, this phenomenological concept enables Peirce to elaborate a characterization of sign that goes far beyond the human dimension, contemplating the rich universe of innumerable beings that are nourished by information, self-organizing their behavior in accordance with sought-after ends.

Perspective (2) of information seems to have emerged towards the end of the nineteenth and beginning of the twentieth centuries, when Peirce's thinking underwent some major changes. During this period, Peirce introduced Phenomenology as the naïve science whose object is everything, without restriction, that appears to the mind. Definitively establishing the categories of firstness, secondness, and thirdness, Peirce unfolds the normative sciences into aesthetics, whose object is the

---

[13] CP 2.418–426.

[14] Cf. CP 2. 418–419.

phenomenon as admirable, and ethics, which studies the phenomenon as an end for the conduct attracted by it. Semiotics then establishes the mediation between the admirable and the sought-for goal, being able to hypothetically explain, in their essential characters, all the signs for a conduct that, by means of the representation they provide, can reach in the future the searched object.

In this new context, the concept of information is considered not only within the realm of formal logic (as a quantity attributed to the concept), but also according to a logic validated by the phenomenological context in which the conduct aimed at desired goals is inserted.

The text Speculative Grammar (1902 and 1903)[15] contains considerations about the classification of signs in ten classes according to the relation of the sign with itself, the sign with its object, and the sign with the interpretant determined by it. In this context, Peirce characterizes information within the semiotic process, clarifying the relations that transform a sign into an icon, an index, or a symbol, and those that in their relation with the interpretant transform it into a rheme, a dicisign, or an argument.

Special attention is given to the presuppositions of the constitution of a dicisign, since to this Peirce attributes the conditions required for the constitution of the arguments. A dicisign is composed of two parts: The first, denoted *subject-terms*, represents an icon of the object (which must have an existence that is independent of the sign in which it is represented), and the second, called the *predicate*, is (or represents) an icon of firstness (which can be a quality or an essence).

The components of the dicisign, thus assigned, address the necessity of the icon to fulfill the conditions for representing an existent object, since it is with this object that the sign must retain (as a necessary but not sufficient condition) some quality in common. Should the sign be a symbol and represent its object as a general class, it would maintain the iconic relation incorporated in this more complex form of representation. It ought also to be noted that the icon should represent the firstness of an idea, as either a simple quality or a general representation in the form of a sign of essence.

The second part of the dicisign consists of a rhematic index, denoting the subject to which the idea conveyed by the rhematic icon should be attributed.

However, the two parts alone are not sufficient for a sign to be interpreted as a dicisign, an effective and existential indicator of the object. A new condition is required, namely the syntax that unites these two parts. Peirce describes this as follows:

> These two parts must be represented as connected and that in such a way that if the Dicisign has any Object, it [the Dicisign] must be an Index of a Secondness subsisting between the Real Object represented in one represented part of the Dicisign to be indicated and a Firstness represented in the other represented part of the Dicisign to be Iconized. [16]

Identifying, in this context, dicisigns with propositions, Peirce affirms that "*every kind of proposition is either meaningless or has a real Secondness as its object.*" [17]

---

[15] CP 2.309–315.

[16] CP 2.312.

[17] CP 2.314.

In the year 1906, Peirce refined the concept of sign by introducing form as that which is transmitted from the object, via the sign or representamen, to the interpreting mind. This refinement was decisive in deepening understanding of the information provided to us about reality by means of signs. Peirce affirms that, in effect, "*a Sign may be defined as a Medium for the communication of a Form*", [18] and reiterates such a conception in a letter to Lady Victoria Welby,[19] stating that "*I use the Word 'Sign' in the widest sense for any medium for the communication or extension of a Form (or feature).*"

Without ever departing from an insistence on the genuinely triadic character of a sign (composed as it is by the correlation between sign and itself, its object, and its interpretant), Peirce explains the communication of the form of the object to the interpretant, with the sign as intermediary, stressing the notion of *conditional proposition.*[20]

In synthesis, in the phenomenological concept of information, which is only outlined here, Peirce differentiates form, as a quality constitutive of the object, and this same form as a relation communicated through the sign to its interpretant. It is by means of this distinction that it is effectively possible to know the object without, *ipso facto*, consuming it. Furthermore, the communicated form of the object is a condition of truth for every subsequent interpretation of that object (here one finds in play all the information that one has about the object and the form that it communicates). We note that in attributing to form "*the truth of a conditional proposition*", conferring on it the function of interpretant, it acquires central importance in the concept of information, since during the course of its constitution in a living thought process it determines the state of information available in that semiotic process.

A further important point is Peirce's preoccupation with the state in which one is, in relation to that which is known, since this state determines the amount of information about the object provided by semiosis. Acknowledging, then, that form consists of the truth of a conditional proposition, Peirce relates the establishment of truth, with respect to the object, to the conditions in which it should, could, or would in effect occur.

In a letter to Lady Welby, after recognizing sign as a means of communicating form, he acknowledges the necessity of distinguishing an *immediate object* and a *dynamic object* of the sign. The first is that which is represented by the sign itself, and the second is that which the object of the sign denotes, and which, independent of the sign, really embodies the form that is communicated through the sign to the interpreter. It is this *dynamic object* that is imposed on the mind as the effective experience during perception, denoted by perceptual judgment, and which, being learned instinctively and immediately, informs the mind about its reality. It is also from it that the information contained in the perceptual judgment proceeds. It seems to be in this sense that Peirce affirms: "*It is an object of actual experience.*" [21]

[18] MS 793:2. EP2:544 n.22.

[19] EP2:477.

[20] See pages 91–92 of this volume for a full quotation on Peirce's view of the being of a form as the truth of a conditional proposition.

[21] Idem, ibidem.

It is the role of the interpretant to represent the object as effectively existing, in the form of the dicent sign, equally representing the reason whereby the object is thus represented, assuming the form that is communicated to it representatively, that is: in the way that the mind is effectively influenced by the object.

In his Prolegomena to the Apology for Pragmaticism,[22] Peirce dedicates a lengthy discussion to the semiotic character of perception, fallible but undeniable, since it invades the mind, secure from any act of will that might wish to dominate it.

In the continuity between perceptual judgment and the abductive processes of concept construction, that which affects us in the instant of perception occurs through a judicative act of recognition (at least potentially). This judgment constitutes a premise, suggested in the form of judgment that exists in the repertory of the percept, belonging to the eidetic domain of the possible. In the same vein, the percept is identified as representing a determined class of objects, even though it may later demand the proof of a perception at the level of attention. In this process, we are faced with the dynamic object as a raw fact, but it is immediately investigated as identifiable with a repertory composed of percepts.

A hypothesis can then be raised that the form entitatively present in the dynamic object is potentially (and even virtually) found in Reason, in the form of percepts. In perceptual judgment, ideas assume the form of predicates; testing the dynamic object in its impact, this judgment acquires the form of a judgment of existence. Viewed abstractly, when predicates assume the function of subjects for new predications, the judgments then denote immediate objects, and in the sign the percepts occupy the logical place of the subjects of predication.

In synthesis, the origin of any informational process can be found in the percept, to be assumed in the perceptual judgment. From a general semiotic perspective, going beyond the limitations of human culture, the gap between percept and mind narrows, since they share the same universal informational instant that precedes all perception (and, consequently, every representation of the dynamic object) and that succeeds all interpretant series in the form of Truth.[23]

From a strictly logical point of view, and hence common not only to human cognition but also to all "*intelligence capable of learning by experience*",[24] all knowledge has as its originating element the sign as a mediating potentiality in the transmission of form. The latter activates the object, from which it is derived, permitting conduct to be once again directed, by means of the interpretant, towards the object.

This non-anthropocentric Peircean hypothesis—that all intelligence interprets percepts in a semiotic process, determining conduct in light of the object that it presents as a goal—will guide the conclusion to this chapter. It seems that here a line of research is opened such that, in the future, the notions of reason and instinct may become closer, both being understood from the level of generality adopted in Peirce's philosophy of expression and action.

---

[22] CP 4.530–571.

[23] CP 4.550–551.

[24] CP 2.227.

Instinct, Abduction, and Information: New Horizons in the Study of Life

So far, we have emphasized the significance of the Peircean phenomenological characterization of information, grounded on the dynamic triadic nature of the sign, which enables an understanding of his claim that abductive reasoning operates in continuity with perceptual judgments. Essentially, in the absence of information embodied in collateral experience, abduction could not even start, because there would be no way of narrowing down the organism's attention, in the right manner, in order to notice the occurrence of a problem. Thus, for example, a Martian who had just arrived on Earth would not have sufficient information to observe recent shifts in climatic conditions; he could not initiate an abductive reasoning and search for a reasonable hypothesis concerning climate change, because the basic information required for diagnosing the situation correctly would be missing from his collateral experience.

In short, information embodied in collateral experience helps with the shaping of perceptual judgments that—by constraining the object of a sign—may initiate abduction. When successful, this brings about new habits or hypotheses (sometimes in a flash, explaining an anomaly responsibly for the disruption of a stable system of habits), which, in turn, affect collateral experiences, and so on.

In his early well-known characterization of abduction, Peirce argues that it resembles instinct:

> This Faculty is at the same time of the general nature of Instinct, resembling the instincts of the animals in its so far surpassing the general powers of our reason and for its directing us as if we were in possession of facts that are entirely beyond the reach of our senses. It resembles instinct too in its small liability to error; for though it goes wrong oftener than right, yet the relative frequency with which it is right is on the whole the most wonderful thing in our constitution.[25]

As we saw in 'Introduction', in maturity Peirce concludes that instinct and reason are not separate, with (as for other animals) our instincts being formed under the influence of the same laws of nature that somehow shaped our existence. However, how can we understand Peirce when he claims "*The word 'instinct' itself is but a generalization of abstractions, one of the brood of language or of thought: there is no great difference between the two*...", and what, after all, is the role of information in instincts?

Our understanding is that Peirce's characterization of information (as a vehicle for the communication of form) goes far beyond the human dimension, contemplating the dynamic universe of myriads of inhabitants, which organize their conduct by means of instinct in order to exist. Information seems to be implicit in his conception of instinct, when he claims: "*...one 'instinct' of an animal may be a special determination of a more general 'instinct' of the same animal...*" In this context, much of human conduct could well be illustrated by the aforementioned passage,

[25] CP 5.173.

where Peirce considers as instinctive the general and voluntary action of an animal, when such action is very similar to the behavior of any other individual of the same species.

As discussed earlier, Peirce argues that when the action of an animal "*...is of such a kind as generally to have a beneficial effect upon the same animal or its progeny, which effect, however, the animals that act so can [...] at any rate not to have ascertained by reasoning from any other facts within their knowledge...*", then the type of action is '*instinctive*'. This action, considered as "*a general habit of behavior, regarded as appertaining to the animal's consciousness...*", he denotes "*an instinct*".

Applying the concepts of breadth and depth (from his formal logic characterization of information, summarized in Part II) now in the semiotic context, Peirce illustrates the above hypothesis with the example of a man of whom may be said, "*... he has a wide acquaintance with animals or plants, however superficial that acquaintance may be; or we may say that a man's linguistic acquirements may be somewhat narrow; yet those languages that [he] pretends to know, he knows to their very bottom...*" (EP2:473)

In this scenario, we can now explain better our hypothesis that the Peircean concept of *information* could help to elucidate not only the notion of instinct, but also the role of instinct in abductive reasoning. By introducing the dynamic triadic relation in his characterization of a symbol (as standing for something other than itself in a process of communicating a form), Peirce recognizes that this triadic relation constitutes the very foundation of any kind of information involved in the evolutionary complex system of habit creation, be it constitutive of the life of humans or any other organism. When stable, this complex system may constitute an instinct in the form of "*a general habit of behavior*" shown to be beneficial for the species. Given the dynamic character of life, abduction will start whenever this habit no longer applies to a present circumstance, and information is available indicating a new path to be initiated.

Furthermore, instinct, grounded in *information*, may not only express a manifestation of abduction, but also allow the generalization of the unfolding semiotic process characterizing innumerable expressions of life beyond human culture. Without departing from the principle of continuity, Peirce agrees with Darwin concerning evolution, with both of them acknowledging the existence of different stages in the plasticity of habit acquisition in living beings. However, it seems that Peirce is a step ahead in offering his semiotic concept of information as a powerful methodological tool for the study of life, as has been developed by contemporary complex system theorists. [26]

The advantage of employing the Peircean semiotic notion of information in the study of life is that it allows us to see new horizons beyond the causal (dyadic), reductionist, paradigm adopted by traditional biology. By dealing with forms, which

---

[26] From the complex system perspective, as pointed out by Merrell, in the study of life no level of information is necessarily more fundamental than any other, given that all levels are dependent on others in this rich, multilayer, interacting system.

convey law-like regularities into the triadic sign relations, Peirce foresees the richness of a scientific semiotic inquiry able to deal not only with data, but also with the effects produced by the data in an interpretant, constraining its conduct and, in a way, anticipating it. As Peirce says:

> A Symbol is a law, or regularity of the indefinite nature. Its interpretant must be of the same description and so must be also its complete immediate Object, or meaning. But a law necessarily governs, or 'is embodied in', individuals and prescribes some of their qualities. (EP2:274)

Since the interpretant of a symbol, being itself a sign and not someone who interprets the symbol, presupposes a mind that is not necessarily human or of any other kind of animal, but could, for example, be embodied in a cell (or a colony of bacteria), the narrowing of the gap between the intellects of humans and other animals, hinted at by Peirce as his writing evolved, could help with a more refined, less anthropocentric (although necessarily anthropomorphic) approach to life.

When applied in the context of life sciences, the three-dimensional aspect of the sign (the body, the medium, and the interpretant of the sign) allows the study of information processes below (and above) the level of human communication. As explained in detail by Robert Lane in his chapter "Peircean Semiotic Indeterminacy and its Relevance to Biosemiotics" (in this volume), the spectrum of possible semiosis involves many levels of complexity, such as animal communication including, naturally, non-human animal signals, animal behavior interpreting environmental phenomena, and sign processes within organs or organ systems, cells, molecular structures, and so forth. As indicated, a dicisign may convey information even in a hypothetical situation in which the interpretant is only realized in a potential interpreter. As Lane suggests, if a tree rots (see page 78) in a forest where there is nobody around to interpret that sign, there could be, nevertheless, meaningful information there ready to make actualized that indeterminate sign of a typical smell.

In short, Peirce's phenomenological concept of information allows us to contemplate the rich universe of innumerable beings that, nourished by information, seem to organize their actions in accordance with law-like regularities and chance, embodied in an ever-growing semiotic process. Furthermore, as signs grow, tracking their developmental trajectory may help with the study of the communication flow of information embodied in the multiple levels of interaction existing in living systems. This multiple (inter-) level communication is nowadays recognized as occurring not only in organisms but also in other types of complex systems whose dynamics the Peircean concept of information may help to understand.

Conclusion

In this chapter, we have presented a basis for our working hypothesis that the Peircean concept of *information* could help to clarify not only the notion of *instinct*, but also the role of instinct in abductive reasoning, narrowing the gap between the human intellect and that of other organisms. To conclude, we are going to outline methodological considerations concerning our assumption that the narrowing of this gap could significantly contribute to the development of biosemiotics.

Following in the tracks of Peirce, and of Jacob von Uexküll (1936)[27], Bateson (1979, 1987, 2000), Merrell (1987), and Rosen (1985), amongst others, we consider that the dynamics of life evolution cannot be properly understood from the mechanistic perspective that focuses only on the strictly causal aspects of the physical and chemical composition of living systems. Criticizing Spencer's mechanistic view of evolution, Peirce argues that the universe "*is not a mere mechanical result of the operation of blind law*". Distinguishing the law of habits from physical laws, he stresses the role of spontaneity, chance, and error in the dynamics of creation in the universe. In particular, chance (that which is not governed by any "deterministic laws," laws constitutes, in general, one of the major possibilities of initiating an abductive reasoning.

Without denying the principle of continuity, Peirce acknowledges different levels of complexity in the process of evolution of the universe, which, we believe, from the point of view of a special science would permit differentiations between living and nonliving beings. In this sense, it seems that he would agree with Jacob von Uexküll, methodological distinction between alive (animate) and inanimate nature, the first being nourished by semiotic meaningful information, available to all organisms, in contrast to a physical environment governed by strict natural laws.

Similar to Peirce, Jacob von Uexküll, investigates the creative nature of organisms, proposing the concept of *Umwelt* to characterize the intricate webs of relations that constitute the meaningful niches in which organisms develop their diverse forms of life. He argues that the main characteristic of living creatures (even the most simple, such as unicellular organisms) is their ability to respond to an external stimulus in a proper, specific, non-mechanical semiotic way. This reaction of all organisms allows them (while they are alive) to respond to signs and not only act according to causal impulses.

Here is not the proper place to reconstruct the very rich argument of Jakob von Uexküll in defense of a semiotic (non-anthropocentric) view of the great diversity of existing *Umwelten*. However, for our present purpose it is worth pointing out that this diversity does not minimize the importance of the common element that unifies us: the sharing of the same instinctive nature. As we saw earlier, this instinctive nature constitutes, according to Peirce, a common basis of the generating process of vitally important inferences that provide us with some grounding in reality, by expressing "*... a beneficial effect upon the same animal or its progeny*". However, as he insists, instincts (in the condition of general habit of behavior) are subject to changes when they lose their original function.

From the contemporary complex systems perspective, small changes (occurring at the microscopic level) in individual activities may produce, under certain unstable conditions, great changes in the organization of well-formed systems at the macroscopic level. In the case of living systems, under unstable conditions changes

[27] The attribution of an Umwelt and an Innenwelt to all the animals is present in work of Jacob Uexküll, 1936 as can be by the homonymous title of one of his first writings.

at the microscopic level may initiate the emergence of new informational patterns that could constrain (at the macroscopic level) their present and future habits.

Thus, consider for example the human instinct, manifested in language, of associating sounds with ideas. This very long-standing habit not only permits communication, but also directs our conduct in general. Strongly supported by linguistic habits, we have developed various types of rational skills that have contributed to a progressive distancing from other non-human animals. This distancing, as suggested by Pennisi and Falzone in *Il Prezzo del linguaggio*, has been responsible for producing a kind of "autism" that makes us insensitive to the recognition of other forms of animal life, sometimes with considerable loss for the environment and even for our own species.

This supposed human autism in relation to the conduct of non-human animals seems to result, in great part, from our habit of overvaluing the rational products of our linguistic abilities, instead of considering them as just one of the many possibilities of self-organization of our species. If this supposition is true, and faced with the problems it may create for individuals (and for our species itself), then abductive reasoning can start, probably grounded in non-linguistic signs, helping us to find habits that support better ways of conducting our present existence in relation to other species.

In this context, one of the main contributions to biosemiotics of the present chapter can be summarized thus: When conceived from the contemporary complex systems perspective, the concept of information developed by Peirce in maturity, and the hypotheses sketched by Jacob von Uexküll, a (concerning the semiotic richness of organisms' existence), could be of great benefit to research on the dynamics of the creation, maintenance, and change of stable habits constitutive of instincts in general. We hope that the relation between instinct, abduction, and information outlined here may also help with Peirce's hypothesis that thought is not in us, as a private property, but we are in thought with all living beings and, as he would say, with the cosmos!

Acknowledgments The authors would like to thank Vinicius Romanini, Mariana Claudia Broens, and José Artur Quilici Gonzalez for their valuable comments on previous versions of this chapter. The work was sponsored by UNESP, CNPq, and Fapesp. Finally, we thank our colleagues from GAEC and CLE for their inspirational research activities.

References

Bateson, G. (1979). *Mind and nature: A necessary unity*. New York: Cambridge Press.

Bateson, G. (1987). Men are like plants. In W. I. Thompson (Ed.), *Gaia—a way of knowing: Political implications of the new biology*. New York: Lindisfarne Press.

Bateson, G. (2000). *Steps to an ecology of mind*. London: The University of Chicago Press.

Gonzalez, M. E. Q., Broens, M. C., & D'Ottaviano, F. L. (2007). Abductive reasoning, information and mechanical systems. In *Studies in computational intelligence* (SCI) (Vol. 64, pp. 91–101). Berlin: Springer-Verlag.

Uexküll, J. von, & Kiszat, G. (1936). *Streifzüge duch die Umwelten von Tieren und Menschen.* Berlin: Springer. (Neudruck Fischer. Frankfurt 1970).
Merrell, F. (1987). *Signs grow: Semiosis and life processes*. University of Toronto Press.
Peirce Edition Project. (Ed.). (1998). *The essential Peirce. Selected philosophical writings Volume 2, 1893–1913*. Bloomington: Indiana University Press.
Rosen, R. (1985). *Anticipatory systems: Philosophical, mathematical & methodological foundations*. New York: Pergamon Press.

The Life of Symbols and Other Legisigns: More than a Mere Metaphor?

Winfried Nöth

> For every symbol is a living thing, in a very strict sense that is no mere figure of speech. The body of the symbol changes slowly, but its meaning inevitably grows, incorporates new elements and throws off old ones. (CP 2.222, 1901)

Life as Semiosis and the Life of Symbols

Biosemiotics, as defined by Jesper Hoffmeyer, assumes as its fundamental tenet that "life is based entirely on semiosis, on sign operations" (1996, p. 24). Major proponents of this approach to the life sciences proclaim that *Life is the Action of Signs* (Emmeche and Kull 2011). They are convinced that semiotic answers to the questions concerning the essence of life are possible and necessary (Emmeche and El-Hani 2000). The determination of the "basic principles of a semiotic study of life" and the demonstration of how "sign processes *per se* and the processes of life" are "intimately and inseparably interconnected" (Kull et al. 2009, p. 168) are among the essential goals of the biosemiotic research program.

This biosemiotic research program has avid supporters, but also critics who reprimand biosemiotics for what they consider to be undue semiotic imperialism (cf. Nöth 2001, p. 74). Some claim that the foundations of biology are in the natural sciences, whereas others believe that semiotics should restrict itself to the study of intentionally produced symbols (Böhme 1996, pp. 20–21). Such criticism is unjustified because it is based on reductionist premises concerning the nature of signs and sign processes. Semiotics is not only the study of intentional and conventional signs. It is also the study of signs in nature. A distinctive feature of signs is their power to make absent things present to an interpreting organism. Peirce shows that this characteristic is shared by signs and living beings. Both have the "active power to establish connections between different objects, especially between objects in different Universes" (CP 6.455; 1908).

W. Nöth (✉)
Catholic University of São Paulo (PUC)
e-mail: noeth@uni-kassel.de

V. Romanini, E. Fernández (eds.), *Peirce and Biosemiotics,* Biosemiotics 11,
DOI 10.1007/978-94-007-7732-3_9, © Springer Science+Business Media Dordrecht 2014

Natural scientists and humanists who resent the claim that the study of life can be approached from a semiotic perspective must find the complementary thesis that *symbols are living beings* even more appalling. Is not a symbol a product of human minds, and if so, how can symbols have a life of their own? Does life not need to be embodied in chemicals, such as amino acids, in molecules, and in real matter, whereas symbols are products of human minds, which are in their essence immaterial? To dualist minds accustomed to dividing the world into matter and mind and into living beings and material things, the suggestion that symbols have life must sound provocative. And yet, it was Charles Sanders Peirce (1839–1914), a natural scientist by training and profession, who put forward this provocative thesis, when he wrote, in 1901, that "every symbol is a living thing, in a very strict sense that is no mere figure of speech" (CP 2.222) or, a year later, when he affirmed that symbols actually live, "that there not only *may be* a living symbol, realizing the full idea of a symbol, but even that there actually *is* one" (CP 2.114; 1902; italics added).

Peirce's assertion that symbols (and other legisigns) live is the topic of the present paper. Since it is an assertion concerning the nature of signs and not a statement about the nature of life, the paper does not directly seem to be a study in biosemiotics. However, if signs are to a certain degree living beings, the scope of biosemiotics must begin with and include the study of the life of signs. Furthermore, biosemiotics is certainly involved in the thesis that symbols are living beings because it presupposes a definition of life, and this definition can only be a biosemiotic one.

Synechistic Premises: Symbols, Things, and Life

What does Peirce mean with his biologically provocative thesis that symbols are living things? Certainly not that symbols have the biochemical or molecular characteristics of organisms. If he nonetheless claims that symbols evince life "in a very strict sense", his words must be examined in light of synechism, his "doctrine of continuity", which teaches "that all that exists is continuous" (CP 1.172, c.1897). Synechism is opposed to dualism, "the philosophy which performs its analyses with an axe, leaving as the ultimate elements, unrelated chunks of being" (CP 7.570, c.1892).

Peirce develops his thesis that symbols are living things under synechistic premises insofar as he presupposes that continuity must not only be assumed very generally between mind and matter, as he postulates (following Schelling) in his evolutionary cosmology (cf. CP 6.158, 1891; Nöth 2004), but that there is also continuity between life and the products of living minds, i.e., symbols. Whether symbols live or not is a matter of the degree to which they have characteristics in common with living beings, but Peirce is convinced that symbols are close to life on the continuum between living beings and lifeless things.

In the light of this premise, Peirce's argument that it is no mere figure of speech to say that symbols evince life must be read as the rhetorical device of a caveat, by which the author positions himself against readers who might be inclined to interpret his argument as a "mere metaphor" and to conclude that symbols are actually only *lifelike,* and in fact *lifeless*, as metaphors are never literally true. Peirce counter-argues that on the life/nonlife continuum, he sees symbols very close to living beings.

It is interesting to note that Peirce uses this rhetorical device against misinterpretations of his arguments as "mere metaphors" in other contexts that deal with his synechistic views concerning the continuity between matter and mind. Another example of this device appears in a manuscript of 1903, in which he develops the argument that the human mind is not only determined by the laws of natural evolution, but also displays similarities, in its modes of operation, with elementary laws of the material world from which it has evolved. His argument is: "It is somehow more than a mere figure of speech to say that nature fecundates the mind of man with ideas which, when those ideas grow up, will resemble their father, Nature" (CP 5.591, 1903).

Habits and Laws, Symbols and Legisigns

The essential criteria which Peirce sets up in defining symbols are neither *codedness* nor *conventionality*, but *generality* and *habit* (cf. Nöth 2010). A symbol is a sign "whose special significance or fitness to represent just what it does represent lies in nothing but the very fact of there being a habit, disposition, or other effective general rule that it will be so interpreted" (CP 4.447, 1903). By *habit*, Peirce means both "acquired" and "inborn" dispositions (CP 2.297, c.1895). Habits in this sense certainly presuppose life. An inborn habit presupposes birth and hence life in a phylogenetic perspective. An acquired habit presupposes learning and hence life in an ontogenetic perspective. But the habit Peirce has in mind is not merely the interpreters' habit of complying with the conventions necessary to make themselves understood; it is neither a "precept" nor does it describe the action of an interpreter "in obedience to a law" (CP 1.586, c. 1903). Instead, the one who interprets a symbol does so according to a general "rule of conduct, including thought under conduct [...] in conformity to [a] norm [...] in the sense of a precept, [...] a pattern which is copied" (CP 2.315, 1902).

All symbols are also legisigns. A sign is a symbol with respect to "the relation of the sign to its object", whereas it is a legisign with respect to "the sign in itself" (CP 2.243, 1903). A legisign is "a sign which is of the nature of a general type" or law (CP 4.414, 1903). Legisigns are *types* which produce *replicas* or *tokens*. Replicas or tokens are sinsigns, signs characterized as singular:

> A Legisign is a law that is a Sign. This law is usually established by men. Every conventional sign is a legisign [but not conversely]. It is not a single object, but a general type which, it has been agreed, shall be significant. Every legisign signifies through an instance of its application, which may be termed a Replica of it. Thus, the word "the" will usually occur from fifteen to twenty-five times on a page. It is in all these occurrences one and the same word, the same legisign. Each single instance of it is a Replica. The Replica is a Sinsign. Thus, every Legisign requires Sinsigns. But these are not ordinary Sinsigns, such as are peculiar occurrences that are regarded as significant. Nor would the Replica be significant if it were not for the law which renders it so. (CP 2.246, 1903)

Cultural anthropologists and humanist philosophers usually define symbols as uniquely human signs (cf. Nöth 2000, pp. 178–184). When Peirce, by contrast, defines the symbol as a sign related to its object by a habit, he means habits which can also be found in nature. Peirce's symbol is a sign "merely or mainly by the fact that it is used and understood as such, whether the habit is natural or conventional, and without regard to the motives which originally governed its selection" (CP 2.307, 1902). Habits in this sense do not presuppose human intentionality. The sign behavior of mating birds and the dance of the honey bee are examples of symbols although they are not conventional signs (cf. Short 1982, pp. 296–297). They are determined by instincts, which are habits in an evolutionary sense.

When Peirce defines the legisign as a sign due to a law "usually established by men", this definition sounds rather similar to the way cultural anthropologists and philosophers of the symbol (such as Cassirer) define the symbol, but in this instance the adverb "usually" is particularly important. Since all symbols are included in the class of legisigns according to his typology, the category of legisigns is no less comprehensive than the one of symbol in the broader Peircean definition. The difference between symbols and legisigns is that all symbols are legisigns, but not all legisigns are symbols. According to Peirce's typology, there are two other types of legisign, the iconic legisign and the indexical legisign (CP 2.258-2.260, 1903). The editors of the *Collected Papers* suggest that a "diagram apart from its factual individuality" (CP 2.258) is an example of an iconic legisign. Short (2007, p. 223) gives the example of a (general) geometrical diagram. In the context of his theory of Existential Graphs, Peirce defines graphs as legisigns and individual instances of graph-replicas (CP 4.414, 1903). Typical examples of indexical legisigns are proper names and demonstrative pronouns. Indexical legisigns differ from symbols because they lack generality and denote individuals. Iconic legisigns differ from symbols since they derive their potential to signify not from a convention, but from the quality which it has in common with the object it denotes.

Symbols may include indexical and iconic features. For example, the dance of the honey bee signaling the direction towards and the distance from the nectar has both indexical (i.e., directional reference) and iconic elements (i.e., similarity of dance angle and proposed flight angle). The reason why symbols are nonetheless symbols, despite their indexical and iconic features, is "that the rules for indexical or iconic legisigns refer interpreters to indexical or iconic grounds, whereas the rules for symbolic legisigns are themselves grounds of significance" (Short 1982, p. 294).

The reason why Peirce often mentions only symbols and no other legisigns when he discusses signs as "living things" is simply that he only introduced the distinction between symbols and legisigns in 1903, so that earlier reflections could not have drawn on this distinction. In short, whenever symbols are discussed in following discussions, it may be assumed that "symbols and other legisigns" are meant according to Peirce's typology of signs after 1903. However, even after 1903, there are contexts in which Peirce restricts himself to speaking merely of symbols, even though his arguments also apply to other legisigns.

It is true that hardly anybody has ever denied that symbols presuppose life, but it is also true that postulating habit as the distinctive feature of the symbol places greater emphasis on the living nature of symbols than on criteria such as codedness and conventionality. Furthermore, postulating that symbols presuppose life is not the same as saying that symbols are living beings. So, what are the really distinctive biological characteristics of symbols?

Purpose and Autonomous Agency

Both symbols and life pursue purposes and act according to a scheme of final causality (Santaella 1999). Purpose is a characteristic of life. The biological purpose of any organism is to survive both individually and as a species. If the final cause of life is self-reproduction and self-replication, symbols are "living realities" (CP 6.152), which have their teleology in self-replication, the creation of interpretants, and in determining future thoughts and interpretations. "The whole purpose of a sign is that it shall be interpreted in another sign" (CP 8.191, 1904). What is characteristic of the teleology of both symbols and biological organisms is that their goals are not predetermined with exact precision. "The law of habit exhibits a striking contrast to all physical laws [...]. A physical law is absolute, [... but] no exact conformity is required by the mental law. [...] The law of mind only makes a given feeling more likely to arise" (CP 6.23, 1891). Only symbols and other legisigns, but not sinsigns and qualisigns, can be said to have the purpose of self-replication. A sinsign cannot replicate since it is "an actual existent thing or event which is a sign" (CP 2.245, 1903), while a qualisign, which is a mere "quality which is a Sign" (CP 2.244, 1903), is too vague for the purpose of replicating in another sign.

What the symbol represents is determined by habits of interpretation, which are not only the habits of individual interpreters; it is not a matter of some individual habit. Symbol users neither are the creators of the symbols they use nor can they do away with any specific symbol. "You can write down the word 'star', but that does not make you the creator of the word, nor if you erase it have you destroyed the word" (CP 2.301, ca. 1895).

Of course, symbols are created by humans, just like children are created by their parents. However, once created, symbols begin to lead a life of their own just like children created by parents do. This is why Peirce says that signs are created *in* and not *by* human minds: "The Sign creates something in the Mind of the Interpreter [...]. And this creature of the sign is called the Interpretant. It is created by the Sign; but not by the Sign quâ member of whichever of the Universes it belongs to; but it has been created by the Sign in its capacity of bearing the determination by the Object. It is created in a Mind (how far this mind must be real we shall see)" (CP 8.179, 1903).

Another characteristic which signs share with biological organisms is that both are agents in processes of semiosis. This may sound paradoxical, for are signs not *produced* and used by biological organisms to serve the purpose of the latter? Undeniably they are, but signs also have lives and purposes of their own independently of the ones of individual sign users. Symbols are not merely the instruments of living agents, but they act in processes of semiosis with purposes of their own: the purpose of the symbol is to represent its object and "to convey some further information concerning it" (CP 2.231, 1910). The autonomy which a symbol develops in relation to its creators raises the question of semiotic agency (cf. Nöth 2009). Agents are by presupposition living beings. To the degree to which symbols act autonomously in the process of semiosis, they act as "living things".

Ransdell (1992) points out that the autonomy of the symbol does not go so far as to make its users blind agents. "To regard semiosis [...] as always due primarily to the agency of the sign itself rather than to the agency of an interpreter, human or otherwise, does not deny that human agency has an important role in the occurrence of meaning phenomena." The determination of the symbol user by the symbol does not exclude the partial determination of the symbol by those who use it. After all, the users' minds, their memories, and their experiences are the loci of the embodiment of the objects of the sign, so that the users are co-agents in the process of semiosis. In this sense, symbols and symbol users are intertwined in a cybernetic master-slave dialectic, which is a synechistic circle of mutual semiotic growth.

Self-Reproduction and Self-Replication of Types in Their Tokens

In organisms, just as in nucleic acid molecules, life shows itself in the form of self-reproduction and self-replication (cf. Kull 2000), features which have more generally been subsumed under the heading of autopoiesis (Maturana and Varela 1972). In 1904, Peirce writes: "A symbol is something which has the power of reproducing itself", which is a reference to the processes of self-replication and self-reproduction in semiosis. In symbols, replication can be found in the process of their embodiment of types in tokens. Peirce describes the process as follows:

> A *Symbol* is a Representamen whose Representative character consists precisely in its being a rule that will determine its Interpretant. All words, sentences, books, and other conventional signs are Symbols. We speak of writing or pronouncing the word "man"; but it is only a replica, or embodiment of the word, that is pronounced or written. [... The] succession of three sounds [...] becomes a sign only in the fact that a habit, or acquired law, will cause replicas of it to be interpreted as meaning a man or men. The word and its meaning are both general rules [...]. A Symbol is a law, or regularity of the indefinite future. [...] But a law necessarily governs, or "is embodied in" individuals, and prescribes some of their qualities. (CP 2.292-93, 1902)

It is not the symbol user, but the symbol itself, *by means of* symbol users, that creates replicas as its embodiments. The three-letter word *man*, which exemplifies the sign type of a symbol, "is not a thing. What is its nature? It consists in the really working general rule that three such patches seen by a person who knows English will effect his conduct and thoughts according to a rule. [...] The being of a symbol consists in the real fact that something surely will be experienced if certain conditions be satisfied. Namely, it will influence the thought and conduct of its interpreter" (CP 4.447, 1903). As a habit, the symbol determines its replicas through the minds of its users in form (pronunciation and spelling) and meaning. As a law, the effects created by the symbol are general, whereas its replicas are the singular instances of its occurrence.

The self-replicative power of the symbol *man* thus consists "in the fact that a habit, or acquired law, will cause replicas of it to be interpreted as meaning a man or men" (CP 2.292, 1902). In sum, the symbol is self-replicative since it has "the power of reproducing itself, and that essentially, since it is constituted as a symbol only by the interpretation" (EP 2: 322, 1904).

Replicas are thus determined by the types they embody, just like the biological phenotype is determined by its genotype. In their singularity, replicas are phenomena of the category of secondness, whereas symbols as types are phenomena of thirdness. Only the replica, not the symbol itself, has a real existence, but although the symbol, because of its generality, cannot be said to exist, its power to produce replicas of itself is a real fact. As types, symbols merely consist in the potential to determine tokens that come into existence and exert their effects: "The word lives in the minds of those who use it. Even if they are all asleep, it exists in their memory" (CP 2.301, ca. 1895). In its genuine thirdness, a symbol is a mere "idea abstracted from all efficiency" (CP 1.213, 1902). The effect a symbol embodies in real tokens is a matter of secondness. This is why it can neither be reduced to any, nor to all of its replicas.

Nevertheless, symbols need replicas to survive. A type "has to be embodied in a Token which shall be a sign of the Type, and thereby of the object the Type signifies" (CP 2.537, 1901). A symbol is the representation of an idea, but even without any real existence, it needs to be embodied, for "an idea without efficiency is something equally absurd and unthinkable" (CP 1.213, 1902). A symbol that is not replicated falls into oblivion and eventually dies out. Peirce compares the symbol as a type to the agency of a court and its replicas to the acts of its sheriff. A symbol is powerless without its replication, just like "a court cannot be imagined without a sheriff" (ibid.).

Procreation and Growth

Symbols procreate symbols insofar as the interpretant they create are (new) symbols, and a symbol "is constituted as a symbol only by the interpretation" (EP 2: 322, 1904). The interpretant of a symbol not only reproduces (or replicates) the symbol as it is, but also displays a proclivity to growth since the interpretation a more complex symbol can offer about its object conveys new information about it. The fact that "symbols grow" (CP 3.302, c. 1895) in time thus means that "the body of the symbol changes slowly, but its meaning inevitably grows, incorporates new elements and throws off old ones" (CP 2.222, 1903).

Growth is an essential characteristic of life. In organisms, the most elementary form of growth is cell division (mitosis), which begins with the egg dividing into "daughter cells" at a ratio of 2, 4, 8, 16, 32 etc. (cf. Schrödinger 1947, p. 8). Symbols grow differently. They are "born", when they are first invented, and they can "die" by falling into oblivion or by being substituted by new symbols. The homologies between the life of words, on the one hand, and biological organisms as well as species, on the other, have been much discussed in the framework of historical and evolutionary linguistics (cf. Driem 2005). It is true that to say that they are born and die makes the argument that symbols are living things somewhat weak. The same can also be said of lifeless objects, such as consumer goods, of which marketers say that they come to life (with their invention) and die (when they become obsolete). Do symbols grow in any stronger sense of the word?

Peirce describes how symbols grow in at least three different ways, (1) the growth of symbols in the process of semiosis, as discussed above, (2) the growth of the meaning of concepts, and (3) the growth of complex ideas and arguments. How concepts grow in meaning is the topic of the following much quoted passage:

> Symbols grow. They come into being by development out of other signs, particularly from icons, or from mixed signs partaking of the nature of icons and symbols. We think only in signs. These mental signs are of mixed nature; the symbol-parts of them are called concepts. If a man makes a new symbol, it is by thoughts involving concepts. So it is only out of symbols that a new symbol can grow. *Omne symbolum de symbolo*. A symbol, once in being, spreads among the peoples. In use and in experience, its meaning grows. Such words as *force*, *law*, *wealth*, *marriage*, bear for us very different meanings from those they bore to our barbarous ancestors. (CP 2.302, c. 1895)

The growth of symbols in this sense is the growth of the meaning of ideas or concepts. Symbols grow as the information stored in them and conveyed by them increases in time (cf. Nöth 2012a). Peirce's favorite examples stem from the natural sciences, where new discoveries lead to redefinitions of key terms: "How much more the word electricity means now than it did in the days of Franklin; how much more the term planet means now than it did in the time [of] Hipparchus. These words have acquired information; just as a man's thought does by further perception" (CP 7.587, 1866).

Words grow in the history of language, acquiring new meanings from generation to generation, but what grows is only the word as a type; word tokens cannot grow, because they exist only in the singular moment of their utterance, never to be replicated as tokens (only as another token replicated from its corresponding type). Consider the example of a novel, which is a complex symbol. An individual copy of the book is a token of this sign; it remains essentially unchanged until it decays or is destroyed. As a sign type, this novel grows with the interpretants it creates, acquires new meanings and loses older ones.

Words do not only grow individually, but the whole vocabulary of a language grows in number. This is so because words have an autopoietic potential, which they share with biological organisms: symbols have a self-regenerative ability that is manifested through the creation of new symbols. Peirce describes this autopoietic potential of symbols as follows: "Perhaps the most marvelous faculty of humanity is one which it possesses in common with all animals and in one sense with all plants, I mean that of procreation. [...] If I write 'Let Kax denote a gas furnace', this sentence is a symbol which is creating another within itself" (CP 3.590, c. 1867).

A different and more general kind of growth takes place as symbols evolve in the semiosphere of ideas. Ideas are symbols. The characteristics of life, which Peirce ascribes to an idea, are three: "The first is its intrinsic quality as a feeling. The second is the energy with which it affects other ideas, an energy which is infinite in the here-and-nowness of immediate sensation, finite and relative in the recency of the past. The third element is the tendency of an idea to bring along other ideas with it" (CP 6.135, 1892).

This kind of growth of symbols is the topic of Peirce's *Law of Mind* (1892). It states that ideas tend to spread continuously, influence other ideas, and thereby become more and more general: "In this spreading they lose intensity, and especially the power of affecting others, but gain generality and become welded with other ideas" (CP 6.104). The continuous growth of symbols in this form is evidence of life, insofar as "general ideas are living feelings spread out" (CP 6.143). Generality, which is a distinguishing feature of all symbols, implies the possibility of determining future events in which the symbol will be used again. "A general idea, living and conscious now, it is already determinative of acts in the future to an extent to which it is not now conscious" (CP 6.156).

Self-Control, the Strength of Symbolic Arguments and the Goal of Revealing the Truth

Anthropologists have claimed that prevarication is at the root of human semiosis (Sebeok 1976). If Peirce says the opposite, namely that "the purpose of signs—which is the purpose of thought—is to bring truth to expression" (CP 2.444 fn, c.1893), he does not mean that humans do not lie. Nor would he deny that the possibility of saying the truth implies the possibility of lying. Instead, Peirce refers to the logical power inherent in symbols of the class of arguments. The argument, he states, is "the only kind of representamen which has a definite professed purpose [… and] the professed purpose of an argument is to determine an acceptance of its conclusion, and it quite accords with general usage to call the conclusion of an argument its meaning" (CP 5.175, 1903). Arguments oblige us by logical necessity to reach the conclusion which derives from their premises. They are, thus. endowed with a capacity for self-control, which consists in their resistance against representations ignoring "that truth and justice are the greatest powers in this world". Although these powers need "defenders to uphold it", they are also able to create their own "defenders and give them strength". After all, "there is efficient causation and there is final, or ideal, causation. If either of them is to be set down as a metaphor, it is rather the former" (CP 8.272, 1897).

The logical force of arguments thus exerts a control over those who use them, and this is why "it is a perfectly intelligible opinion that ideas are not all mere creations of this or that mind, but on the contrary have a power of finding or creating their vehicles, and having found them, of conferring upon them the ability to transform the face of the earth" (CP 1.217, 1902). Therefore, the revelation of truth is not only a purpose of individuals; arguments have this purpose in themselves. In the long run, the correct arguments cannot be ignored.

Strong ideas and arguments have a strength that resides in their evidence and not in the minds of their proponents. Not only proponents create good ideas, but good ideas create their own proponents since the very "ideas somehow manage to grow their machinery, and their supporters, and their facts, and to render the machinery, the supporters, and the facts strong" (CP 2.149, 1902). It is important to emphasize that not only words (rhemes) are signs in the framework of Peirce's semiotics, but also propositions (dicisigns) and arguments. Only propositions and arguments can be true or false, valid or invalid, but they are so independently of what a sign producer claims. Ideas have their own strengths or weaknesses, the power "to grow their machinery and their supporters". Although "most of us [...] look askance at the notion that ideas have any power [...], that some power they have we cannot but admit" (CP 2.149, 1902).

On the Parasitic Nature of Symbols

Parasites are organisms which live as uninvited tenants at the expense of their host organism, a life from which only the tenant benefits, whereas the host is mostly harmed. In a weak sense, all symbols live in symbiosis. Words live in symbiosis when they acquire signification in and from sentences and texts. Like symbionts that depend on other species for their survival, symbols live as agents connecting themselves with objects of a different kind, *viz.,* with the objects they represent and the interpretants they create. In the morphology of inflected languages, affixes (prefixes and suffixes) live a vicarious life. Like parasites, that can only live attached to their hosts, they need *stems* (or *roots*) to live and to survive in language and discourse (cf. Nöth 2012b).

What is true of morphemes and words is no less true of *texts*. How they live at the expense of other texts has been studied in the theory of *intertextuality*, according to which no text can live on its own. In fact, there are no texts, but only *intertexts*, which, like parasites feeding on hosts, nourish themselves from other texts since intertexts, like symbionts, are made up of "citations, references, echoes, antecedent or contemporary cultural languages (what language is not?), which cut across it through and through in a vast stereophony" as Roland Barthes (1977, p. 160) puts it.

To the degree that signs live in symbiosis, they have an agency of their own. Can this be true in a more than metaphorical sense? Skeptics will object that a symbol is a mere mark on paper or a *flatus vocis*, a mere external tool of its user, who is the only and real semiotic agent. However, in alignment with Peirce's argumentation, it must be objected that this is to confound the replica of a symbol with the symbol as type. Replicas of symbols in their acoustic or written form are indeed lifeless things (phenomena of secondness), but symbols as genuine thirdnesses live on as self-replicative beings. It is true that symbols cannot replicate themselves without the agency of symbol users, but neither can parasites live without their hosts. Like a parasite that only reproduces in the organism of another species, a symbol needs a symbol user to replicate. We, as symbol users, are of course living semiotic beings, too, but not the creators of the symbols we use. We are not the copyright owners of our own words since we did not invent them. In much the same vein we can say, in extension of Peirce and Serres (1982), that *we* are the parasites of the symbols we use and on which we feed—quite to the contrary of those who maintain the opposite, namely that the symbol (resp. language) is the parasite (e.g., Salverda 1998; Kortlandt 2002).

References

Barthes, R. (1977). *Image—Music—Text* (trans: Heath, S.). London: Fontana Press.

Böhme, H. (1996). Über vormoderne Naturkonzepte. *Zeitschrift für Semiotik, 18*(1), 19–22 (in German).

Driem, G. V. (2005). The language organism: The Leiden theory of language evolution. In J. W. Minett & W. S.-Y. Wang (Eds.), *Language acquisition, change and emergence: Essays in evolutionary linguistics* (pp. 331–340). Hong Kong: City University of Hong Kong Press. http://www.himalayanlanguages.org/files/driem/pdfs/prague.pdf.

Emmeche, C., & El-Hani, C. N. (2000). Definindo vida. In C. Niño El-Hani, & A. A. Passos Videira (Eds.), *O que é vida?* (pp. 31–56). Rio de Janeiro: Relume Dumará.

Hoffmeyer, J. (1996). *Signs of meaning*. Bloomington: Indiana University Press.

Kortlandt, F. (2003). The origin and nature of the linguistic parasite. In B. L. M. Bauer G.-J. *Pinault (Eds.), Language in time and space: A Festschrift for Werner Winter on the occasion of his 80th birthday*, (pp. 241–244). Berlin: Mouton de Gruyter.

Kull, K. (2000). Copy *versus* Translate, Meme *versus* Sign: Development of biological textuality. *European Journal for Semiotic Studies, 12*(1), 101–120.

Kull, K., et al. (2009). Theses on biosemiotics: Prolegomena to a theoretical biology. *Biological Theory, 4*(2), 167–173.

Maturana, H. R., & Varela, F. J. (1972). *Autopoiesis and cognition*. Dordrecht: Reidel.

Nöth, W. (2000). *Handbuch der Semiotik* (2nd rev. ed.). Stuttgart: Metzler.

Nöth, W. (2001). Ecosemiotics and the semiotics of nature. *Sign Systems Studies, 29*(1), 71–82.

Nöth, W. (2004). Semiogenesis in the evolution from nature to culture. In M. Alač, & P. Violi (Eds.), *In the beginning: Origins of semiosis* (pp. 69–82). Turnhout: Brepols.

Nöth, W. (2009). On the instrumentality and semiotic autonomy of signs, tools, and intelligent machines. *Cybernetics & Human Knowing, 16*, 11–36.

Nöth, W. (2010). The criterion of habit in Peirce's definitions of the symbol. *Transactions of the Charles S. Peirce Society, 46*(1), 82–93.

Nöth, W. (2012a). Charles S. Peirce's theory of information: A theory of the growth of symbols and of knowledge. *Cybernetics & Human Knowing, 19*(1–2), 99–123.

Nöth, W. (2012b). Signs from the life of organisms, species, languages, and the media. In T. Maran, K. Lindström, R. Magnus, & M. Tønnessen (Eds.), *Semiotics in the wild: Essays in honour of Kalevi Kull on the occasion of his 60th birthday* (pp. 123–130). Tartu: Tartu University Press.

Peirce, C. S. (1931–1958). *Collected Papers*, vols. 1–6, ed. C. Hartshorne and P. Weiss, vols. 7–8, ed. A. W. Burks. Cambridge, MA: Harvard University Press (quoted as CP).

Ransdell, J. M. (1992). Teleology and the autonomy of the semiosis process. In M. Balat, & J. Deledalle-Rhodes (Eds.), *Signs of humanity/L'homme et ses signes* (Vol. 1, pp. 43–48). Berlin: Mouton de Gruyter. http://www.cspeirce.com/menu/library/aboutcsp/ransdell/autonomy.htm.

Salverda, R. (1998). Is language a virus? Reflections on the use of biological metaphors in the study of language. In M. Janse (Ed.), *Productivity and creativity. Studies in general and descriptive linguistics in honor of E. M. Uhlenbeck* (pp. 191–209). Berlin: Mouton de Gruyter.

Santaella, L. (1999). A new causality for the understanding of the living. *Semiotica, 127*(1/4), 497–519.

Schrödinger, E. (1947). *What is life*? Cambridge: University Press. Online: http://whatislife.stanford.edu/LoCo_files/What-is-Life.pdf.

Sebeok, T. A. (1976). Notes on lying and prevarication. In T. A. Sebeok (Eds.), *Contributions to the doctrine of signs* (pp. 143–147). Lanham: University Press of America.

Serres, M. (1982). *The parasite*, L. R. Schehr, transl. Baltimore: Johns Hopkins University Press.

Short, T. L. (1982). Life among the legisigns. *Transactions of the Charles S. Peirce Society, 18*(4), 285–310.

Short, T. (2007). *Peirce's theory of signs*. New York: Cambridge University Press.

Signs Without Minds

John Collier

A ***Sign,*** or ***Representamen,*** is a First which stands in such a genuine triadic relation to a Second, called its ***Object,*** as to be capable of determining a Third, called its ***Interpretant,*** to assume the same triadic relation to its Object in which it stands itself to the same Object. The triadic relation is ***genuine,*** that is its three members are bound together by it in a way that does not consist in any complexus of dyadic relations. That is the reason the Interpretant, or Third, cannot stand in a mere dyadic relation to the Object, but must stand in such a relation to it as the Representamen itself does. Nor can the triadic relation in which the Third stands be merely similar to that in which the First stands, for this would make the relation of the Third to the First a degenerate Secondness merely. The Third must indeed stand in such a relation, and thus must be capable of determining a Third of its own; but besides that, it must have a second triadic relation in which the Representamen, or rather the relation thereof to its Object, shall be its own (the Third's) Object, and must be capable of determining a Third to this relation. All this must equally be true of the Third's Thirds and so on endlessly; and this, and more, is involved in the familiar idea of a Sign; and as the term Representamen is here used, nothing more is implied. A ***Sign*** is a Representamen with a mental Interpretant. Possibly there may be Representamens that are not Signs. Thus, if a sunflower, in turning towards the sun, becomes by that very act fully capable, without further condition, of reproducing a sunflower which turns in precisely corresponding ways toward the sun, and of doing so with the same reproductive power, the sunflower would become a Representamen of the sun. But ***thought*** is the chief, if not the only, mode of representation. (CP 2.274, 1897)

Introduction

In a number of places, such as in the above quote, Peirce argues that anything with the necessary triadic structure of sign, object and interpretant can be a proper sign, even if there is no mental representation involved, though he says in the same passage that signs usually, if not always, have a mental interpretant (Peirce 1940, p. 100). He notes that the notion of interpretant does not logically require consciousness, but since we have no clear cases that do not, so we must at least fix our understanding of semiosis with these clear cases (1940, p. 282). The advent of biosemiotics has extended the notion of semiosis well into the non-mental sphere.

J. Collier (✉)
University of Kwazulu-Natal, Durban, South Africa
e-mail: collierj@ukzn.ac.za

V. Romanini, E. Fernández (eds.), *Peirce and Biosemiotics,* Biosemiotics 11, DOI 10.1007/978-94-007-7732-3_10,

In some cases in biosemiotics the signs are similar to human cognitive signs by involving perceptions and possibly even deliberate action, but many proposed cases within biosemiotics do not involve anything that might be considered to be mental, especially within endobiosemiotics, which involves at its lowest level chemical processes. If these extensions of Peircean semiotics are sound we need a clear idea of what it is to be a sign when minds are not involved. Peirce gives us some hints about how the notion of semiosis might be extended, and what the limits of its extension might be. These come from both his paradigmatically mental signs as well as what he says about non-mental signs, including signs in biological systems.

It is useful to make a number of distinctions in order to get a grip on increasingly difficult aspects of the problem of delimiting non-mental signs. First, we have to distinguish between non-mental things that are signs for some mind or minds and non-mental things that are signs for non-mental things. Each case further breaks down into cases that involve communication (or some sort of transmission of information between bearers and receivers of the signs) and those that do not, in which any transmission and interpretation is internal. In the latter case there is a logically possible distinction between cases that require some sort of internal transmission, and those that do not. The purpose of this chapter is to explore and explain each of these cases. The internal cases are especially important to endobiosemiotics, whose Peircean version has been derided as "unscientific." By examining the stronger cases in order to see what is essential to their being semiotic, the understanding gleaned can be extended to the more questionable cases. This will allow us to put some limits on what can be understood semiotically, contrary to movements towards pansemioticism according to which semiotics pervades the entire Universe.

Peirce's famous case of the sunflower, and whether it is a sign of the direction of the sun, is a case in point. Since sunflowers face the sun reliably, they can serve as a sign of the sun for someone that observes them. Peirce, though was interested in whether the sunflower's facing the sun could serve as a sign for the sunflower itself. The very fact that he considered the possibility indicates that Peirce did not rule out the possibility. However, he thought that facing the sun did not serve as a sign for the sunflower itself. Looking at this and other examples, this chapter will try to determine what properties something non-mental must have in order to serve as a sign for its bearer. The answer lies in teleology, specifically in functionality. I will argue that candidate non-mental signs must be functional for their bearer, and that the grounds of functionality ensure that there is a Peircean interpretant, which completes the triad, which becomes a sign with a further interpretant, and so on, much as is the case for minds.

Signs with Minds

Peirce clearly thought that mental signs were the paradigmatic case (Peirce 1940, p. 100). If we are going to look at signs without minds, it is best to look at this case so that we can then subtract the mental aspect and see if we have enough left to fit a semiotic model. First of all, mental signs have meaning. They are interpreted to at least some degree, though it is quite possible that some representation we have is puzzling, and we don't know how to identify it. Consider, for example, a glimpse of movement in the shadows late at dusk. It might be a large animal, or a person, or even an illusion of the lighting. We are unsure of what the sign indicates, let alone what it means. However the very fact that we have the sign within our experience implies that taken as it is, it is definite. This "as it is" is an iconic aspect of representation (Firstness). But the glimpse is more than this. It is an indication that something is going on in our visual sensory apparatus: it is a visual experience. This takes us beyond the experience in itself to the kind of experience that produces it, an index of visual goings on (Secondness). But it is more than that. The glimpse is a particular kind of visual experience that we habitually (Thirdness) connect to a range of possible interpretations, as well as to certain emotions (surprise, edginess, perhaps startle). This is the meaning of the glimpse as we experience it. These make the three components of anything that could be a sign: icon, index and interpretant. However there is something else lurking in the background, which is the reference of the sign. As described, the reference of the glimpse is nothing but itself; however we know that it might be a sign of something else, perhaps outside the mind (whence much of the emotional component). This uncertainty gives the sign, the glimpse, a meaning beyond itself, leaving it open, but only in certain ways compatible with its nature. It couldn't be a sign of a bright light, for example. It is the character of most signs to be open in this respect; their connection to other signs that might enhance their meaning is not fully circumscribed by the sign itself. Closing part of this openness produces new signs with a broader (beyond the sign itself) but at the same time more specific interpretant. It is important to recognize that the reference of a sign is not part of the sign itself, but the result of an abductive hypothesis of something which the sign is of (the exception being purely iconic signs, called qualisigns, like the bare glimpse itself, whose object and interpretant is themselves, allowing no external reference).

A glimpse is first of all a mental experience, a visual experience. Its interpretant is also mental. In order to get to the external world we need to hypothesize first that there is such a world, and that some experiences are signs of it. Peirce thought that we came to realize that there was a world outside of ourselves as very young children, when some things did not go as we expected or wished (CP 5.51). He thought that we distinguish a certain set (revisable) of experiences as being of the external world. In particular, these would include sensory experiences. Barring some defeater, a glimpse is of something external. This then raises a further question of what the glimpse refers to, what it is of. Experience, and thus its signs, has two aspects, internal and external. Only the latter involves reference, where a sign indicates something outside of our immediate thoughts. This external part is not required; thoughts can be interpreted internally, and related to other thoughts, and so on, without specific end. Of course, we may be mistaken about any part of this (CP 1.171), except for the brute experience itself (CP 6.454).

Peirce's pragmatic theory of meaning comes in a number of different versions, but one of his latest, from his *Harvard Lectures* is:

> For the maxim of pragmatism is that a conception can have no logical effect or import differing from that of a second conception except so far as, taken in connection with other conceptions and intentions, it might conceivably modify our practical conduct differently from that second conception. (CP 5.196)

In other words, any difference in meaning entails a difference in some conceivable difference in our practical conduct. The last requires (if we are to be conceived as the least bit rational) a difference in our possible expectations about the world. These are given by our possible expected set of signs within the (revisable) set that we take to be of the external world.

In Peirce's pragmatic view of meaning a mere correlation or correspondence between a sign and its reference does not in itself produce a meaning. Meaning requires an irreducibly triadic production of what Peirce calls the *interpretant*, a relation in which the sign (*representamen*) bears some variety of correspondence to its reference through the immediate object of the sign (*ground*), which is an "idea" corresponding to the object not in all its respects, but only under certain considerations (Peirce CP 2:228, 1940 p. 275). A sign must be taken as a sign in a context supporting interpretation in order to be interpreted. Mere function is not sufficient (Peirce argues) since, for example, a

> … thermometer dynamically […] connected to a heating and cooling apparatus, so as to check either effect, we do not, in ordinary parlance, speak of there being any *semiosy*, […], but, on the contrary, say that there is an 'automatic regulation', an idea opposed, in our minds, to that of semiosy. (Peirce 1940, p. 275, italics in original)

We should, in parallel, say the same thing about automatic elements of thought. Sensations and feelings, inasmuch as they are brute experiences, are not in themselves semiotic, though they can be classified in semiotic terms. They are pure firsts, not existing in themselves, but only in relation to (seconds) other thoughts, and interpreted through (thirds) the way they are integrated into our thought. Likewise pure mimicry, a causal process (second), the only pure form of memes, is not in itself semiosic. If integrated into thought, say we wear jeans to fit in with our crowd, this is not merely mimicry, but connected to other thoughts, desires and actions.

Semiosy proper, then, has four elements: the sign (representamen), reference, object and the essentially triadic interpretant, the reference being something external (extensional in the language of modern logic, which is perhaps determined by the dynamical interpretant—see below) determined only indirectly through the mediation of the full triadic sign. It is the last which is irreducibly pragmatic and contextual in nature, involving anticipated (expected) behavior of the object in open-ended (and possibly frustrated) interactions with the world. It should be noted that the mediation through signs is typically through a network of signs, in which each sign serves as an icon for another, so that the interpretant of a particular sign, say a glimpse, is mediated by this network. The reference of the glimpse, though very vague because of its nature, is the hypothesized product of this network, which becomes increasingly sophisticated and refined as we learn more about both the world and how our mind works.

One of the peculiarities of the human mind is that we are able to partially consider (as Peirce says, "prescind") aspects of full blown signs, so we can consider just the interpretation in the form of a proposition or thing or other sign, or the raw experience, or the connection to other thoughts, or some combination of these. Each of these possible combinations presents a semiotic category within Peirce's thinking. However in a full sign all of the elements are present equally, whether we are focusing on them or not. Peirce called such a "perfect sign." As T. Short (2007, p. 227) remarks, it is not entirely clear what Peirce meant by this, with CP 4.448 suggesting that it bears on what the fullest function of a sign is. Later (EP 2:545n25) Peirce says that a perfect sign is one that "involves the present existence of no other sign except that are ingredients of itself" (quoted in Short 2007, p. 227). Peirce is clear in the cited passages and context that the perfect sign is dynamic, it ages, but is being constantly renewed by its object; in a sense it grows. Peirce calls this, therefore, a "quasi-mind" though his complete meaning is unclear. What does seem clear is that the perfect sign is self-generating (he says "spontaneous") and has a susceptibility to determination. It is therefore not closed, but open, growing or evolving as it tracks its object. Some observers (Lyris 2006) have suggested that this is related to entelechy (the actualization of form-giving or final cause, that which makes actual what is merely potential), identifying the perfect sign with entelechy in Peirce. If this is correct, then perfect signs are not merely functional or even merely meaningful, but they are the realization of function and meaning. I will return to this later when I deal with signs without minds. I should mention before going on, however, that the notion in this paragraph suggests that a sign can change. This is a tricky issue. As generals we can talk about the same sign over time and space, but it seems to me that if the interpretant changes, then the sign itself must change, but then it is not clear that it is the same sign any more. I think it is safer to refer to temporal lineages of signs, in which later versions inherit some properties of their predecessors. This would be analogous to the way we talk about genes and inherited traits. Perhaps it is more than an analogy.

How do we get from a specific sign in our mind to the external world? As indicated above but worth repeating, there must be a chain of interpretations leading to a set of expectations of possible behaviour within our set of signs of the external world. Each sign in the chain can be a sign for another sign (with the sign, especially its interpretant playing the role of an icon for the next sign), forming a net of signs that includes expectations of the results of possible behavior. Following the pragmatic maxim this gives our meaning for the sign, so the sign of a glimpse, for example, means that we have expectations of possible experiences (that given the vagueness of a glimpse would be vague themselves) that might follow if certain further events should occur, or, importantly, if we were to follow certain actions, like shining a light in the direction associated with the glimpse, or approaching closer.

Peirce's overall approach to semiotics matured and changed somewhat with time, and became more articulated, with many finer divisions. I won't have space to go into all the details here, especially since much of this is at its best arcane (see Marty and Lang 2012). The important elements of Peirce's classification of signs, grounded in the logic of signs, are (1) as mentioned above all signs are irreducibly triadic, with representamen, object and interpretant being the three relata, (2) signs must be of something; in the mind they are of signs, giving 27 possible combinations which are given by the crossproduct of icon, index and symbol, so that we can have an icon of an icon, and so on, of which only 10 can be occupied, so we can, for example, have a common noun or a proposition considered as a symbol, but only proposition can represent a fact (CP2.254–263). These basic kinds of signs are laid out with each aspect as below:

	Phenomenological category	Relation to object	Relation to interpretant	Specificational redundancies in parentheses	Some examples
(I)	Qualisign	Icon	Rheme	(Rhematic Iconic) Qualisign	A feeling of "red"
(II)	Sinsign	Icon	Rheme	(Rhematic) Iconic Sinsign	An individual diagram
(III)		Index	Rheme	Rhematic Indexical Sinsign	A spontaneous cry
(IV)			Dicisign	Dicent (Indexical) Sinsign	A weathercock or photograph
(V)	Legisign	Icon	Rheme	(Rhematic) Iconic Legisign	A diagram, apart from its factual individuality
(VI)		Index	Rheme	Rhematic Indexical Legisign	A demonstrative pronoun
(VII)			Dicisign	Dicent Indexical Legisign	A street cry (identifying the individual by tone, theme)
(VIII)		Symbol	Rheme	Rhematic Symbol (-ic Legisign)	A common noun
(IX)			Dicisign	Dicent Symbol (-ic Legisign)	A proposition (in the conventional sense)
(X)			Argument	Argument (-ative Symbolic Legisign)	A syllogism

There are three other aspects of Peirce's thinking about mental signs that need to be considered before going on.

First, repeatability: from the discussion of perfect signs above, it is evident that at least these are self-producing. However, this is a property of all signs, if not in quite the same way, since in general the production is not spontaneous. It is the nature of signs that they are general. A single instance of an isolated relation between an icon and an object could not be a sign as there would be no basis for connecting the two. Furthermore, there needs to be some way of associating the two according to some habit. In order for smoke to be a sign of fire either there must be some experience of the association of the two, or else one must have background knowledge that the two should be associated. It is the very generality of signs that requires that being a sign is ensured if the correct triadic relation between icon, index and interpretant holds. Being a sign implies repeatability, and the nature of the sign itself implies its reproduction, if the conditions for it hold. That holds not only for the icon-index aspect of a sign, but also for the other possible combinations.

Second, reference: Peirce is clear that signs have a physical basis. Even an icon in the form of a glimpse has a material basis that must be compatible with its iconic nature, even if we cannot be directly aware of this. This applies equally for indexical relations (with typically some sort of causal material basis, whether in the present or the past) and interpretants. Knowing the causal properties alone, however, does not give the semiotic properties. The brief explanation is that full signs are irreducible triadic relations, whereas all causal relations are either binary or composable from binary relations. The interpretant in particular, necessary for meaning, cannot be recovered from binary relations and their composition because meaning requires integration of immediate sensation, thoughts and the interpretive context. Any full sign is an irreducible triad of icon, index and interpretant. These are not separate entities but are aspects of a sign being a sign. The physical embodiment of a sign, then, will also have an irreducible triadic form, compatible with each aspect of the sign. It would be a mistake to assume, for example, that because a weathervane is a sign of the direction of the wind that as a sign it is nothing but the material that makes up the weathervane. This would be to ignore its interpretant, and it would be deficient as a full sign (though we can talk about the aspects in the abstract). For mental signs, with our power of abstraction (prescision) it is possible for us to not only consider aspects of signs independently and in various incomplete combinations, but also in complete negligence of their material embodiment. In fact much mental activity ignores physicality, which is only required when we interact with the world. This connection is given only as a hypothesis, or abduction, as described above.

Signs may also have a referent outside of themselves, but this is not necessary. If they do, and the referent is physical (as opposed, say, to mathematical objects), then there must be some causal connection between the referent and the sign, and this must be of a general nature to ensure the reproducibility of the sign. This causal relation is not sufficient for reference, though, as causal relations are dyadic, and a sign gets reference only through its interpretation, which is triadic. This is not to say that reference cannot be given a causal basis, just that this cannot be the whole story, as it is in theories of perception originating with John Locke and popular within analytic philosophy (Grice 1961; Goldman 1967; Dretske 1981).

Third, finality: The third and last aspect that needs consideration is the evolution of signs towards greater accuracy (clarity). Peirce discusses this in detail in his earlier work (Peirce 1878), but it is only later that he gives the projected end result of this process the name 'final interpretant'. In a letter to William James in 1909 (EP 2:496-7) he says, "…there is certainly a third kind of Interpretant, which I call the Final Interpretant, because it is that which would finally be decided to be the true interpretation if consideration of the matter were carried so far that an ultimate opinion were reached." (italics in original). He goes on to refer to his 1878 work. Earlier (1906) he admitted that the idea was a bit unclear, and he gave a somewhat different interpretation:

> In regard to the Interpretant we have […] to distinguish, in the first place, the Immediate Interpretant, which is the interpretant as it is revealed in the right understanding of the Sign itself, and is ordinarily called the meaning of the sign; while in the second place, we have to

> take note of the Dynamical Interpretant which is the actual effect which the Sign, as a Sign, really determines. Finally there is what I provisionally term the Final Interpretant, which refers to the manner in which the Sign tends to represent itself to be related to its Object. I confess that my own conception of this third interpretant is not yet quite free from mist. (CP 4.536)

I will not try to explain in any detail what he meant in the mental case, since it involves his views about convergence on the truth, the communal nature of inquiry, and idealist aspects of his metaphysics, not to mention his views on universal love and evolution that are "not yet quite free from mist." However it is clear that any account of sign will need to give an account of either its final interpretant, or else of why the idea is not relevant. I will argue later that the idea is relevant to biosemiotics in a fairly straight-forward and fundamental way.

The essential elements of mental signs, then, are (1) an icon, index and interpretant in an irreducible triadic relation, best symbolized by a centered triple, somewhat like this: ⅄. As it is symmetrical, the labels can be any of the three points. The symbol indicates that the three meet together, and cannot be broken into dyadic relations. It is, however, possible for our minds to abstract or prescind aspects of signs, and consider them separately as further signs, just as a whole or perfect sign can be sign of something else, taking the role of an icon for the further sign, and so on. (2) Meaning arises through the interpretant and through its connections to other signs, determined by expectations about what differences meanings would make to the world, according to the Pragmatic Maxim. Our connection to the world is through a (revisable) class of signs that we can call experiences of the external world. We know the external world to exist because our signs do not always meet our expectations. This last is an important aspect of representations in general; they are fallible. If not I think it does not make sense to call them representations. Because we can be mistaken about what our signs represent we must distinguish between the immediate interpretant and the dynamical interpretant, which is "the actual effect that the sign determines." Our immediate interpretant is corrigible as we learn more what effect(s) it actually determines. This gives a connection to the physical embodiment of signs for signs of the external, and derivatively for internal signs. As described above simpler signs are interpreted through a chain or network of signs grounded in expectations of possible experiences. Lastly, (3) signs reproduce themselves, and they can grow (improve in clarity) as we are corrected; signs evolve. This should not be taken as implying that the same sign persists through changes in its interpretant, but that there is a lineage of signs (perhaps splitting) that inherit some properties of earlier signs, but are clearer in the sense that Peirce discusses in his paper "How to make our ideas clear" (Peirce 1878). In the process of inquiry, which is a mind dependent process, the final interpretant not only represents its object, but represents it as related to its object, if I read him right. As Peirce says, this is a bit misty. Provisionally I would say that a final interpretant for mental signs is an ideal state (that may or may not be achieved in finite time) in which there are no expectations of the object of the sign that are to be contradicted by experience. Of course we cannot know for certain that we have achieved a final interpretant for a sign (more properly, of a sign lineage). The world can always surprise in ways that we might not even be able to understand fully unless it happens (Collier 1990).

In summary, if there are signs without minds they must, if mental signs are a paradigm, have irreducible triadicity, something equivalent to meaning, and reproduction leading to some sort of improvement or at the very least directional change. They must show the properties of repeatability, reference and finality in achieving these properties.

Signs Without Minds

If there are signs without minds, then there is no partial consideration, no prescinding or abstracting, no expectations, and no meaning in the usual sense. We need to find satisfactory analogues for these, or at least show why they are not required. Furthermore, we need to do this in a way that satisfies the triadicity, meaning, and reproduction requirements, while embodying the repeatability, reference and finality properties.

The idea that there are signs without minds is belied by the following quote:

> A Representamen is the First Correlate of a triadic relation, the Second Correlate being termed its Object, and the possible Third Correlate being termed its Interpretant, by which triadic relation the possible Interpretant is determined to be the First Correlate of the same triadic relation to the same Object, and for some possible Interpretant. A Sign is a representamen of which some interpretant is a cognition of a mind. Signs are the only representamens that have been much studied. (2.242–Syllabus, 1902)

This implies that there are triadic relations that are not signs because their interpretant is not a cognition. However it also implies that all signs must involve cognitions, and hence minds. If so, there are no signs without minds even if we can satisfy (or explain away) the three requirements and three properties.

On the other hand, Peirce sometimes appears to say (as in EP:455)[1] that there are thoughts without minds, which would require signs without minds, since, at first glance at least, thoughts are signs. Peirce certainly toyed with objective idealism, according to which the world is made up only of mind. If this were correct, it would solve a problem for him, which is that he thought that signs could be related only to other signs (e.g., Peirce 1868).

If the world is made up of signs, then a weathervane can itself be a sign of the direction of the wind independently of our thoughts. This would make our internal signs continuous with signs in the world. When we become conscious of it, and understand its reliability (if not how it works) then we can use it as a sign through the manipulation of our internal signs. But in itself it is a sign, we might say, composed of the icon of the shape of the weathervane, the index of the wind direction, and the interpretant which is the thought that the wind is in the direction in which it is pointing. This gives triadicity, meaning and reproduction, with the properties of repeatability and, perhaps, reference. Finality, however, seems to be lacking. It might be suggested that a weathervane can be better or worse in indicating the direction of the wind, and some might not do so well at all. However there is no inevitable

[1] See full quotation on page 118 of this volume.

progress towards better indication of wind direction (in fact, without external maintenance they get steadily worse at this function). Furthermore, and relevantly, it is only through the needs of someone who uses the weathervane that the difference between accuracy and inaccuracy matters. It seems then that the weathervane, as Peirce said of the thermostat in the quote above, is not fully a sign. Its action is automatic, necessitated by its nature, and there is nothing in it itself to determine whether or not it is successful. Because of this, it is not really clear that the weathervane even represents anything. The basic problem is that the weathervane is not a sign for itself or anything else unless minds that use it as a sign are invoked.

As part of our built environment, our extended phenotype, weathervanes do, like thermostats, perform a function. We might then argue that they are constructed by us as semiotic entities, and that they are therefore in themselves signs. I have my doubts, though. We could just as easily use the direction in which smoke, dust or leaves blow, all natural objects, to determine wind direction. But these would not be artefacts, part of our built environment, part of our extended phenotype. I don't see that the difference of being built should affect their intrinsic nature as signs. Their being signs depends on our thinking in certain ways about them. They are signs only with minds.

There are some approaches to pansemiosis that try to resolve these problems through evolutionary ideas of natural laws as fixations of some sort of material habit, making them thirds that is some sense interpret the indicating activity of, for example, weathervanes. Whether or not this can be successful I will not pursue here. I am not sympathetic to the idealism that makes it possible, and I am unclear what value it would have even if it could be worked out adequately. If we except the Peirce quote about palaeontologists, this would leave us with cognitions without signs. But this would be consistent with the first quote in this section. In biology, however, there are non-cognitive things that do appear to have finality and the properties of representations. They use signs for their own purposes, but not through conscious activity. I will turn to these next.

Biosemiotics

Jesper Hoffmeyer states:

> In the biological world, certainly, signs incite the generation of interpretants in the form of actions which are future-oriented, inasmuch as living beings always seek signs for survival and for reproduction. (Hoffmeyer 2008, p. 65).

In other words in the biological world there are signs that contribute to biological function, i.e., to biological fitness. Fitness implies function, and this implies finality. Understanding the nature of biological function is nontrivial. I have argued elsewhere that we can go only so far with information theory (Collier 2003, 2008) and control theory (Collier 2011), arguing that we need to look to biosemiotics to give a complete account of function. As Terrence Deacon (2012b) has argued, function, like intention, has the peculiar character of invoking something that can be absent. He uses the neutral term "ententional" to refer to this property, presumably on analogy to intentional. In order for there to be reference of some sort to something absent, we need something like representation, so we need at least an analogue to meaning in the mental case. There is also a sense of finality involved, since function can be better or worse. If the other aspects of mental semiosis are required, then we have a good case for full semiosis in biological systems.

Some biological systems use external signs in much the same way as conscious minds do. In some cases similar enough to human cases we can assume that these signs are conscious. However in other cases it is not so clear. It is also too easy to mistake who the sign serves. I noted:

> Brooks and McLennan [8–9] discussed this issue with respect to biological signaling. They concluded that most signaling done by organisms is signaling to oneself, both about itself and about its conditions. Whether this is intentional or not is moot. Some of this self-conversation may produce changes in the organism detectable by other organisms. The meaning that those other organisms place on the "signaler" is not caused by the intentions of the signaler, relieving us of the burden of having to postulate a causal link between intention and meaning, or function of the sign. Breaking this link permits signals and meanings to evolve in a purely Darwinian manner (i.e., accidentally, with both costs and benefits, so long as the benefits outweigh the costs by at least a tiny bit). So, for example, a male stickleback turns red as a result of biochemical changes related to testosterone levels. The color change is a by-product of an internal chemical signal from the animal to itself, telling it that it is ready to breed. In that sense, the color change is completely unintentional. However, the color change does occur and as a sign may have different functions for different receivers in the environment ("mate" to a female stickleback, "dinner" to a heron). And so long as "mate" benefits are slightly greater than "dinner" costs, the system will continue. (Collier 2003, p. 74–75)

The point here is that the "sign" produced by the male is not immediately a sign for the male, though it serves an evolutionary purpose for the male. It is a side-effect of other processes. It is only because the female can recognize it as a sign that it becomes a sign for the male as well. But it is not a conscious sign for the male. Nonetheless, it serves an evolutionary function for the male by making it attractive to the female. It seems then, that it is a sign without a mind. I would assume that humans also have many unconscious external signs. The stickleback case is interesting because although the female may be conscious of the sign, the male is not, and there is no reason that both sides in such a relation need be conscious. There also seems to be no reason to deny the sign aspect of the relation merely because there is no conscious experience. So it seems that there may be functional biological signs without minds.

From the above it is evident that there can be external signs in biological systems that are not conscious. Are there internal signs? There has been a developing split between exobiosemiotics and endobiosemiotics, with some holding that endobiosemiotics can be understood entirely in terms of codes, e.g., Marcello Barbieri. I will not go into this dispute in depth here, though I have taken the side that endobiosemiotics can be Peircean, and that codes are not sufficient (Collier 2008, 2012). Hoffmeyer points out that there is no clear distinction between endo and exobiosemiotics (Hoffmeyer 2008, p. 213 ff.). I think that functionality in biosemiotics applies equally internally as externally, and if the reasons for applying it externally are sound, then it is also reasonable to apply it internally. This argument is made in some detail with examples in (Arnellos et al. 2012),[2] and I will not repeat it here. Instead I would like to look in more detail at the implications of the Peircean view, as developed in (Collier 2011, 2012).

I think the differences between the Peircean and the code only views are obscured when focusing on details of individual signs and sign systems. Peirce's semiotics allows for systems of signs in which one sign can serve as an icon for another sign, which gives a further interpretant. This implies a network of signs, as described for the mental case above, with interpretation becoming more general as we move upwards and outwards. A Peircean sign is an indecomposable triad of icon, object and interpretant. The icon is the "bare feel," or presentation; the object is picked indexically, and the interpretant brings them together in a unified context. This triad can then be an icon for a further sign that is more general. Several signs together can provide the presentation for an even more general sign. This regress (or recursion) comes to an end in some most general sign, with its interpretant being ultimate.

What are the ultimate interpretants in biology? Suppose we have smell A (icon) of something dangerous (interpretant), then it is incumbent to avoid (object).

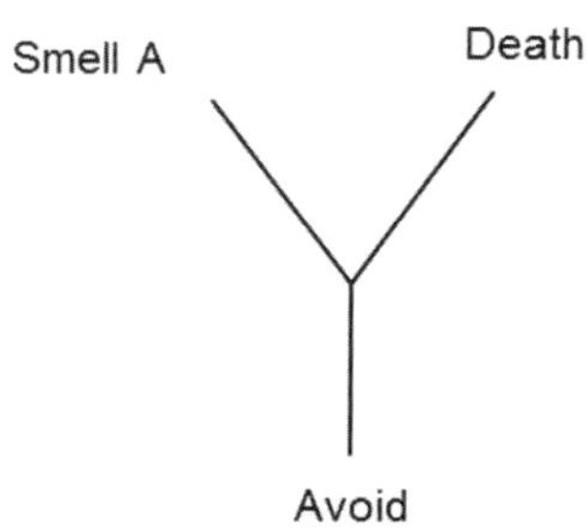

For good biological reasons, this sort of interpretation has very short chains. Chains related to survival are typically longer. A slightly longer chain: Suppose we have smell B (icon) that indicates food (interpretant) that can be eaten (object). This itself is a sign (icon) that falls under survival (interpretant) indicating it should be accepted (object).

[2] Although we used 'meaning' in a paper on anticipatory functions (Arnellos et al. 2012), I was never happy with the term, thinking it was not quite the correct concept, and was potentially misleading. However there was no available term to invoke, so I went along with our use of 'meaning' in our discussion of endobiosemiotics. Deacon's term 'ententional' is better due to its neutrality, though the idea is analogous to concepts of intentionality and meaning.

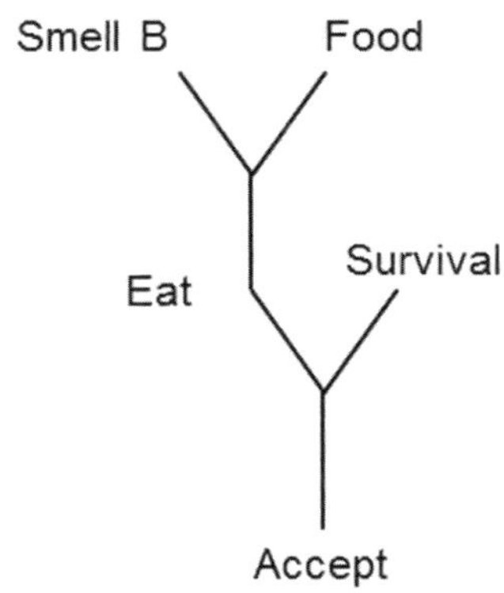

Typically there will be longer chains both for the interpretant and often the icon. Often these will involve chemical processes that can act as signs. What gives them their semiotic quality, ultimately, is their interpretation in terms of survival. This is not an accidental property, but one that is central to biology, especially the biological autonomy of organisms and their reproductive lineages. I suppose that there could be biosemiotic conditions that don't contribute to survival, but they are very likely to be weeded out by evolution, so they would be rare and temporary at best. This justifies the quote from Hoffmeyer that opened this section (Collier 2012).

There are a few things to clear up at this point. On the surface it seems that there is nothing like the capacity to prescind without the abstracting ability of consciousness. If so, then the table of Peircean signs in the second section are not filled out. Each sign can only be taken as an icon by another biological sign without mind, which might seem to limit us to class I in the diagram. We can infer other sign classes abstractly, but these, it might be argued, are not in the organism itself. I think this would be a mistake. It is the whole triadic sign that is taken as a sign for the next sign in line, which is also produced in its fullness. In other words, all three parts of the triad are there equally in each biological sign. The inability to prescind does not reduce the complexity to just iconic aspects, but ensures that every biological sign is a perfect sign in Peirce's sense. This means that every non-mental biological sign is an entelechy, which is directly commensurate with its necessarily being functional. It further means that we can make sense of each biological sign as any of the ten categories. It conveys an icon, an index and a proposition, in the role of a feeling, an indicator and a reason, each being part of its functionality together.

One last issue that should be addressed (there are many ramifications to be worked out, but I focus on just the barest outlines, details to be filled in) is finality. What is the final interpretant of biological signs, if any? The idea of functionality suggests that there can be better or worse function, which suggests that the final interpretant would be the best function possible. According to the widely adopted optimality theory in neo-Darwinism, every trait is already optimally adapted, so it could not be better. On this account every biological sign except those involving minds is already at its final interpretant. The problem with this view is that it assumes that genetic and environmental changes are slow enough to allow selection to produce optimal adaptation. Biologists I work with tell me that this is seldom the case. If genetic and environmental changes are too fast, then there is room for self-organization within the genetic and trait space and the information of adaptation can increase indefinitely (Collier 1998). This is still a controversial area, with traditional neo-Darwinists holding that progress in evolution is a mistaken idea, whereas self-organization theorists see it as both possible and actual. If the latter are correct, then the idea of a better interpretant is on the table. But since the target is always moving, it seems that final interpretant is not, unless there is some ideal evolved being, analogous to the ideal theory of the world. I will leave the issue there.

Further Remarks on Differences Between Mental and Bio-Semiotics

Unless we favor some version of pansemiosis and the accompanying objective idealism, non-mental signs without function seem to be implausible. The main things lacking are not triadicity, repeatability and reproduction, but reference and finality. These, however, are found in biology, and are characteristic of biology. Repeatability and reproduction are a given in biology, so the only missing element is triadicity, which is essential for Peircean signs. I have argued that we can tie together aspects of exo and endobiosemiotics by assuming a similar sign process in both cases, and that there is no clear place where functional traits used for passing information in biological systems cease to be signs. So the Peircean account of signs fills in the story, giving us genuine signs without minds. This is an abductive argument of some complexity, and one could resist it at any particular step. I would argue that there is no good reason or boundary at which this resistance would be reasonable.

Conclusions

Biological systems are capable of having signs in the Peircean sense. This is obvious in cases in which we have conscious minds, but there is no clear place where the application of Peircean semiotics fails to be explanatory where we have functional information transfer. There are analogues to all of the properties and conditions for signs with minds, with intentionality being taken over by the weaker notion of ententionality, characteristic of functionality, which is manifest throughout biology. Finality is guaranteed by the same conditions. Once we have introduced Peircean semiotics to fill a very obvious place in this biological story, reference is determined for each sign by way of its object. Essential to being a sign is the possibility of the absence of an actual referent, but this is just a consequence of the possibility of function going wrong. Most importantly, perhaps, biological signs are signs for those that have them.

On the other hand, non-biological non-mental things do not exhibit reference or finality, and consequently they do not have signs that are signs for them. I think then that it is proper to say that they do not have signs as such, but only derivatively from things that have signs for themselves.

References

Arnellos, A., Bruni, L. E., Niño El-Hani, C., & Collier, J. D. (2012). Anticipatory functions, digital-analog forms and biosemiotics: Integrating the tools to model information and normativity in autonomous biological agents. *Biosemiotics*, 5, 331–367.

Collier, J. (1990). Could I conceive being a brain in a vat? *Australasian Journal of Philosophy, 68,* 413–419.

Collier, J. (1998). Information increase in biological systems: How does adaptation fit? In van der V. Gertrudis & Stanley N. Salthe (Eds.), *Evolutionary systems* (pp. 129–140). Dordrecht: Kluwer.

Collier, J. (2003). Hierarchical dynamical information systems with a focus on biology. *Entropy, 5,* 57–78.

Collier, J. (2008). Information in biological systems. In P. Adriaans & J. van Benthem (Eds.), *Handbook of philosophy of science, volume 8: Philosophy of information* (Chap. 5 f., pp 763ap.). Dordrecht: North-Holland.

Collier, J. (2011). Explaining biological functionality: Is control theory enough? *South African Journal of Philosophy, 30*(4), 53–62.

Collier, J. (2012). *Interpretants*. In D. Favereau, P. Cobley, & K. Kull (Eds.), *A more developed sign: Interpreting the work of Jesper Hoffmeyer* (pp. 175. f.). Tartu: Tartu Semiotics Library 10, Tartu University Press.

Deacon, T. W. (2012a). *Information*. In D. Favereau, P. Cobley, & K. Kull (Eds.). pp. 161–64.

Deacon, T. W. (2012b). *Incomplete nature: How mind emerged from matter*. London: Norton.

Dretske, F. (1981). *Knowledge and the flow of information*. Cambridge: MIT Press.

Goldman, A. I. (1967). A causal theory of knowing. *The Journal of Philosophy, 64*(12), 357–372.

Grice, H. P. (1961). The causal theory of perception. *Proceedings of the Aristotelian Society Supplementary, 35,* 121–152.

Hoffmeyer, J. (2008). *Biosemiotics: An examination into the signs of life and the life of signs*. Scranton: University of Scranton Press.

Lyris. (2006). Discussion of the philosophy of Charles Peirce, Entelechy 21:39. http://permalink.gmane.org/gmane.science.philosophy.peirce/786. Accessed 9 May 2006.

Marty, R., & Lang, A. (2012). 76 Definitions of The Sign by C. S. Peirce, with 12 Further Definitions or Equivalents. http://www.cspeirce.com/rsources/76DEFS/76defs.htm. Accessed 19 Feb 2013.

Peirce, C. S. (1868). Questions concerning certain faculties claimed for man. *Journal of Speculative Philosophy, 2,* 103–114.

Peirce, C. S. (1878). How to make our ideas clear. *Popular Science Monthly,* 12 (January), pp. 286–302.

Peirce, C. S. (1940). *The philosophy of Peirce, selected writings*. In J. Buchler. London: Kegan Paul, Trench, Tubner and Company.

Short, T. (2007). *Peirce's theory of signs*. New York: Cambridge University Press.

Dicent Symbols and Proto-propositions in Biological Mimicry

João Queiroz, Frederik Stjernfelt and Charbel Niño El-Hani

> (…) A Dicent Symbol, or ordinary Proposition, is a sign connected with its object by an association of general ideas, and acting like a Rhematic Symbol, except that its intended interpretant represents the Dicent Symbol as being, in respect to what it signifies, really affected by its Object, so that the existence or law which it calls to mind must be actually connected with the indicated Object. Thus, the intended Interpretant looks upon the Dicent Symbol as a Dicent Indexical Legisign; and if it be true, it does partake of this nature, although this does not represent its whole nature. Like the Rhematic Symbol, it is necessarily a Legisign. Like the Dicent Sinsign it is composite inasmuch as it necessarily involves a Rhematic Symbol (and thus is for its Interpretant an Iconic Legisign) to express its information and a Rhematic Indexical Legisign to indicate the subject of that information. But its Syntax of these is significant. The Replica of the Dicent Symbol is a Dicent Sinsign of a peculiar kind. This is easily seen to be true when the information the Dicent Symbol conveys is of actual fact. When that information is of a real law, it is not true in the same fullness. For a Dicent Sinsign cannot convey information of law. It is, therefore, true of the Replica of such a Dicent Symbol only in so far as the law has its being in instances. (CP 2.262)

Introduction

Sign-mediated processes show a remarkable variety. The construction of appropriate typologies of these processes is a requisite for a deeper and more refined understanding of animal communication. In an attempt to advance in the understanding of sign-mediated, i.e., semiotic processes, Peirce proposed several typologies, with different degrees of refinement and diverse relationships to one another. A basic

J. Queiroz (✉)
Federal University of Juiz de Fora, Juiz de Fora, Brazil
e-mail: queirozj@pq.cnpq.br

F. Stjernfelt
University of Aarhus, Aarhus, Denmark
e-mail: semfelt@hum.au.dk

C. N. El-Hani
Federal University of Bahia, Salvador, Brazil
e-mail: charbel.elhani@pq.cnpq.br

V. Romanini, E. Fernández (eds.), *Peirce and Biosemiotics,* Biosemiotics 11,
DOI 10.1007/978-94-007-7732-3_11,

typology in his framework differentiates between iconic, indexical, and symbolic processes. An attempt to classify the signs mediating mating, courtship, competition, and predation in fireflies in accordance with the Peircean framework shows, however, that this basic typology is not sufficient and, moreover, leads to unexpected findings with regard to firefly communication, as we will see below. The fact that we need to go beyond this basic typology is in itself telling.

Peirce developed several classifications of signs based on different trichotomies (see MS 540; L 463: 132–6). One of them concerns specifically the 10-fold classification as described in the 1903 Syllabus (MS 540, EP 2: 289–299), while the others deal with the deep structure of Peirce's various trichotomic classifications—six trichotomies (28 classes), and ten trichotomies (66 classes) (see Farias and Queiroz 2000, 2003; Freadman 2001, 2004). This implies a deep level of detail in the description of the relations observed in semiosis, as a process involving the interaction of sign, object, and interpretant.

Here we are especially interested in how Peirce's 10-fold classification of signs can contribute to the construction of models that may serve as tools for the investigation of biological mimicry. As a corollary to our analysis of firefly signaling (El-Hani et al. 2010), we analyze the capacity of producing propositions (i.e., dicent symbols) as a general requisite for a semiotic system to act as a mimic. We intend to show that Peirce's mature theory of signs brings an important contribution to the building of a general semiotic theory of mimicry, since it is quite helpful in addressing semantic and pragmatic aspects of biological information. As is well known, the semiotic processes involved in biological mimicry most often do not result from learning processes taking place in the individual semiotic system, but from the fine-tuning of inherited capacities by natural selection among variants over thousands or millions of generations. Still, the concrete sign exchange takes place within the lifetime of a single individual, and those signals, indicating and describing at the same time, should be conceived of as dicent symbols or dicisigns. This calls for an investigation of the Peircean notion of the dicisign, which is a generalization of the notion of proposition. A key feature of this Peircean notion is that it liberates assertions from the confines of language and points to their appearance also in pictures, gestures, mixed media, etc. That is, it generalizes propositions from being a human privilege so as to also embrace simpler dicisigns found in the biological realm.

Some Basics of Peirce's Model of Sign-Action

Peirce's semiotics is defined as the doctrine of the essential and fundamental nature of all possible varieties of sign-mediated processes. From this perspective, we can speak of such processes at several biological levels of organization, and, certainly, a treatment of animal communication from a semiotic perspective is worthwhile. Living beings are generally treated, according to this perspective, as semiotic systems, i.e., systems that produce, communicate, and interpret signs. No anthropomorphism is necessarily involved in this approach, and the presence of consciousness needs not to be assumed in talking about signs. We can think of semiotic modeling as an attempt to build general, abstract models of sign-mediated processes, which select among the several features involved in such processes some which are taken to be fundamental elements to be modeled in a general manner, and some which are regarded as peculiar elements of some kinds of sign processes.

In this model, we find several clues to understand how signs act. The model says that any sign is something that stands for something else, its object, in such a way that it ends up producing a third relational entity/process, an interpretant, which is the effect a sign produces on an interpreter (in our case, an animal). In many sign-mediated processes, sign interpretation results in a new sign within the interpreter, which refers to the same object to which the former sign refers, forming a chain of inferences, or ultimately to result in an action, which can lead to the termination of the process.

A sign is also pragmatically defined as a medium for the communication to the interpretant of a form embodied in the object, so as to constrain, in general, the interpreter's behavior (Peirce MS 793:1-3. See EP 2.544, n.22, for a slightly different version).[1]

The object of sign transmission is a habit (a regularity, or a 'pattern of constraints') embodied as a constraining factor of interpretative behavior—a logically 'would be' response. The form is something that is embodied in the object as a 'regularity', a 'habit', a 'rule of action', or a 'disposition'. Form is defined as having the 'being of predicate' (EP 2.544) and it is also pragmatically formulated as a 'conditional proposition' stating that certain things would happen under specific circumstances (EP 2.388). For Peirce, it is nothing like a 'thing' (De Tienne 2003), but something that is embodied in the object (EP 2.544, n. 22) as a habit, a 'rule of action' (CP 5.397, CP 2.643), a 'disposition' (CP 5.495, CP 2.170), a 'real potential' (EP 2.388) or, simply, a 'permanence of some relation' (CP 1.415).

Fundamental Classes of Signs

As it is well-known, sign-mediated processes show a notable variety. According to Peirce, there are three fundamental kinds of signs underlying meaning processes—icons, indexes, and symbols (CP 2.275). Respectively, a sign may be analogous to its object, spatiotemporally connected to it, or might represent it by means of a law, rule or norm. These classes correspond, then, to relations of similarity, contiguity, and law between sign and object.

Icons are signs that stand for their objects through similarity or resemblance, no matter if they show any spatiotemporal physical correlation with an existent object (CP 2.276). In this case, a sign refers to an object in virtue of a certain quality that sign and object share. When the interpreting system receives an effect of the sign (i.e., when an interpretant takes place within it or when it acts due to the detection

[1] See pages 91–92 of this volume for full quotation.

of the sign), this is due to the communication of a quality of the object to the system through the sign. An icon can refer to an object independently of the spatiotemporal presence of the latter because it denotes the object merely by virtue of characters of its own, and which it possesses no matter if the object is present or not, and, in fact, no matter if the object actually exists or not. Icons play a central role in sensory tasks since they are associated with the qualities of objects. Thus, they are present in the sensorial recognition of external stimuli of any modality. Sebeok (1989) called attention to the fact that iconic signs are found throughout the phylogenetic series, in all sensory modalities, and, furthermore, that iconicity is crucially involved in the phenomenon of mimicry. Through a sign-mediated process with an iconic nature, an animal can modulate the behavior of another animal by communicating through the sign a certain quality shared with an object.

Indices are signs that refer to an object due to a direct physical connection between them. Since in this case the sign should be determined by the object, for instance, through a causal relationship, both must exist as actual events. Thus, in order to have actual effects, icons must be supported by indices. Accordingly, spatiotemporal covariation is the most characteristic property of indexical processes. For instance, if we see an English pointer pointing out the location of a bird, we can analyze the pointing paw as an indexical sign of the bird, its object. This sign has an effect on the human hunter, modulating his behavior so that he looks for the bird, due to the correlation in space and time between the sign—the pointing paw—and the object—the bird. Notice that in this case the sign will misinform the interpreter, the human hunter, if it shows no spatiotemporal correlation with its object. Icons, in turn, do not misinform when the object is not present in the same spatiotemporal frame. But notice that icons, in order to be efficient signs, must appear in connection with indices and their physical pointing to the object of the sign; if this is not the case, icons remain vague possibilities. When associated with indices, however, icons may be part of a process leading the interpreter to be misinformed about an object.

Finally, symbols are signs that are related to their object through a determinative relation of law, habit, or convention. A symbol becomes a sign of some object merely or mainly by the fact that it is used and understood as such, due to some of the kind of relations mentioned above. When we are speaking about the emergence of a symbol, the community of sign-users has a crucial role in establishing the conventions, habits, or laws on which the sign-object relationship is based. Furthermore, differently from indices, symbols can refer to objects not present in the same spatiotemporal frame, and, moreover, they can refer to abstract objects, such as classes or types of entities/processes. It is not necessary, however, that the sign-users have any awareness of the nature of the symbolic relation, or, else, that this relation emerges through sign use and learning during the lifetime of individual organisms. Symbol-mediated processes can emerge throughout phylogenetic time as a result of selective processes operating on a population.

Symbols can be found in non-human animals, such as African vervet monkeys (*Chlorocebus aethiops*, MAMMALIA: PRIMATES), whose alarm-call system is a particularly well-known case of vocal communication in non human primates (see Ribeiro et al. 2007; Queiroz 2003; Queiroz and Ribeiro 2002). The object of the sign, in the case of the vervet monkeys alarm-call system, is not an object-token but rather a class of objects, i.e., an object-type and, therefore, does not need to exist as a singular event only. This is yet another feature expected in communication by means of symbols. We can analyze the semiotic processes involved in this alarm-call system as involving signs that communicate to (i.e., have an effect on) the interpreting animal a lawful relationship between kinds of calls (signs) and kinds of objects (predators). It allows, thus, the emitter of the call to repeatedly modulate the behavior of its conspecifics in the same, general way so that it is more likely that they successfully avoid a potential predator.

Icons, indices, and symbols are differentiated by Peirce based on how the sign relates to its object, which may be defined, in turn, as the item to which the interpretants are related by the mediation of sign. In an attempt to advance in classifying semiotic processes, Peirce proposed several typologies, with different degrees of refinement. Around 1903, he developed a division into ten classes. According to this typology, symbols can be further analyzed in three subclasses.

The Extended Theory of Signs—Ten Classes of Signs

Impelled by discoveries in different domains (existential graphs, phenomenology, pragmatism), Peirce developed several classifications of signs based on several trichotomies (see MS 540; L 463:132-60; CP 2.233-72 and 8.342-76; EP 2:289-99 and 478-91; Lieb 1977, pp. 80–85).[2] The consequence is an enormous accuracy concerning the relations observed within S-O-I, including the differentiations between Immediate and Dynamical Object, and between Immediate, Dynamical, and Final Interpretant, and the conception of 10 trichotomies, resulting by combination in 66 classes of signs. The trichotomies are aspects according to which the semiosis can be observed and can be translated into questions (cf. Houser 1991). In order to understand the design of the ten classes of signs based on the first three trichotomies, three questions should be formulated: (i) "What is the relation of the Sign with itself?" (1st trichotomy); (ii) "What is the relation between the Sign and its Object?" (2nd trichotomy); (iii) "What is the relation between the Sign and its Object for its Interpretant?" (3rd trichotomy). Following a suggestion given by Rescher (1996, p. 36), we can understand these questions as "oriented problems" about semiosis. For each question, there are three kinds of relation as answers. As an example, for the first trichotomy we have a "monadic relation answer." In this case, this relation

[2] For an introduction to Peirce's extended classifications of signs, see Savan (1977), Lizska (1996), Parker (1998), Houser (2010); on the classes and their compositions, Weiss and Burks (1945), Lieb (1977), Sanders (1970), Müller (1993); on the theoretical aspects involved on this issue, Savan (1977b), Short (2007); on its genealogy, Freadman (2004, 2001); on its structural organization, Jappy (1989), Serson (1997), Farias and Queiroz (2000, 2003, 2004), Hoffman (2001); on the description of proposition, Hilpinen (1992), Houser (1992), Stjernfelt (2011); on the modeling of biosemiotic phenomena, Queiroz (2012a, b).

	1st Trichotomy	2nd Trichotomy	3rd Trichotomy
Monadic relation	QUALISIGN in itself, the sign is of the nature of an appearance.	ICON a sign which refers to the object merely by virtue of characters of its own (CP 2.247).	RHEME a sign which, for its Interpretant, is a Sign of possibility.
Dyadic relation	SINSIGN in itself, the sign is of the nature of an individual object or fact.	INDEX a sign which refers to the object by virtue of some existential relation.	DICENT a sign which, for its Interpretant, is a Sign of actual existence.
Triadic relation	LEGISIGN in itself, the sign is of the nature of a general type (CP 8.334).	SYMBOL a sign which refers to the object by virtue of some kind of convention.	ARGUMENT a sign which, for its Interpretant, is a Sign of law (CP 2.252).

Table 1 Three trichotomies and three kinds of relation. (see CP 2.243)

is described as a qualisign, which notation is designated by the integer 1—it "is any quality in so far as it is a sign" (CP 2.254). If the answer is a dyadic relation, it is a sinsign, which "is an actual existent thing or event which is a sign" (CP 2.245); if it is triadic, it is a legisign, which is "a law that is a Sign" (CP 2.246). The second and the third trichotomies are described, respectively by: icon (1), index (2) and symbol (3); rheme (1), dicent (2) and argument (3). A rheme is "a Sign which, for its Interpretant, is a Sign of qualitative Possibility, that is, is understood as representing such and such a kind of possible Object" (CP 2.250); a "Dicent Sign is a Sign, which, for its Interpretant, is a Sign of actual existence" (CP 2.251); and "An Argument is a Sign which, for its Interpretant, is a Sign of law (CP 2.252).

The results of the trichotomic questions (Table 1) may be combined, in a certain way, by building up a system of cross-relations (see Freadman 2001, 2004; Fig. 1). The kind of relation that answers the first question qualifies (cf. Savan 1977, p. 14) the second, which qualifies the third. A class of sign can be described as a *complex* of relations based on the notion of *logical constraints* (cf. *praecisio*) operating between the categories (monadic, dyadic, triadic relations).

The cross-relations that satisfy the *constraints* are:

(I) 111: a Qualisign is a quality "in so far as it is a sign"; its object is interpreted as being of the same nature—"a feeling of red" (CP 2.254); (II) 211: an Iconic Sinsign is a sign-event interpreted as possibly (rheme) standing for its object (icon)—"an individual diagram" (CP 2.255); (III) 221: a Rhematic Indexical Sinsign is a sign-event interpreted as possibly standing for another event (index)—"a spontaneous cry" (CP 2.256); (IV) 222: a Dicent Sinsign is an sign-event interpreted as

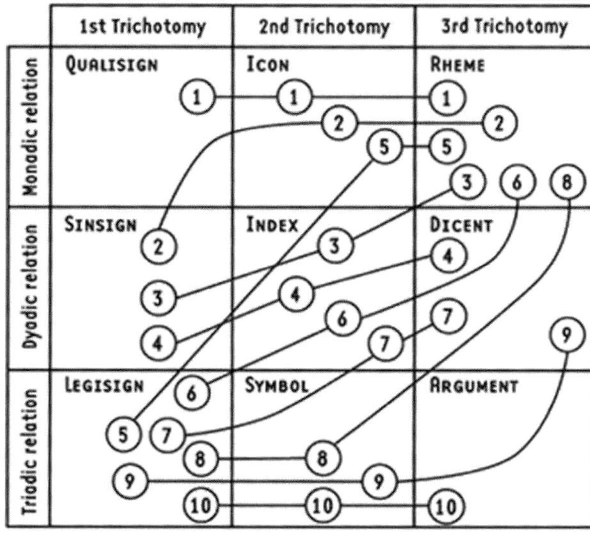

Fig. 1 The 10 classes of signs as a system of cross-relational classes. The paths correspond to the possible compounds of relations. (Figure based on 'Table 1', Merrell 1996, p. 8)

spatio-temporally standing for another event (index)—"a weathercock" (CP 2.257); (V) 311: an Iconic Legisign is a *type*, or a law, a "regularity of the indefinite future" (CP 2.293), interpreted as possibly standing for its object (icon)—"a diagram, apart from its factual individuality" (CP 2.258); (VI) 321: a Rhematic Indexical Legisign is a *type* interpreted as possibly standing for its object (another event)—"a demonstrative pronoun" (CP 2.259); (VII) 322: a Dicent Indexical Legisign is a *type* interpreted as spatio-temporally reacting with its object (another event)—"a street cry" (CP 2.260); (VIII) 331: a Rhematic Symbol is a *type* interpreted as possibly standing for its object (law)—"a common noun" (CP 2.261); (IX) 332: a Dicent Symbol is a *type* interpreted as physically standing for its object (law)—an "ordinary Proposition" (CP 2.262); (X) 333: an Argument is a *type* interpreted as semiotically standing for its object (law) (CP 2.263).

Our interest here is the sub-division of the symbols. According to this typology, there are three classes of symbols, called rhematic, dicent and arguments. A symbol is a general *type* (1st trichotomy) and its object can only be a general (2nd trichotomy). But symbols can also be interpreted as "qualities" or "events" (3rd trichotomy). In other words, a symbol, which is a general, and represents its object as a category of objects, can be interpreted as a possibility (e.g., rheme, hypothesis), as an existing fact (e.g. dicent, proposition) and as a law (e.g. argument, declarative rule). There are many examples of rhematic symbols. Unsaturated mathematical entities, general diagrams, and predicates are usually described as structures of this type. In natural languages, the onomatopoeic words are good examples of symbolic analogical signs. They are dependent on the properties (e.g. phonetic, prosodic) interpreted as shared by signs and objects. For Peirce, "Many words, though strictly symbols, are so far iconic that they are apt to determine iconic interpretants, or as we say, call up lively images" (NEM 4: 243).

Other symbols are interpreted as "existents", or dicent signs, for instance, demonstrative pronouns, nouns, quantifiers, as they appear in logical propositions (CP 2.262)—"A Dicent Sign is a sign, which, for its Interpretant, is a Sign of actual existence" (EP 2:292). The indexes and their fundamental properties are well-known (see Atkin 2005). Thermometers, for instance, represent the "temperature" due the covariation between the height of the mercury column (sign) and the energy of the represented system (object). Importantly, this covariation is not mediated by the action of the interpretant. Of course, it cannot work as a sign without the existence of such an interpreter. But the process is primarily dependent on the fact that two existent events have their natures (topology, shape, dynamics) connected. The example of the thermometer is interesting because there are several rules involved in the interpretation of the device. But it works semiotically mainly because a variation on the dynamics of its object physically *corresponds* to the variations in sign form and structure. This is an indexical legisign. A dicent symbol is a sign interpreted as connected with its object, but its relation with its object depends on a law, rule or acquired habit. Dicent signs in general are able to take truth values—thus conveying information about states-of-affairs (Stjernfelt 2011).

Field research about the vervet monkey alarm-call system revealed that infantile and young adult vervets do not have the competence of either interpreting or emitting these calls efficiently (Cheney and Seyfarth 1990). Learning is involved in vocal production, in using the calls for specific events and responding to the calls. Infant vervets already babble alarms for broad and mutually exclusive categories like "flying birds", but they are unable to recognize whether the birds are predators of their group or not (Seyfarth and Cheney 1986). Although vervet monkeys appear to have an innate predisposition to vocalize calls which are similar to alarm calls for predator-like objects, they have to learn to recognize and respond to those calls. Moreover, the assumption that the mapping between calls and predators can be learned is supported by the observation that cross-fostered macaques, although unable to modify their call production, "did learn to recognize and respond to their adoptive mothers' calls, and vice versa (Cheney and Seyfarth 1998).". In our approach, we assumed that an associative learning competence is used for the acquisition and response to all alarm calls (Ribeiro et al. 2007; Queiroz and Ribeiro 2002). As we have pointed, alarm calls vocalized by vervet monkeys are signs of classes of objects that exist in the real world. These signs are symbols interpreted as indices of the presence of the predator. In Peirce's mature semiotic terminology, alarm-calls are dicent symbols, for the object of a dicent symbol is a general interpreted as an existent. It is exactly this class of processes that some authors have identified as proto-propositional (see Hurford 2007). The importance of this class has been strongly emphasized, since it seems to work between two semiotic classes (index—symbol).

If symbols can be analyzed in three subclasses, only one, namely, argument, possesses metasemiotic properties. Arguments are genuine symbols, types interpreted as generals—"An Argument is a Sign which, for its Interpretant, is a sign of law. Or we may say [...] that an Argument is a Sign which is understood to represent its Object in its character as Sign" (EP 2:292). Several anthroposemiotic phenomena are examples of arguments. Complex chains of reasoning are based on metasemiotic properties, as metalinguistic activities and description of codes and messages. Intersemiotic translations from a semiotic system to another one (e.g., speech > writing; writing > formal languages) only appeared once the material was interpreted as semiotically manipulable, based on laws, or rules. It is to this aspect that what Tomasello et al. (2005) called the "ratchet effect" must be attributed, since it permits accumulation and explosive growth of information. The emergence of symbolic artifacts (semiotic structures physically and culturally available in the environment) produces a very distinct type of cognitive niche (see Clark 2006). They make complex semiotic processes available to attention, perception, etc., opening up a new range of semiotic operations, allowing self-inspection of semiotic complex processes. They make it possible for semiotic systems to think about their own dynamics. What was unexpected to us when performing a semiotic analysis of firefly communication was to find that arguments were also relevant to models of this phenomenon.

Firefly Propositions: An Unexpected Finding

When we try to interpret the firefly flashes involved in courtship or predation based on the basic categories above, we find that three fundamental classes are not sufficient to account for those signs. They seem to combine different aspects of the basic categories. Consider, for instance, that, on the one hand, the flashes point to the presence and location of the emitting firefly, just like the pointing paw of the English pointer indicates the presence and location of a bird. In these terms, the firefly flashes, with their acute figure/ground distinction, can be said to be indexical in nature. However, their pattern also refers to a particular species and gender of firefly, and, for this reason, can be said to be iconic signs, referring to certain qualities of the emitter. It is true that the flashes only refer to a very specific and small part of the iconic qualities of the emitter, but (in the dark) they are very significant, pregnant signs. This is precisely what facilitates their being copied by other species, as it indeed happened in the case of *Photuris* femme fatale fireflies, which signal courtship signs for males of other fireflies, such as *Photinus* males, and when they come to reach what was supposed to be a female of its species, what they find is a voracious predator (Lloyd 1965, 1975, 1986). Indeed, an arms race seem to be taking place between *Photinus* and *Photuris* fireflies, with more complex signalling being selected for in the former, while correspondingly more complex (aggressive) mimicry being selected in the later. This 'arms race' between *Photuris* females and their prey, say, *Photinus* males, was made possible by the fact that the firefly flashes became so conventionalized, in their iconic aspect, that they became stable symbols of the species in question, which are also readable by members of other species.

In order to account for the sign-mediated processes involved in firefly deception, we need to go beyond the basic classification of signs as icons, indices, and symbols.

We should take into account that icons, indices, and symbols are just sign aspects, not mutually exclusive classes of empirical, observable signs. To use a familiar example, photographs are both icons (resembling their objects to different degrees) and indices (being physically connected to their object by the photographic process). The possession of one aspect—e.g. iconicity—does not preclude, thus, the same sign from being also indexical or symbolic.

But notice that, in the case of the fireflies, the sign—the species-specific flash—is referring to its object—the very firefly—in two different senses, combining information about its presence and location, and about its species and gender. These flashes are more complex than simple signs; they are composite signs. We come, thus, to a finding which was unexpected to ourselves: according to Peirce's theory of signs, fireflies are producing propositions!

The situation is more general, however. It may be that, in all cases of deception, we need to turn to Peirce's account of propositions, since propositions are the (composite) signs which may be true or false, and deception is obviously a case of a false proposition.

Robert Mitchell formalizes the concept of deception as follows:

(i) An organism R registers (or believes) something Y from some organism E, where E can be described as benefiting when (or desiring that)
(iia) R acts appropriately toward Y, because
(iib) Y means X; and
(iii) it is untrue that X is the case.

In order to be explicit about the role of signs in the mediation of deception, El-Hani et al. (2010) propose the following modified version of Mitchell's definition:

(i) An organism R registers a sign Y emitted by organism E, and E can be described as benefiting when
(iia) R behaves toward Y, as if
(iib) Y means that X is the case; but
(iii) it is untrue that X is the case.

It is clear, then, that deception demands a situation in which a sign emitted by a given organism and registered by another organism conveys a false information. Only (proto-)propositions can be true or false, and this is one of the reasons why we argue that we should say that the composite signs used by fireflies in deception should be treated as a proposition. Is this too far-fetched?

To answer this worry, let us examine what is a proposition in terms of Peirce's theory. Usually, a proposition is taken to mean linguistically represented signs claiming something about an object (e.g., "S is P"). But, for our arguments to be properly understood, it is important to take into account that Peirce generalized the notion of proposition in his theory of dicisigns (i.e., signs-which-say-something; propositions) to encompass also non-linguistic cases. The proposition forms part of the concept triad rheme-proposition-argument, as discussed above. The rheme is the skeleton of a proposition—but with one or more of its subject slots left blank. Thus "_is blue" is a rheme just like "_ gives _ to _". When one or more slots in a rheme are filled in by an index (a pointing finger, a demonstrative pronoun, a proper name, etc.), it becomes a proposition. When a proposition is inferred from one or more other propositions in a process of inference, it is an argument.

The crucial structure uniting all propositions, linguistic or not, is that they refer twice to the same object, by means of indices and icons, respectively. In the proposition, the object dealt with must be indicated by means of an index (in the linguistic case: a proper name, a class name, a pronoun, etc.), thus constituting the subject of the proposition, while the predicate of the proposition is presented by means of an icon (in the linguistic case: an adjective, a class name, or verb describing aspects of the object indicated by the index). Thus, a proposition forms a special type of syntactic combination between icon and index, claiming that the two refer to one and the same object.

In order to appreciate how the proposition does not need to be linguistically represented, consider the case of a caricature accompanied by a pointing gesture towards a person. This is a non-linguistic proposition: the pointing functions as the index indicating the subject; the caricature drawing as the icon predicate part. This double reference of the proposition is the reason why it may claim something (iconic) about something (indexical)—and this is why a proposition may be true or false, depending on whether the iconic quality claimed actually exists in the object referred to or not. And this is why deceptive signs must necessarily be propositions. Notice, further, that it is not the indexical part of these signs that lies: in our firefly example, the signs correctly indicate the firefly femme fatale at a specified point in space and time. It is the symbolic-iconic code indicating the character of the animal emitting the sign that does the lying.

From the several complex features of fireflies' communication-mediated interactions, Lloyd (1986) concludes that deception can act as a triggering factor in the evolution of communication complexity. We agree with him. Lies are only possible at the level of—ever so primitive—propositions, claiming that something is the case when it is not. Simple propositions can lie and, by lying, they can trigger increasingly more complex sign processes. This complexity scale should not be mapped, however, onto the distinction icon-index-symbol. It should be rather conceived by other means, e.g., the number of nested levels of communication, the complexity of the sign vehicles, the degree of explicit articulation of the different semiotic functions involved, the complexity of the perceptive and cognitive processes necessary to interpret these signs, etc. (Stjernfelt 2012).

The evolution of communication complexity in fireflies occurs at a very slow pace, since the evolution of the next level of complexity is likely to take an immense amount of generations. While the single game is constantly acted out between particular insects out there, the 'intelligent' step taking us to the next level of deception in this fly-over-fly game is performed by the whole firefly population subjected to natural selection.

We consider these findings, although initially unexpected, to be a crucial part of a proper semiotic modelling of deception in animal communication. It is worth summarizing the basic ideas, thus, in the end of this paper:

1. Our semiotic analysis of firefly communication—in particular, deception—shows the necessity of considering even simple sign uses in deceptive strategies in biology as dicisigns, proto-propositions, displaying the duplicity of referring indexically and signifying iconically. These signs can only be used to deceive because they propositionally claim that something is the case, when it is untrue that it is indeed the case.
2. This analysis casts doubt upon the widespread assumption that the distinction between icons, indices, and symbols can be mapped directly onto biological evolution so as to form three distinct, consecutive phases. Rather, all three aspects of object reference are present in different combinations already in relatively simple biological sign uses. Firefly signal patterns are a case in point, hinting at a continuous scale between simple icons at one end and symbolical stylizations of icons at the other end.

If we consider the following continuum—the flash pattern as an icon of a firefly, the stripe pattern as an icon of a zebra, two half circles of the breasts as an icon of a woman, the outline of the body shape as an icon of a man—, we can see that, in all these cases, the patterns mentioned have iconic qualities allowing conspecifics to recognize each other. But they do not display the same amount of stylization nor the same degree of arbitrariness in relation to other aspects of the animals' appearance.

The flash pattern is evidently the most symbolic of these cases, because it is extremely stylized, can be repeated identically with a stable meaning, is used only for signifying mating behavior and only at a specific time (night), and could, in theory, be substituted by another pattern. Moreover, it is the only sign facilitating the meeting between the firefly males and females in the dark. By contrast, each of the other three signs mentioned above are only one among many other possible iconic signs among conspecifics (smell, sound, other visual icons, etc.). It is the specific signification of the firefly flashes that make it possible that they are used in such a manner that fireflies are deceived by apt mimics of the species-specific flashes. They are deceived by different manners of using these proto-propositions that both refer indexically and signify iconically.

3. Our analysis also shows the need of considering two timescales to understand the evolution of deception (in the case of fireflies, and, certainly, in the case of many other deceptive mimicry phenomena): one of individual, highly routinized sign exchanges involving particular firefly specimens with relatively small interpretation plasticity, and one of the arms race between species leading to the evolution of nested deception strategies. Such deceptive strategies in general—between predator and prey, between species competing for related ecological niches, between individuals or groups within the same species—may constitute a major motor in the semiotic evolution of complex signs, to some extent analogous to the code maker-code breaker arms race in human military intelligence.

In summary, the study of firefly signal patterns seems very promising for further investigation, both empirical and theoretical, of the role of biological deception in the evolution of animal communication.

Discussion

The vast morphological variety of empirically observed semiotic processes is usually reduced to three (non-exclusive and hierarchically organized) classes of signs based on the sign-object relation (icon, index, symbol). But there are a number of semiotic phenomena which correspond to "intermediate" classes of symbols such as linguistic deictic, proper names, logical quantifiers, which are cases of "indexical symbols", or musical onomatopoeia, which are "iconic symbols". In order to more accurately describe these classes, Peirce developed several classifications. According to the ten-fold classification, a sign is grounded in some property, event, or regular pattern, by virtue of which it stands for some quality, occurrence, or law to a third element, an interpretation of possibility, physical connection or rule-based tendency. Through such classification, it becomes possible to represent more accurately several morphological variations that we can empirically find.

In the symbolic sign process described above, what is communicated from the object to the interpretant through the sign is a lawful relationship between a given type of alarm and a given type of predator. Generally speaking, a symbolic sign communicates a habit embodied in an object to the interpretant as a result of regularity in the relationship between sign and object (see Queiroz and El-Hani 2006). If iconic signs inform a certain quality shared by signs and objects and indexical signs inform a direct physical connection between signs and objects, then symbolic signs inform a habit embodied in the object to the interpretant as a result of a law-like regularity in the relationship between sign and object. Constraints imposed by spatio-temporal presence of an existent object are not functional in symbolic processes.

Acknowledgements C.N.E. thanks The State of Bahia Research Foundation (FAPESB) for research funds and the Brazilian National Research Council (CNPq) for a productivy in research grant, F.S. thanks Aarhus University and the The Danish Council for Independent Research/ Humanities.

References

Atkin, A. (2005). Peirce on the index and indexical reference. *Transactions of The Charles S. Peirce Society, 41*(1), 161–188.

Cheney, D. L. & Seyfarth, R. M. (1990). *How Monkeys See the World.* Chicago, University of Chicago Press.

Cheney, D. L. & Seyfarth, R. M. (1998). Why monkeys don't have language. *The Tanner Lectures on Human Values.* G. Petersen. Salt Lake City, University of Utah Press. 19, 173–210.

Clark, A. (2006). Language, embodiment, and the cognitive niche. *Trends in Cognitive Sciences, 10*(8), 370–374.

De Tienne, A. (2003). Learning qua semiosis. S.E.E.D. *Journal–Semiotics, Evolution, Energy, and Development, 3,* 37–53.

El-Hani, C., et al. (2010). Firefly femmes fatales: A case study in the semiotics of deception. *Biosemiotics, 3,* 33–55.

Farias, P., & Queiroz, J. (2000). Notes for a dynamic diagram of Charles Peirce's classifications of signs. *Semiotica, 131*(1/2), 19–44.

Farias, P., & Queiroz, J. (2003). On diagrams for Peirce's 10, 28, and 66 classes of signs. *Semiotica, 147*(1/4), 165–184.

Farias, P., & Queiroz, J. (2004). *10cubes* and *3N3*: Using interactive diagrams to investigate Charles Peirce's classifications of signs. *Semiotica, 151*(1/4), 41–63.
Freadman, A. (2001). The classifications of signs (II): 1903. In J. Queiroz & R. Gudwin (Eds.), *Digital encyclopedia of C. S.Peirce*. http://www.digitalpeirce.fee.unicamp.br. Accessed 1 Jan 2012.
Freadman, A. (2004). *The machinery of talk—Charles Peirce and the sign hypothesis*. Stanford: Stanford University Press.
Hilpinen, R. (1992). On Peirce's philosophical logic: Propositions and their objects. *Transactions of the Charles Sanders Peirce Society, 28*(3), 467–488.
Hoffman, M. (2001). The 1903 classification of triadic sign-relations. In J. Queiroz & R. Gudwin (Eds.), *Digital encyclopedia of C.S.Peirce*. http://www.digitalpeirce.fee.unicamp.br. Accessed 1 Jan 2012.
Houser, N. (1991). A Peircean classification of models. In M. Anderson e F. Merrell (eds.). On Semiotic Modeling. Mouton de Gruyter, pp. 431–439.
Houser, N. (1992). On Peirce's theory of propositions: A response to Hilpinen. *Transactions of the Charles Sanders Peirce Society,* XXIII(3), 489–504.
Houser, N. (2010). Peirce, phenomenology, and semiotics. In Paul Cobley (Ed.), *The routledge companion to semiotics* (pp. 89–100). London: Routledge.
Hurford, J. (2007). The origins of meaning. Oxford: Oxford University Press.
Jappy, A. (1989). Peirce's sixty-six signs revisited. In G. Deledalle (Ed.), *Semiotics and pragmatics proceedings of the perpignan symposium on semiotics and pragmatics* (pp. 143–153). Amsterdam: John Benjamins.
Lieb, I. C. (1977). Appendix B. In C. S. Hardwick (Ed.), *Semiotics and significs: The correspondence between Charles S. Peirce and Victoria Lady Welby* (pp. 161–166). Indiana: Indiana University Press.
Lizska, J. (1996). A general introduction to the semeiotic of Charles Sanders Peirce. Indiana: Indiana University Press.
Lloyd, J. E. (1965). Aggressive mimicry in Photuris: Firefly femmes fatales. *Science, 149,* 653–654.
Lloyd, J. E. (1975). Aggressive mimicry in Photuris fireflies: Signal repertoires by femmes fatales. *Science, 187,* 452–453.
Lloyd, J. E. (1986). Firefly communication and deception: "Oh, what a tangled web". In R. W. Mitchell & N. S. Thompson (Eds.), *Deception: Perspectives on human and nonhuman deceit* (pp. 113–128). Albany: SUNY Press.
Merrell, F. (1996). Signs grow. Toronto: University of Toronto.
Parker, K. (1998). *The continuity of Peirce's thought*. Nashville: Vanderbilt University Press.
Queiroz, J. (2003). Comunicação simbólica em primatas não-humanos: uma análise baseada na semiótica de C.S.Peirce. Revista Brasileira de Psiquiatria, *25*(Supl II), 2–5.
Queiroz, J. (2012a). Dicent symbols in non-human semiotic processes. *Biosemiotics, 5,* 1–11.
Queiroz, J. (2012b). Peirce's ten classes of signs: Modeling biosemiotic processes and systems. In T. Maran, K. Lindström, R. Magnus, & M. Tønnessen (Eds.), *Semiotics in the wild—Essays in honour of Kalevi Kull on the occasion of his 60th birthday* (pp. 55–62). Tartu: Tartu University Press.
Queiroz, J., & El-Hani, C. (2006). Towards a multi-level approach to the emergence of meaning in living systems. *Acta Biotheoretica, 54,* 179–206.
Queiroz, J., & Ribeiro, S. (2002). The biological substrate of icons, indexes and symbols in animal communication: A neurosemiotic analysis of Vervet monkey alarm-calls. In M. Shapiro (Ed.), *Peirce seminar V* (pp. 69–78). Oxford: Berghahn Books.
Rescher, N. (1996). Process Metaphysics: An Introduction to Process Philosophy. New York: State University of New York Press.
Ribeiro, S., et al. (2007). Symbols are not uniquely human. *Bio Systems, 90,* 263–272.
Sanders, G. (1970). Peirce sixty-six signs? *Transactions of Charles Sanders Peirce Society, 6*(1), 3–16.

Savan, D. (1977). Questions concerning certain classifications claimed for signs. *Semiotica, The Hague*, *19*(3/4), 179–196.
Sebeok, T. A. (1989). *The sign and its masters*. New York: University Press of America.
Serson, B. (1997). On Peirce's pure grammar as a general theory of cognition: From the thought-sign of 1868 to the semeiotic theory of assertion. *Semiotica, 113*(1/2), 107–157.
Seyfarth, R. M. and Cheney D. L. (1986). "Vocal development in vervet monkeys." *Animal Behavior*, *34*, 1640–1658.
Short, T. (2007). *Peirce's theory of signs*. New York: Cambridge University Press.
Stjernfelt, F. (2011). Signs conveying information: On the range of Peirce's notion of propositions: Dicisigns. *International Journal of Signs and Semiotic Systems, 1*(2), 40–52.
Stjernfelt, F. (2012). The evolution of semiotic self-control: Sign evolution as the ongoing refinement of the basic argument structure of biological metabolism. In T. Schilhab, F. Stjernfelt, & T. Deacon (Eds.), *The symbolic species evolved* (pp. 39–63). Dordrecht: Springer.
Tomasello, M., et al. (2005). Understanding and sharing intentions: The origins of cultural cognition. *Behavioral and Brain Sciences*, *28*(5), 675–735.
Weiss, P., & Burks, A. (1945). Peirce's sixty-six signs. *Journal of Philosophy, XLII,* 383–388.

Semeiosis as a Living Process

Vinicius Romanini

> We must distinguish between the Immediate Object, —i.e. the Object as represented in the sign, —and the Real (no, because perhaps the Object is altogether fictive, I must choose a different term, therefore), say rather the Dynamical Object, which, from the nature of things, the Sign *cannot* express, which it can only *indicate* and leave the interpreter to find out by *collateral experience.* For instance, I point my finger to what I mean, but I can't make my companion know what I mean, if he can't see it, or if seeing it, it does not, to his mind, separate itself from the surrounding objects in the field of vision. It is useless to attempt to discuss the genuineness and possession of a personality beneath the histrionic presentation of Theodore Roosevelt with a person who recently has come from Mars and never heard of Theodore before. A similar distinction must be made as to the Interpretant. But in respect to *that* Interpretant, the dichotomy is not enough by any means. For instance, suppose I awake in the morning before my wife, and that afterwards she wakes up and inquires, "What sort of a day is it?" *This* is a sign, whose Object, as expressed, is the weather at that time, but whose Dynamical Object is the *impression which I have presumably derived from peeping between the window-curtains.* Whose Interpretant, as expressed, is the quality of the weather, but whose Dynamical Interpretant, is *my answering her question.* But beyond that, there is a third Interpretant. The *Immediate Interpretant* is what the Question expresses, *all* that it immediately expresses, which I have imperfectly restated above. The *Dynamical Interpretant* is the actual effect that it has upon me, its interpreter. But the Significance of it, the *Ultimate,* or *Final, Interpretant* is her *purpose* in asking it, what effect its answer will have as to her plans for the ensuing day. I reply, let us suppose: "It is a stormy day." Here is another sign. Its *Immediate Object* is the notion of the present weather so far as this is common to her mind and mine—not the *character* of it, but the *identity* of it. The *Dynamical Object* is the *identity* of the actual or Real meteorological conditions at the moment. The *Immediate Interpretant* is the *schema* in her imagination, i.e. the vague Image or what there is in common to the different Images of a stormy day. The *Dynamical Interpretant* is the disappointment or whatever actual effect it at once has upon her. The *Final Interpretant* is the sum of the *Lessons* of the reply, Moral, Scientific, etc. Now it is easy to see that my attempt to draw this three-way, "trivialis" distinction, relates to a real and important three-way distinction, and yet that it is quite hazy and needs a vast deal of study before it is rendered perfect. Lady Welby has got hold of the same real distinction in her "Sense, Meaning, Significance," but conceives it as imperfectly as I do, but imperfectly in other ways. Her *Sense* is the *Impression* made or normally to be made. Her *meaning* is what is intended, its purpose. Her Significance is the real upshot. (CP 8.314)

V. Romanini (✉)
University of São Paulo, São Paulo, Brazil
e-mail: vinicius.romanini@usp.br

V. Romanini, E. Fernández (eds.), *Peirce and Biosemiotics,* Biosemiotics 11,
DOI 10.1007/978-94-007-7732-3_12, © Springer Science+Business Media Dordrecht 2014

Introduction

Peirce's theory of signs is not easy to grasp and there are a number of reasons for that. The most important is that it is unfinished. For nearly 40 years, Peirce actively worked on his system of logic he regarded to be the same as semeiotic. In these four decades, Peirce produced dozens of different definitions for the term "sign" and its fundamental aspects (which he called respects, probably meaning that they were always respective to one another, as in CP 8.343). The simple ones are quite similar because they involve only the three basic aspects and their correlation. So we can say without fear of mistake that a sign is anything that represents its object as to produce an effect, which is its interpretant.

But how these three aspects should be understood, eventually further divided and correlated to produce a whole system of logic, changed a lot in Peirce's works. During his youthful years, his semeiotic was intended to cover primarily the relation between concepts—which he probably conceived just as symbols produced by human minds—and their objects. In his mature years, mostly after 1900, Peirce's definition of *sign* was being continuously reformulated to incorporate his logic of relatives and his metaphysical musings about cosmology and a semeiotic theory of reality.

Furthermore, Peirce found out that the three basic aspects described above in the basic definition should be decomposed into more subtle ones. The aspect of the object is then split into ***immediate*** and ***dynamic*** objects; and the aspect of the interpretant is thrice divided in ***immediate***, ***dynamic*** and ***final***. The myriad of possible relations among these now six aspects, as well as their ordering in a general classification of all possible classes of signs occupied Peirce's mind until near his death in 1914.

We can clearly see from sketches drawn in his notebooks, as well as in his letters to Lady Welby and William James in the 1900's, that Peirce spent the last years of his life working strenuously to reach the broadest possible definition for the word "sign." In 1906 he even considered dropping the word "sign" and adopting the word **medium**, which seemed to him less contaminated with historical confusions. A "medium for the communication of a form" is then one of his preferred encapsulated definition for sign (EP II: 544, n. 22).

Another reason for the difficulty in understanding Peirce's semeiotic is that it cannot be isolated from the rest of his philosophy. Quite the contrary, he considered his semeiotic as the cornerstone of his lifelong study of the scientific method to pursue the truth. Pragmatism or—as he sometimes preferred—"pragmaticism", was his scientific method to clarify ideas by their consequences. A concept, or symbol, is a mental habit functioning as a **conditional tendency** to produce effects accordingly, and its meaning is the whole of the conceivable effects that it **would produce** in *general* conceivable situations. Since symbols must embody both indices and icons to produce reasoning, pragmatism should be considered a corollary of his semeiotic.

To catch the complexity of Peirce's theory of sign, then, one needs to dive into the deep waters of this general, unfinished and many times idiosyncratic philosophical system in which metaphysics, phenomenology, philosophy of mind and logic would converge harmonically under his architectonic system. In the dawn of the twentieth century, his theory of signs left the threshold of the human mind and culture to recognize that there is semeiosis whenever and wherever we find living systems. Putting together Plato's idealism, Aristotle's concepts of forms and final causes, Duns Scott's medieval realism and Darwinian evolutionism, Peirce announces his "extreme realism" by saying that the whole of reality has the nature of sign, and semeiosis is akin to a fundamental theory of continuum (which he names synechism).

Peirce was particularly worried about the direction science was going under mechanistic positivism, by which the explanation about observed phenomena was into discrete interactions of matter,[1] instead of searching the *final* causes. In a letter of 1905, for instance, he claims that:

> to try to peel off signs & get down to the real thing is like trying to peel an onion and get down to onion itself, the onion per se, the onion an sich'. (MS L387)

Some critics say that Peirce was advocating a sort of religious vitalism. Or that he was letting metaphysics dominate his thoughts as he got old, sick and worried about his death. Let's not underestimate Peirce, though. He was a man of science, worked as chemist and physicist, knew mathematics as few of his time did, and was well read in philosophy of science. But first of all, Peirce was a logician and this grand vision of semeiosis should be understood as a statement about the foundations of logic and the pursuit of truth. His contribution to biosemiotics cannot be detached from these more general purposes.

A Peircean Approach to Biosemiotics

It has been often said that semeiosis is coextensive with life. I take this for granted, but also stress that this idea needs further development if we want to build a logical foundation for the field of biosemiotics. We can start by asking about the nature of such coextension. Viewing from a particular side, one can answer that semeiosis is a phenomenon emerging from life processes[2] which are considered primary. Taking the opposite perspective, another can say that life is the emergent phenomenon and semeiosis is the most fundamental.

A third and more profound hypothesis is that coextension might mean a continuous gradient by which life and semeiosis are two different words to signify the same

---

[1] This led Peirce to an interesting semeiotic version of the quantum uncertainty principle: "But as soon as a man is fully impressed with the fact that absolute exactitude never can be known, he naturally asks whether there are any facts to show that hard discrete exactitude really exists. That suggestion lifts the edge of that curtain and he begins to see the clear daylight shining in from behind it" (CP 1.172). He would have regarded as pointless the search for fundamental quantum particles and would have instead exhorted physicists to search for the logical causational principles governing reality.

[2] Sebeok (2001), for instance, says that "semiosis presupposes life".

thing: the creation of meaning, a process by which the real is produced as general representation developed by the continuous convergence of a multitude of interpretants (a concept which should no be equated to the action of interpreters made of flesh and blood, but more generally as a correlate of a logical triadic relation that can even be a simple possible effect of a sign.).

This last approach is the one I will develop here, which prompts me to add another concept that is coextensive both to life and semeiosis: mind.[3] Putting it in a nutshell: life, mind and semeiosis might be just different words to describe the action of signs (or a genuine triadic relation, which amounts to the same). If that is the case, a truly Peircean approach to biosemiotics cannot escape from these considerations, although such a hypothesis must be also studied by the pragmatic method. That is, we should start by asking: What would be the conceivable normative practical consequences that we should establish in the scientific community of biosemioticians if we were to adopt the belief that mind, life and semeiosis are the same phenomenon viewed from different perspectives?

The first consequence is that we should not try escape from a metaphysical inquiry about the foundations of our concept of reality. Another consequence is that we should try to understand the minute aspects of the sign relations and how they would help us to understand biosemiotics as fundamentally the study of symbols as living signs.[4] A taxonomy of all classes of signs and their logical relation is the necessary scaffolding for wrapping around the three concepts we are trying to weld into a single general and more powerful conception of reality. In the following pages we will give some hints about how we think these consequences could be taken into account, although much yet will be left for future research.

Peirce's Metaphysics

Following Kant's Copernican revolution, Peirce early understood that our concept of reality is the result of a logical process by which the multitude of sense impressions given by perception are subsumed under a general concept produced *a posteriori*, or after the experience. This is done by the work of our mind, which is able to synthesize percepts due to the application of the concepts of space and time. For Kant, the thing in itself is not cognizable and reality appears as the arrangement of what is given during experience into space (giving us the outer sense of the world) and time (the inner sense of change and modality). The categories of space and time must then be *a priori*, which means that the conditions of possibility of any intelligible experience are built into our own human minds.[5]

[3] My frame of interpretation is similar to Ivo Ibri's definition of connaturality of mind and matter. See Chapter "The continuity of life: on Peirce's objective idealism."

[4] See Chapter "The life of symbols and other legisigns: More than a mere metaphor." by Winfried Nöth for an explanation of the life of symbols.

[5] Jakob von Uexküll used this same line of argument to create his concept of Umwelt, which is basically a generalization of the Kantian system to all living beings: "No matter what kind of qual-

Having solved the enigma of the origin of our ideas in experience, Kant surrenders to an even greater mystery: how are pure sciences, such as mathematics and theoretical physics, possible? No one can deny that the theorems of geometry, the operations of arithmetic or the inductive method applied to physics bring us important practical results, but none of these are synthesized *a posteriori*. Neither are they analytical. They all seem to depend on a quite strange *a priori* synthesis. Without explaining this kind of knowledge we cannot justify causality, for instance. To say that the sun causes rocks to heat, (in general, that is, not only this or that rock but any conceivable one), would be meaningless, because we could only be sure of the heating of the particular rocks effectively subjected to our perception. We would never be able to suppose rocks in Mars are heated by the Sun unless we somehow actually measure them. Neither that a diamond is hard unless we actually scratch it.[6]

Nevertheless, we all reasonably guess that any rock, be it on Earth or Mars, get heated when illuminated by the sun, or that a diamond will not be scratched by a knife. How can we justify such guesses—which are pretty much the kind of guesses that are behind any new discovery in science, the work of a geometer or of an artist?[7]In other words, from the very beginning of his studies in logic Peirce was interested in finding out how the human mind got its surprising ability to guess the laws of nature. He concludes that Kant had put a lock on the door of philosophy when he formulated the most important epistemological question ever made:

> Kant declares that the question of his great work is "How are synthetical judgments a priori possible?" By a priori he means universal; by synthetical, experiential (i.e., relating to experience, not necessarily derived wholly from experience). The true question for him should have been, "How are universal propositions relating to experience to be justified?" (CP 4.92, 1893)

The answer Peirce attempts to give involves, in the first place, the denying of any uncognizable ***thing-in-itself***. He assumes that everything is fundamentally cognizable because it participates in the whole of experience that springs from real possibilities. The "ding an sich" does not have a place in Peirce's philosophical system because semeiosis is naturalized to explain mental and living processes, which are then considered to be of the same nature as a symbol. Only symbols can be used

---

ity it may be, all perceptual signs have always the form of a command or impulse. If I claim that the sky is blue, I am doing so because the perceptual signs projected by myself give the command to the farthest level: Be blue! The sensations of the mind become, during the construction of our worlds, the qualities of the objects, or, as we can put it in other words, the subjective qualities are building up the objective world. If we, instead of sensation or subjective quality, say perceptual sign, we can also say: the perceptual signs of our attention become the perceptual cues (properties) of the world." (Uexküll 1973/1928). Peirce would regard this as an unattainable nominalistic view of perception.

[6] Peirce himself fell into this sort of nominalism in his early writings, as he himself admits in CP 5.453 and later on, in CP 5.457, correcting his own mistake, he explicitly declares that "we must dismiss the idea that the occult state of things (be it a relation among atoms or something else), which constitutes the reality of diamond's hardness can possibly consist in anything but in the truth of a general conditional proposition."

[7] See the quote from CP 1.383 that illustrates Vincent Colapietro's Chapter "The ineffable, the individual, and the intelligible:Peircean reflections on the innate ingenuity of the human animal."

to express general conditional propositions and communicate the modal law they embody. Being indeterminate by their generality and bearing vague pregnancy,[8] symbols have a tendency to develop towards complexity by gaining information as they get more determined.

This is precisely the general mode of reality, as Peirce sees it: from the continuous flow of qualities of feelings underlying the real (with no clear cuts among them) unfolds the ever more clear and distinct attributes of reality. As they get determined, the work of mind (not this or that mind, but the communion of all possible minds) ***would*** eventually combine them by association so as to synthetically produce all types of complex beings.[9]

> Kant gives the erroneous view that ideas are presented separated and then thought together by the mind. This is his doctrine that a mental synthesis precedes every analysis. What really happens is that something is presented which in itself has no parts, but which nevertheless is analyzed by the mind, that is to say, its having parts consists in this, that the mind afterward recognizes those parts in it. Those partial ideas are really not in the first idea, in itself, though they are separated out from it. It is a case of destructive distillation. When, having thus separated them, we think over them, we are carried in spite of ourselves from one thought to another, and therein lies the first real synthesis. An earlier synthesis than that is a fiction. (CP 1.384)

This implies that thought is a pervasive constituent of reality from which all kinds of predication are distilled, beginning at the level of sensation. The common mind of a community of interpretants then associate these seeds of ideas, or "semes,"[10] to produce cognitions and to represent them in general propositions which are then continuously reshaped and communicated to strengthen our common concept of truth:

> This theory of reality is instantly fatal to the idea of a thing in itself, —a thing existing independent of all relation to the mind's conception of it. Yet it would by no means forbid, but rather encourage us, to regard the appearances of sense as only signs of the realities. Only, the realities which they represent would not be the unknowable cause of sensation, but ***noumena,*** or intelligible conceptions which are the last products of the mental action which is set in motion by sensation. The matter of sensation is altogether accidental; precisely the same information, practically, being capable of communication through different senses. And the catholic consent which constitutes the truth is by no means to be limited to men in this earthly life or to the human race, but extends to the whole communion of minds to which we belong, including some probably whose senses are very different from ours, so that in that consent no predication of a sensible quality can enter, except as an admission that so certain sorts of senses are affected. This theory is also highly favorable to a belief in

---

[8] See Robert Lane's Chapter "Peircean semiotic indeterminacy and its relevance for Biosemiotics." for an excellent treatment of vagueness and indeterminacy in Peirce's philosophy.

[9] See CP 1.383, as well as Vincent Colapietro's explanation about this synthesis in Chapter "The ineffable, the individual, and the intelligible:Peircean reflections on the innate ingenuity of the human animal."

[10] CP 4.538–9: "Of course, I must be understood as talking not psychology, but the logic of mental operations. Subsequent Interpretants furnish new Semes of Universes resulting from various adjunctions to the Perceptual Universe. They are, however, all of them, Interpretants of Percepts. Finally, and in particular, we get a Seme of that highest of all Universes which is regarded as the Object of every true Proposition, and which, if we name it [at] all, we call by the somewhat misleading title of 'The Truth.'"

> external realities. It will, to be sure, deny that there is any reality which is absolutely incognizable in itself, so that it cannot be taken into the mind. But observing that "the external" means simply that which is independent of what phenomenon is immediately present, that is of how we may think or feel; just as "the real" means that which is independent of how we may think or feel ***about it;*** it must be granted that there are many objects of true science which are external, because there are many objects of thought which, if they are independent of that thinking whereby they are thought (that is, if they are real), are indisputably independent of all ***other*** thoughts and feelings. (CP 8.13)

For the universe to be intelligible, generality must be a fundamental condition of reality, for only regularities are intelligible. These are what Peirce calls "thought," his third category, or Thirdness. But no pattern is in a completely static and immutable situation. On the contrary, everywhere we perceive changes and the influx of novelty, which is Peirce's first category, or Firstness. Experience, which is always *hic et nunc*, must involve intrinsic qualities which render it unrepeatable but must also be governed by some general patterns. The friction of experience is Peirce's second category, or Secondness.

> Now in genuine Thirdness, the first, the second, and the third are all three of the nature of thirds, or thought, while in respect to one another they are first, second, and third. The first is thought in its capacity as mere possibility; that is, mere mind capable of thinking, or a mere vague idea. The second is thought playing the role of a Secondness, or event. That is, it is of the general nature of experience or information. The third is thought in its role as governing Secondness. It brings the information into the mind, or determines the idea and gives it body. It is informing thought, or cognition. **But take away the psychological or accidental human element, and in this genuine Thirdness we see the operation of a sign.** (Peirce, CP 1.537, emphasis added)

Peirce does not rule out that there might be universes, even an infinite number of them, which are in a stage of pure firstness or eventually producing infinitesimal flashes of secondness. They might be all around us and have no consequence at all because they would not be reals. Continuity is what grants reality, which means that some sort of habit or law must be taken into account, although never so endured as to freeze changes and make growth impossible. Our Universe must have acquired the necessary generalizing tendency capable of governing replicas and growing towards complexity whenever novelty is internalized.[11]

[11] "The tendency to obey laws has always been and always will be growing. We look back toward a point in the infinitely distant past when there was no law but mere indeterminacy; we look forward to a point in the infinitely distant future when there will be no indeterminacy or chance but a complete reign of law. But at any assignable date in the past, however early, there was already some tendency toward uniformity; and at any assignable date in the future there will be some slight aberrancy from law. Moreover, all things have a tendency to take habits. For atoms and their parts, molecules and groups of molecules, and in short every conceivable real object, there is a greater probability of acting as on a former like occasion than otherwise. This tendency itself constitutes a regularity, and is continually on the increase. In looking back into the past we are looking toward periods when it was a less and less decided tendency. But its own essential nature is to grow. It is a generalizing tendency; it causes actions in the future to follow some generalization of past actions; and this tendency is itself something capable of similar generalizations; and thus, it is self-generative. We have therefore only to suppose the smallest spoor of it in the past, and that germ would have been bound to develop into a mighty and over-ruling principle, until it supersedes itself by strengthening habits into absolute laws regulating the action of all things in every respect in the indefinite future" (CP 1.408).

If we want a scale, we could say that the more inveterate generals are the laws of nature that govern the so-called physical phenomena, followed by the much more plastic patterns of biological behavior of protoplasm, such as instinctive habits of actions and reactions. The more flexible habits are those of mental action, which are able to produce self-corrective habit-changes. Mental habits govern the inferences of reasoning by bringing forth their own possible future laws, which are the hypotheses that we create and entertain while shaping our conduct as to bring a harmonic sentiment of identity between the creations of our minds and the aesthetical feeling we gather from experience. That's why minds are so much driven by iconic representation of possible relations, where mostly associations take place.

This finally solves the Kantian question of the possibility of every pure science, since all of them are based on diagrammatic representation of the laws of nature—that is, pure sciences represent a seme of the true functioning of the real, although never a complete truth about it. Reality, Peirce explains, has the form of a general capable of growth and development because animated by final causes, which are the general laws they embody and communicate. The synthesis of the multitude of percepts given by experience in the unity of a symbol is possible because the *schema* of time is not a transcendental entity, as Kant suggested, but is a regularity transforming and developing itself inside a living ***symbol***:

> The reality only exists as an element of the regularity. And the regularity is the symbol. (…) A symbol is an embryonic reality endowed with the power of growth into the very truth, the very entelechy of reality. This appears mystical and mysterious simply because we insist on remaining blind to what is plain, that there can be no reality which has not the life of a symbol. (EP2: 323-324)

Some might think all this is an anthropomorphic view of Nature, which Peirce would promptly agree, for he considers this kind of anthropomorphism much better than anthropocentric illusion that reality is what it seems to be only because there are humans to observe it:

> I hear you say: "This smacks too much of an anthropomorphic conception." I reply that every scientific explanation of a natural phenomenon is a hypothesis that there is something in nature to which the human reason is analogous; and that it really is so all the successes of science in its applications to human convenience are witnesses. (CP 1.316)

Putting it in simple words, when developing our science as an interpretation of natural processes we can only hope that the way our minds work has something in common with the way reality develops independently of ourselves. From this hope, which is a primordial hypothesis, we extract possible consequences and proceed to act accordingly. This means that the *rationale* of all reasoning is deductive at its bottom, although every ampliative inference produced during experience is either abductive or inductive.[12]

---

[12] "(…) although Abductive and Inductive reasoning are distinctly not reducible to Deductive reasoning, nor either to the other, nor Deductive reasoning to either, yet the rationale of Abduction and of Induction must itself be Deductive. All my reflections and self-criticisms have only served to strengthen me in this opinion. But if this be so, to state wherein the validity of mathematical reasoning consists is to state the ultimate ground on which any reasoning must rest." (Peirce in Turrisi, 1997, p. 276–277)

Moreover, Peirce denies Cartesian intuition and Cogito to proclaim that perceptual judgments, in the guise of unconscious abductions, are responsible for producing fallible but self-corrective hypotheses that pave our path to the truth. Experience is the mother of all possible information and true knowledge is the information contained in that symbol which **would be** shared and agreed upon by an ideal community of interpretants.

This would-be is the conditional law, or form, embodied by every symbol. This information might not be manifested to any actual mind but it is nevertheless real. This implicit or 'unfolded' form[13] is the foundation of Peirce's so called scholastic realism and objective idealism. What is the meaning he gives, for instance, to the word "hard" when he describes the hardness as the predicate of a rock? Peirce explains:

> (…) if he thinks that, whether the word "hard" itself be real or not, the property, character, the predicate, *hardness* is not invented by men, as the word is, but is really and truly in the hard things and is one in them all, as a description of habit, disposition, or behaviour, *then* he *is* a realist. (CP 1.27, n. 1)

Peirce conceived the whole of reality as a conditional argument from which general modality any number of particular world views could be derived, once more circumscribed communities of minds would develop, with their particular perceptive frames. As thirdness, the real is vague and indeterminate, but different degrees of determination can be accomplished by different living species, as the queries of reality are opened to their communicating minds.

Peirce's Logical Conception of Mind

One of the reasons some biosemioticians do not agree with Peirce's metaphysics is that they understand thought and mind in their traditional Cartesian and folk psychological definitions. Peirce, quite differently, saw them as logical entities, as in this important quote from 1909:

> All the various meanings of the word "Mind," Logical, Metaphysical, and Psychological, are apt to be confounded more or less, partly because considerable logical acumen is required to distinguish some of them, and because of the lack of any machinery to support the thought in doing so, partly because they are so many, and partly because (owing to these causes), they are all called by one word, "Mind." In one of the narrowest and most concrete of its logical meanings, a Mind is that Seme of The Truth, whose determinations become Immediate Interpretants of all other Signs whose Dynamical Interpretants are dynamically connected. In our Diagram the same thing which represents The Truth must be regarded as in another way representing the Mind, and indeed, as being the Quasi-mind of all the Signs represented on the Diagram. For any set of Signs which are so connected that a complex of two of them can have one interpretant, must be Determinations of one Sign which is a ***Quasi-mind***.(CP 4.550)

[13] As David Bohm (2000) puts it.

It is a mistake of folk psychology to suppose that biosemeiosis depends upon individual agency, or upon a brain doted of some inexplicable intentionality. The semeiotic mind is not a product of brain activity, although a large brain might intensify and speed the process. Mind is simply a logical triadic relation producing and partaking of meaning, such as when we draw a diagram representing the relations conceived about the object of our inquiry and then proceed to transform these relations so as to reveal facts that were not noted before. For example, every living being participates in the instinctual search for the food that it needs to survive, and any particular hungry animal should be considered a particular instantiation of this very general and instinctual process happening throughout the living world.

The worldview of the species (what Uexküll called its Umwelt) is the diagram of all the relations concerned with its survival and permanence, while the actual manipulation of these relations is performed by particular specimens, or group of specimens living in community in a specific time and space. Whenever new hypothetic relations are discovered by chance during ontogenesis and internalized by the species as phylogenesis, we have the evolutionary working of mind reshaping the diagram of relations, producing co-evolution of both the living species and their own significant environment. Their immediate interpretants are deeply welded as their dynamic interpretants become dynamically connected in a habitual manner.

If it is true that any hungry living being feels the hunger and emotionally responds to this feeling, this only indicates that semeiosis is happening at a level near immediate perception, when feelings have not yet been generalized into cognitive processes capable of being shared. Emotions are insipient cognitions, produced inferentially to reduce the manifold of sensations gathered through perception into a single predicate—one capable of allowing a quick response to some sensitive situation.

> (…) when our nervous system is excited in a complicated way, there being a relation between the elements of the excitation, the result is a single harmonious disturbance which I call an emotion. Thus, the various sounds made by the instruments of an orchestra strike upon the ear, and the result is a peculiar musical emotion, quite distinct from the sounds themselves. This emotion is essentially the same thing as a hypothetic inference, and every hypothetic inference involves the formation of such an emotion. (CP 2. 643)

Peirce's view of emotion connects it to organisms and their accidental situations (the ontogeny of individuals), while cognitions and sentiments are much more general and dependent on the development of instinct and the phylogenic development of species:

> That which makes us look upon the emotions more as affections of self than other cognitions, is that we have found them more dependent upon our accidental situation at the moment than other cognitions; but that is only to say that they are cognitions too narrow to be useful. The emotions, as little observation will show, arise when our attention is strongly drawn to complex and inconceivable circumstances. Fear arises when we cannot predict our fate; joy, in the case of certain indescribable and peculiarly complex sensations. (CP 5.292)

Following this line of argument, our individual "self" must be considered the result of a flow of feelings, sensations and emotions, but which must sooner or later give place to a more general and mediated concept to be shared by a community of interpretants—a shared sentiment, which is the basis of Peirce's sentimentalism: an instinct-like disposition to act in accordance to our past—or "collateral," as we will see below—experience, without which no living being could ever communicate anything to his peers, his enemies or to any other participant of his ground.

In fact, without this continuous spreading of emotions into more community-like sentiments, no logical conception of reality would ever be possible, no learning and no memory would ever be produced. What we call instinct is just the repertory of all the past experiences lived by a community of living beings (a species, for example, but also any number of species that have a common past of shared experiences):

> (…) every phenomenon of our mental life is more or less like cognition. Every emotion, every burst of passion, every exercise of will, is like cognition. But modifications of consciousness which are alike have some element in common. Cognition, therefore, has nothing distinctive and cannot be regarded as a fundamental faculty. If however, we ask whether there be an element in cognition which is neither feeling, sense, nor activity, we do find something, the faculty of learning, acquisition, memory and inference, synthesis. (CP 1. 376)

As much as we see growth and information in every corner of the Universe, so we notice that even the most ingrained instincts can develop as living species experience the flow of reality:

> Instinct is capable of development and growth—though by a movement which is slow in the proportion in which it is vital; and this development takes place upon lines which are altogether parallel to those of reasoning. And just as reasoning springs from experience, so the development of sentiment arises from the soul's Inward and Outward Experiences. Not only is it of the same nature as the development of cognition; but it chiefly takes place through the instrumentality of cognition. The soul's deeper parts can only be reached through its surface. In this way the eternal forms, that mathematics and philosophy and the other sciences make us acquainted with, will by slow percolation gradually reach the very core of one's being; and will come to influence our lives; and this they will do, not because they involve truths of merely vital importance, but because they are ideal and eternal verities. (CP 1.637)

To sum up, meaning is a matter of general relations that evolve on the level of species or, more generally yet, on the level of the whole biosphere. That's why Peirce says "even plants make their living… by uttering signs." (MSS 205, 318). Every living species must be considered, as a whole, a cognitive repository of learned experiences, and its metabolism—as well as any of his more complex functions—are communicative utterings necessarily attuned to the patterns and laws that govern reality. Here echoes again the Kantian challenge about the possibility of *a priori* synthetic inferences and Peirce's answer: the inward and outward senses are indeed the ground of all instinctual inference, but there are not *a priori*. They are the ground of the intelligible reality, the very common ground of every living species.

A white blood cell of our immune system moving after bacteria in a way that resembles a cat chasing a mouse[14] is not a solitary bunch of blind physical and chemical reactions, but a marvelous example of semeiotic conduct towards a definite, if only instinctual, purpose. Without this broad conception of mind, we would have to bring into the action of the white cell the finger of a magical divinity, or disguise biosemeiosis in some strange theory similar to the Cartesian homunculus.

[14] I owe this example to Daniel Meyers (University of San Diego).

Take it All or Leave Him Alone

Biosemioticians embarrassed by Peirce's extreme realism and synechism prefer to skip such ideas and use only the excerpts where Peirce deals only with semeiotic terminology. Others prefer to learn Peirce's doctrine of signs second-handedly through commentators that have hidden Peirce's objective idealism under the carpet. In both cases, the pervasive mentality that bathes reality becomes the hidden variable of Peirce's theory of signs. Both options lead invariably to huge mistakes. I will give you a concrete example. One of the most famous and quoted definitions of the "necessary and sufficient condition for something to be a semiosis"—found in dozens of articles, books and papers on biosemiotics—was given by Posner, Roberick and Sebeok:

> A interprets B as representing C.
> In this relational characterization of semiosis, A is the interpreter, B is some object, property, relation, event, or state of affairs, and C is the meaning that A assigns to B. (Posner et al. 1997, p. 4)

I am not saying the above is necessarily wrong. The authors in question have all the right to propose their own definition of sign and semeiosis. My question is: is this definition in accordance with Peirce's semeiotic? Let's see.

In Peirce's terminology of the triadic relation among sign, object and interpretant, when the above authors say that "A interprets B as representing C" we must understand that A is the place of the **interpretant**, B is "some object, property, relation, event, or state of affairs" that functions as the **sign** and C is the **object** being represented. Paraphrasing in a somewhat tautological expression we have kept the original definition in italics and added Peirce's three basic elements of his triadic definition in capital letters:

INTERPRETANTS *(A) interprets* SIGNS *(B) as representing* THEIR OBJECTS *(C)*.

So far so good. But we get into muddy waters when the authors define C as "the meaning that A assigns to B." That is wrong in Peircean semeiotics.

If by C we understand the **dynamic object**, that is, the object which the sign professes to represent because it conveys some information about it, then C cannot be the **meaning** of semeiosis in Peircean terms because meaning is for Peirce just another name for the interpretant of the sign:

> A 'sign', I say, shall be understood as anything which represents itself to convey an influence from an Object, so that this may intelligently determine a 'meaning', or 'interpretant'. (MS 318 1907)

The problem I am raising is **not** a mere technicality. Putting the meaning of a sign under the realm of the interpretant is the kernel of Peirce's theory of signs and pragmatism. Indeed, the pragmatic maxim states that the 'entire meaning and significance of any conception lies in its conceivably practical bearings' (EP: 145), that is, the effects and consequences of accepting the truth of such sign under conceivable circumstances.

If we simply equate meaning and the object of the sign, we slip into nominalism. Peirce would turn over in his grave if we were to transform his semeiotic into a nominalistic enterprise. What is missing, then, in the definition given above? Final causation, for interpretation is a teleological process of production of effects through the action of signs, or semeiosis.[15] Putting this in semeiotic terms, we cannot explain reality, or mind, or life or any other process in nature if we do not admit final causation based on the process of the transmission of forms from the dynamic object to interpretant through the sign.

The quoted authors put the action of semeiosis in the interpreter, when it should be in the sign. Semeiosis means "action of the sign" and not "action of the interpreter."

But how is such active transmission of form actually done?

Collateral Experience

The answer is given by Peirce himself in the lengthy quote at the head of this chapter: the action of the sign in any cognitive semeiosis is responsible for calling the attention to the form of the dynamic object, but leaving to the interpreter the work of finding out what is being signified.

> (…) the Dynamical Object (…) the Sign *cannot* express, (…) it can only *indicate* and leave the interpreter to find out by *collateral experience*.(A Letter to William James, EP 2:498, 1909)

Here we come to a central feature of Peirce's theory: collateral experience, through perception, grounds semeiosis.[16] The sign can only denote its dynamic object by making evident some relation that, although already familiar in a very vague and general way, was kept hidden in the background until now. But what is precisely this collateral experience?

> (…) by collateral observation, I mean previous acquaintance with what the sign denotes. Thus if the Sign be the sentence "Hamlet was mad," to understand what this means one must know that men are sometimes in that strange state; one must have seen madmen or read about them; and it will be all the better if one specifically knows (and need not be driven to presume) what Shakespeare's notion of insanity was. All that is collateral observation and is no part of the Interpretant. But to put together the different subjects as the sign represents them as related—that is the main [i.e., force] of the Interpretant-forming. (CP 8.179)

Collateral experience is responsible for producing the necessary familiarity with the dynamic object for the sign function as such. The form that inhabits the sign,

[15] For a better understanding of Peirce's semeiotic theory of causation, see Chapter "Semeiotic causation and the breath of life" by Hulswit and myself.

[16] See Chapter "Instinct and abduction in the Peircean informational perspective: contributions to biosemiotics." by Quilici and Silveira for a relation between collateral experience, information and instinct.

or the immediate object, not only grounds semeiosis but is also responsible for transforming pure blind indexes into cognitions. Through knowledge accumulated by perception and stored in memory, semeiosis offers the necessary predicate to ground the meaning of immediate interpretants, putting semeiosis into movement and producing information.[17]

A common example used by Peirce to illustrate this process is the weather-cock: as it rotates driven by the wind, it blindly points to some direction, but this is a blind index that brings no information in itself. Familiarity about the way a weather-cock works, gathered by collateral experience, will enable the association between the form of the movement made by the weather-cock (an iconic representation of the direction the wind is blowing), and its pointing to a specific direction. Information is then produced (see CP 5. 138 and CP 5.287, for instance).

The Solenoid of Semeiosis

The usual representation of the sign relations simply puts the aspects of sign, object and interpretant at the vertices of a triangular figure. To my wit, Peirce has never drawn such triangles and their appearance was probably due to Ogden and Richards (1923). The insistence in applying this triangle to represent semeiosis is the cause of great mistakes, and biosemiotics has become contaminated with them. In this section I provide my own model for the flow of information as the sign develops toward its final interpretants.

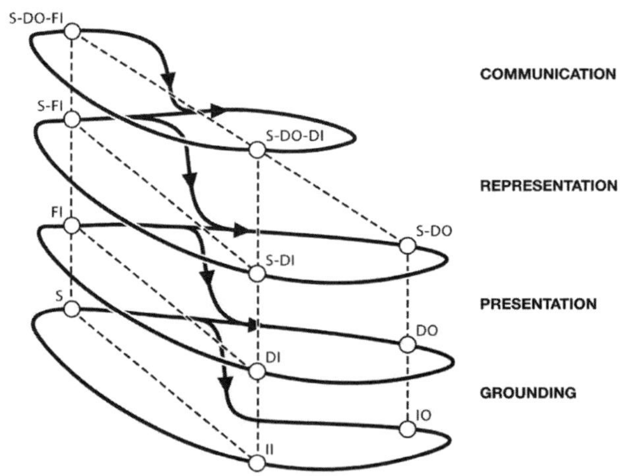

[17] The semeiotic information could be identified with Fisher's information, which is dependent on an amplitude of probabilities. In Frieden and Romanini (2008), we show that this amplitude might express the resultant of a community of interpretants and is the effect of measurements made (and registered as signs) about a parameter (the dynamic object).

The solenoid of semeiosis (see figure above) is a recursive torus organizing the sign aspects and their possible relations into four periods. Its recursive logic resembles the Smale-Williams attractor, the Moebius strip, the Klein bottle and other autopoietic-like, recursive, flows of information.[18] The solenoid is (Romanini 2006):

- **Dynamic**: it behaves like a dynamic system built from recursive interaction. The great system of semeiosis can be divided into smaller sub-systems. This nesting of systems creates dynamical hierarchies. (Collier 1999)
- **Periodic**: it shows a periodical flux. By periodicity we mean the phenomenon of repetition of a group of properties at steady intervals (Scerri 1998), although there is an increase of complexity in the whole.
- **Autopoietic**: it is *autopoietic* (Maturana and Varela 1980), that is, it produces itself from a fundamental complementarity between structure and function.
- **Ampliative**: it develops from the simple and goes towards the varied and complex, that is, it moves towards the increase of information.

One continuous arrow goes bottom-up and shows the order of determination among 11 aspects[19] of the sign: this means that a genuine and habitual triadic relation (generality in the relation sign, dynamic object and final interpretant: S-DO-FI) implies that the immediate object (IO) is also a general (this is so as well with every other intermediary aspect of the solenoid). Conversely, if the immediate object is a mere possibility, then the relation S-DO-FI will necessarily be a mere possibility (as well as every other intermediary aspect of the solenoid of semeiosis).

Aspects of the Grounding Period

IO (Immediate Object): In living systems, it is the non-conscious flow of quality of feelings that grounds the real; in mental semeiosis, it is the "idea," "form" or "seme of the truth" about the dynamic object that any sign must embody to be able to represent the dynamic object. In the final true opinion held by an ideal community, such immediate object would become "the real."[20] Since fallibilism forbids truth from being ever fully reached, the immediate object is always somewhat metaphoric, i.e., an approximation. In physical systems, it can be considered the multiple entangled states expressed by an amplitude of probabilities. It is internal to the sign.

[18] See, for instance, Roy Frieden (2004/2000).

[19] Peirce identified 10 aspects of the sign, but did not ruled out the possibility of more (CP 8.343: "... I do not say that these divisions are enough..."). In fact, I found out the need for at least one more: the relation among sign, dynamic object and dynamic interpretant, which is important to logically differentiate between assertions (what is effectively said, as a question, order, doubt etc) and propositions (the informed relation between subjects and predicates, which is the intellectual pattern expressed by a Dicisign). Assertions are dynamic utterances dependent of contextual and emotional accidents, while propositions are general conditionals expressing universality.

[20] As Peirce explains, "the immediate object of thought in a true judgment is the reality" (CP 8.16).

II (Immediate Interpretant): the non-conscious effect of a sign. In living systems, it ranges from emotions to quasi-automatic bodily effects. In mental semeiosis, it is perceptual judgment and the set of already established meaning of a symbol (as expressed in a dictionary, for instance). In physical systems, it can be the meaningful "decoherence" of multiple possible states of the immediate object. It is also internal to the sign. It is a symmetry-breaking that produces habits.

S (Sign itself or representamen): it is the phaneron, or what appears as phenomena to the mind, before it can represent anything else; "a medium for the communication of a form"; whatever can be noted by living systems; the "furniture" of reality as the result of perception; the manifested reality resulting from decoherence; the present as it appears to be related to the past and to produce the future. Its tendency is to develop toward the final interpretant, which is its intended future. A habit governing the representamen, and consequently the period of grounding, produces the worldview (not only of individuals, but of communities and species).

Aspects of the Presentation Period

DO (Dynamic Object): it is whatever the sign professes to represent, but can only imperfectly do. In physical systems, it can be regarded as the "past" information; in living systems, it is the pattern or information that must be gathered, or the "objective" environment. This bears some further explanation: we know that the light coming from a distant star tell us about its past, but the same must be said of the environment around a living being or species: an animal smelling food is getting information from its surrounding ***past***. In any informational flow, the dynamic object is the "utterer" of the information (form) to be communicated by the sign.

DI (Dynamic Interpretant): the actual effect produced by the action of a sign; in living systems, it is the manifestation of its behavior, for instance; in physical systems, the sequential change of states of any dynamic system.

FI (Final Interpretant): what every sign would become if all its possibilities of signification were totally fulfilled. The ideal future of the sign, its entelechy, the apprehension of all the information the sign is capable of communicating; the final general state towards which any system tends. As the final interpretant is a normative parameter, but never actually fulfilled, semeiosis is necessarily fallible. A habit governing the final interpretant, and consequently the period of presentation, produces **legisigns**. In genuine classes of signs, if the sign (S) is thirdness and the final interpretant (FI) is secondness, we have **replicas**. If both S and FI are secondness, we have **sinsigns**. When they are both firstness, we have **qualisigns**.

Aspects of the Representation Period

S-DO (relation between the sign and its dynamic object): the nature of the relation between a sign and its object; as the sign develops, its power to function as a medium improves.

S-DI (relation between the sign and its dynamic interpretant): the nature of the relation between a sign and its dynamic interpretant.

S-FI (relation between the sign and its final interpretant): the nature of the relation between the sign and the final interpretant. In genuine classes of signs, if there is a habit governing this aspect, and consequently the whole period of representation, we have **symbols**. If the aspect S-DO is governed by thirdness and the aspect S-FI is secondness, we then have **replicas**. If both the aspect S-DO and S-FI are both secondness, we have **indexes**. If they are both firstness, we have **icons.**

Aspects of the Communication Period

S-DO-DI (relation among the sign, its dynamic object and its dynamic interpretant): the actual dynamic communicative action of the sign; the realm of assertions and of all kinds of performative speech acts. The actual sharing of the form of the dynamic object as to produce a communicative effect. The actual performance of a mating ritual, for instance.

S-DO-FI (relation among the sign, its dynamic object and its final interpretant): the logical realm of propositions and arguments; the normative final upshot of any communicative enterprise. The final purpose (reproduction) guiding a mating ritual, for instance. A habit governing this aspect, and consequently the period of communication, produces **arguments**. In genuine classes of signs, if there is thirdness in the relation S-DO-FI and secondness in the relation S-DO-FI, we have **inductions**. If both S-DO-DI and S-DO-FI are secondness, we have **dicisigns**. If they are both firstness, we have **rhemas**.

These four periods are closed when there is habit dominating all the aspects they involve, which means that novelty and new information becomes increasingly irrelevant. In the limiting case, if these resulting habits are so rigid as to block the possibility of novelty in the whole solenoid, we have what is usually called the laws of nature.[21] In this case, the habitual forms governing semeiosis are those we identify by geometry and which allow the deduction of mathematical theorems. This is the kind of causation we experience in phenomena such as gravitation and

[21] CP 6. 25: "The one intelligible theory of the universe is that of objective idealism, that matter is effete mind, inveterate habits becoming physical laws. But before this can be accepted it must show itself capable of explaining the tri-dimensionality of space, the laws of motion, and the general characteristics of the universe, with mathematical clearness and precision; for no less should be demanded of every philosophy.".See Ivo Ibri's Chapter "The continuity of life: on Peirce's objective idealism." of this volume for a full exposition of this doctrine.

acceleration,[22] where feelings are almost totally absent. If the lawfulness of the generated habits is less rigid, novelty pops up as feelings and we have life-like purposeful behavior shown by protoplasm (CP 6.259 and ff.), such as metabolism and instinct.[23] Eventually, vague and general habits become smooth and fluid by information generated during sensations, which we then identify as mental semeiosis.[24] In actual semeiosis, though, we have none of these extreme limits, for no law of nature is so determined as to eliminate every possibility of unpredictable results (which would amount to a probability of 1/1), nor is any real phenomenon so free from regularity as to be absolute novelty (which would then amount to a probability of 0/1).

The grounding period, which is the most fundamental one, is closed into recursive loops only when there is generality in the aspects of the immediate object (IO), immediate interpretant (II) and the representamen or sign in its materiality (S). Whenever habit is not totally established in this period, we experience active ***perception***, which is the non-conscious production of general predicates by collateral experience. When habit becomes inveterate in the grounding period, perception is not the relevant activity anymore, and knowledge production decreases. It is what happens when we get bored by something and stop paying perceptive attention to it. The grounding period produces whatever is immediately present to the mind: the non-conscious phaneron, the flow of feelings, the perceptive, and active, continuum of information that connects the real (which is very similar to David Bohm's non-local, implicit although active information, real but nonetheless waiting to be manifested during experience).

The presentation period, which is the second of the solenoid from the bottom-up, is closed only when there is habit dominating the aspects of sign (S), dynamic object (DO) and final interpretant (FI). If habit is ruling this period (which implies that habit is also ruling the more basic grounding period), when the manifestation of the phaneron as consciousness: the world is disentangled from our subjectivity and becomes "objective" due to the habitual manifestation of the dynamic object. The sign itself (S) develops into the general final interpretant (FI), creating legisigns, which are the laws that effectively rule the manifested world. Inveterate habits in this period produce a complete regularity of manifested phenomena, which is commonly taken as the "furniture" of the world: rocks, rivers, clouds, wind-blows, animals, people. If the grounding period produces perception, the presentation period

[22] In fact, Peirce puts both gravitation and acceleration under the general law of causation, as Peirce explains in CP 1.270 and CP 6.68.

[23] CP 2.170: "If I may be allowed to use the word 'habit,' without any implication as to the time or manner in which it took birth, so as to be equivalent to the corrected phrase 'habit or disposition,' that is, as some general principle working in a man's nature to determine how he will act, then an instinct, in the proper sense of the word, is an inherited habit, or in more accurate language, an inherited disposition. But since it is difficult to make sure whether a habit is inherited or is due to infantile training and tradition, I shall ask leave to employ the word 'instinct' to cover both cases."

[24] For an excellent treatment of Peirce's concept of habit, see Chapter "Peircean habits, broken symmetries, and biosemiotics." by Eliseo Fernandez.

produces ***inquiry***, which is simply the act of questioning the rules that govern the manifested reality.

The representation period, which is the third of the solenoid from bottom-up, is closed when habit dominates the aspects of the relation between the sign and the dynamic object (S-DO), the relation between the sign and the dynamic interpretant (S-DI) and the relation between the sign and final interpretant (S-FI). If rigid and inveterate habit dominates this period (which means that it also dominates the two previous ones), attention is not paid to the manifested "objective" reality anymore, but to what it represents. An actor on stage representing Hamlet presents his own body to the public but represents the fictional character created by Shakespeare. But to be effective such representation must count on the public collateral experience of what is being represented in a very general and vague way. That is to say, the public must have a grounding (a familiarity gained by a continuous flow of feelings) about Hamlet in such a way that what is being staged is welded to what the public already know about kings, madness, vendetta, cruelty etc.[25] In this period we have ***deliberative*** semeiosis, because the act of representation encompasses a great degree of freedom, since collateral experience varies greatly among different minds or communities of minds. We can easily prove this by the many possible ways political representation can be established in different countries and in different historical periods according to different culture, influences, etc.

The communication period, which is the forth and last of the solenoid of semeiosis from the bottom-up, closes when we have habit ruling the relation among sign, dynamic object and final interpretant (S-DO-FI). This necessarily implies that the relation among sign, dynamic object and dynamic interpretant (S-DO-DI) is also habitual, as well as all previous periods of the solenoid. If rigid habit rules this period, we have law-like communication. The most rigid habitual communication is the physical transmission of information from past to future (the so-called light cone), but the gradient of possible flexibility of communication is enormous, as the chemical communication of substances, the organic communication of living organs, the instinctual communication among species. If communication is freely self-organized, we have the ***scientific*** phase of semeiosis: through a rhetoric based on the scientific inquiry, a community of minds welded by a common purpose, or *commens*, endeavors to share meaning and reach mutual persuasion on the basis of evidence gathered through perception. Peirce calls it methodeutics, the science of communication when the pursuit of truth and the revelation of reality is the only guiding desire. This commens is a logical entity and must not be confounded with only the union of all human minds at every time. It is the communion of all possible minds active in the Universe, of which we might be a tiny sample.

---

[25] "When the universe of discourse relates to a common experience, but this experience is of something imaginary, as when we discuss the world of Shakespeare's creation in the play of Hamlet, we find individual distinction existing so far as the work of imagination has carried it, while beyond that point there is vagueness and generality" (CP 4.172).

The Solenoid Applied to Biosemiotics

One interesting feature is that the *rationale* of the solenoid of semeiosis could be used to build a taxonomy of the living being based purely on their cognitive capabilities. The lower levels of mental complexity exhibit habit-taking only in the grounding period, where perception is active. The next level, one period above, shows habit-taking in the presentation level, where the world appears as a regular pattern of events. Only the most complex types of living-beings show habit formation in the representation and communication periods, where the work of abstractly produced signs, capable of being shared among living and freely cooperative specimens, become predominant. That would mean that not only ontogenesis and phylogenesis are acting to create the complexity of the living, but also the more fundamental process of semeiosis, which could here be called semiogenesis.[26]

Genuine triadic relations as expressed by the solenoid of semeiosis are easily found in living systems. A hungry dog sniffs the smell of food in the air and follows its path. The dog is a specimen of a species, which means that much of his behavior is an expression of general instinctual modes of living developed during the phylogenetic evolution of its species. The smell is the *sign* that denotes the existence of food (its *dynamic object*) and immediately produces an effect (an emotion, such as an excitation and desire to follow the smell), which is the *immediate interpretant*. This happens only when its concentration on the environment reaches a level sufficiently high as to be perceptible by the olfactory apparatus of the dog, becoming then able to indicate to the animal that food is nearby.

If this kind of smell embodies mainly a quality of the food, its concentration must be a fact of perception, a difference in the environment that the dog must be able to note. On the other hand, the smell can only function as a sign because the dog has some habitual or pre-dispositional familiarity, usually instinctual, with the fact that eating that food represented by the smell-sign will satisfy him. This familiarity must be gotten through previous experience by which a particular *kind* of smell becomes significant to a canine species during its evolution. This is collateral experience on the level of evolution, a phylogenetic collateral experience. In conclusion: qualitative properties working as icon, an environmental concentration working as index, and a previous familiarity or habitual instinct working as symbol are all intertwined in semeiosis to produce a purposeful action.

But the dog is not after the smell. It is after the food or, more correctly yet, he is after the feeling of satisfaction that he would get after eating the food. This is the *final interpretant* or purpose of his following the path of the smell. The smell, as a sign, is just a medium by which the food—represented in the smell-sign in its pattern of qualitative odor—has the potentiality of affecting the dog's behavior: the action of moving and sniffing around, always correcting its trajectory to keep himself in the direction from which the smell is coming—which are all *dynamic interpretant*s.

[26] In fact, Peirce says that "Genesis is production from ideas. It may be difficult to understand how this is true in the biological world, though there is proof enough that it is so" (EP2:127).

The habitual relation between the appearance of smell-sign in the dog's phaneron and its dynamic object (the relation *sign-dynamic object*) is granted by habit: in order to survive every dog must be able to relate a *kind* of smell to the *kind* of food he needs to eat. His actually accomplishing this relation in order to act accordingly is granted by habit in the relation between the sign and the dynamic interpretant (*sign-dynamic interpretant*), and the security that the same would happen whenever the dog is hungry and is exposed to the same kind of smell is granted by habit in the relation between the smell-sign and the fulfillment of the purpose of getting food (relation *sign-final interpretant*).

The whole process involving the dog searching and getting the food it needs to survive is a semeiotic communication between the dog and the reality its species is determining as it evolves. This continuous flow of communicative actions, by which every dog of this species gets food while co-influencing the significant environment where it lives and aims to continue to live, is the relation sign-dynamic object-final interpretant (S-DO-FI). Each of these acts performed by a particular dog is expressed by the relation sign-dynamic object-dynamic interpretant. This means that every species is in continuous communication with its significant "Umwelt," even if only instinctually, when the purpose is survival.

There is no doubt that signs get their meaning by deep and mostly unconscious processes of perception, but semeiosis can also produce conscious and rationally self-controlled purposeful actions. A particular dog cannot prevent himself from feeling hungry (which is the result of unconscious mental activities), although it can be taught not to jump on a table to get the food there but wait until the food is served in the pot by the master. Learning is a marvelous example of ***purposeful*** semeiosis because it involves habit-taking or the modification of old habits, even deep instinctual ones, by the introduction of self-controlled behavior expressed by new habits. This is possible only when true reasoning of a diagrammatic type takes place to produce a breakdown of inveterate symmetries and the production of new ones.

In fact, a diagrammatic process must take place by which the master makes evident to the dog the habitual bad consequences of jumping on the table, as well as the habitual good ones of waiting to be served with food in his pot. This diagram must embody a set of syntactical general relations involving the dog, the master, the food, the table, the pot and the treats and punishments the dog effectively gets (*dynamic interpretants* coordinating the actions of the master and the dog) according to the purpose of the learning (*final interpretants*). To be effective, learning must be a process involving the unconscious generation of a new hypothesis (perceptual judgments or habitual *immediate interpretants*—the pleasures and displeasures at stake), the evidence of some of the consequences extracted from the new hypothesis (deduction) and the continuous satisfaction of acting according (*dynamic interpretants*) to the new hypothesis, producing an induction leading to *final interpretants*. Peirce gives us an example from daily experience with his dog *Zola* where a similar process is described:

> I tell my dog to go upstairs and fetch me my book, which he does. Here is a fact about three things, myself, the dog, and the book, which is no mere sum of facts relating to pairs,

> nor even a pairing of such pairs. I speak to the dog. I mention the book. I do those things together. The dog fetches the book. He does it in consequence of what I did. That is not the whole story. I not only simultaneously spoke to the dog and mentioned the book, but I mentioned the book to the dog; that is, I caused him to think of the book and to bring it. My relation to the book was that I uttered certain sounds which were understood by the dog to have reference to the book. What I did to the dog, beyond exciting his auditory nerve, was merely to induce him to fetch the book. The dog's relation to the book was more prominently dualistic; yet the whole significance and intention of his fetching it was to obey me. In all action governed by reason such genuine triplicity will be found (CP 2.86).

In the above example of semeiosis given by Peirce, we easily notice that the periods of representation and communication are not instinctual but contingently produced by the relation between the dog and his master and dependent on diagrammatic reasoning involving abductions, deductions and inductions.

An Example From David Bohm

I want to finish this chapter with an example of learning extracted from Bohm and Peat (1987) that might offer some important lessons in this epoch of too much audiovisual semeiotic analysis. Although starting from a different road, the quantum physicist David Bohm came to similar conclusions about the ontological status of reality.[27] As is known, Bohm's interpretation of quantum mechanics proposes the reality of a peculiar implicit, non-manifested field of information bathing the whole of reality, and which becomes manifested or explicit as particles of matter during perceptive processes such as measurements. Bohm describes this kind of active information as being like the *form* of the streams of oceans and rivers in the manner in which they make particles behave the way they do. Like Peirce's information that brings life to the real, Bohm's form has mind like properties and depends on the wholeness of a communicative process carried out during the experiment. And like Peirce's participatory theory of communication, Bohm's theory of meaning also depends on perception and communication.

Bohm illustrates his own hypothesis with the case of Helen Keller (1880–1968), a deaf and blind girl from early age who not only was contemporaneous with Peirce, but whose process of learning how to communicate happened precisely in the same years Peirce was developing his semeiotic theory of meaning and communication. When Peirce died in 1914, Helen was 34 and had already earned a degree of Bachelor of Arts, had become an author of books, lecturer and political activist. Helen's destiny changed when Anne Sullivan accepted the job of privately teaching the little girl. She did not realize that she was to find a "wild animal" that could not communicate. Bohm explains how Sullivan managed to teach Helen Keller:

> The key step was to teach Helen to form a communicable concept. This she could never have learned before, because she had not been able to communicate with other people to any significant extent. Sullivan, therefore, caused Helen, as if in a game, to come into contact

[27] Brent (1993) was one of the first Peirce scholars to point out the similarity of views between them.

> with water in a wide variety of different forms and contexts, each time scratching the word water on the palm of her hand. For a long time, Helen did not grasp what all this was about. But suddenly, she realized that all these different experiences referred to one substance in many aspects, which was symbolized by the word water on the palm of her hand. (…) Thus, the different experiences were implied in some sense as being equal by the common experience of the word water being scratched on her hand. (…) Up to that moment, Helen Keller had perhaps been able to form concepts of some kind, but she could not symbolize them in a way that was communicable and subject to linguistic organization. The constant scratching of the word water on her palm, in connection with the many apparently radically different experiences was suddenly perceived as meaning that, in some fundamental sense, these experiences were essentially the same. (Bohm and Peat 1987, pp. 36–37)

Here is another wonderful example of semeiosis bringing life to a concept. The patterns we usually associate with water (which are the *dynamic object* to be represented) have a number of regular characters, such as fluidness in its liquid state, a certain average temperature ideal for drinking, an evocative sensation of pleasure when we drink being thirsty, another sensation when we bathe etc. These general properties are as real as the hardness of a rock we previously discussed. They can be learned by non-conscious perceptual judgments if we are exposed to them through our senses. The possible cognitive patterns of predicates (the general feeling of fluidness of water, the general cooling temperature when drinking it etc) are the *immediate objects* of the sensation. The possible emotional effects produced by the sensation of the qualities of water are their *immediate interpretants*.

The English word "water" was invented by men. It is a legisign grounded by the information brought by the community of its users—including the teacher Ann Sullivan—and animated by a form of significance, which is a teleological component. But how can we manage to relate the word water scratched on the palm of the hand of someone who does not know its meaning to the bunch of real properties held by the substance water? Simply by putting them into continuous association, side by side, in a multitude of occasions, until a synthetic inference is created.

Guided by Ann Sullivan, Helen produces a series of collateral experiences initially unrelated but that are mentally associated by the scratching of the same word "water" on the palm of her hand. The scratching itself produces some sensations as a regular form is drawn, when *immediate objects* and *immediate interpretants* connected to the scratching are also unconsciously gathered. The continuity of experience between the sensation of water and the sensation of scratching produces a common ground between them both, while bringing them into relation. Not only the scratching becomes the *sign* of the flow of qualities of feelings we usually identify with water, but the familiar patterns of those ***past*** qualities of feelings stored in memory becomes the *dynamic object* to be represented. They are then welded by the tendency of mind to produce a more general pattern resulting from their unification.

When the sequence of scratchings is felt as a regular pattern by Helen Keller (a legisign), it becomes for her a symbol of the general qualities of feelings she experienced collaterally when in perceptive contact with the substance water. Here begins the relation *sign-dynamic object* (S-OD), which is consolidated at the almost magical moment when the girl Helen, feeling thirsty, takes the hand of her teacher

Ann and scratches a similar form on the palm, putting the relation S-DO to an actual use (the relation *sign- dynamic interpretant*, or S-DI). This is done with a clear purpose: asking for those pleasant qualities of feelings that will bring her painful thirst to an end, which is described by the relation *sign-final interpretant,* or S-FI, when purposeful representation between Ann Sullivan and Helen Keller becomes habitual. A concept is born.

As we have already seen, a concept, or symbol, is precisely a mental habit capable of communicating the form of the dynamic object so as to produce general interpretants in a conditional future. This communication happens whenever Ann, Helen or any other person of a community of users of the word "water" effectively uses it to reach pragmatic effects. Utterances made in particular contexts (contingent and therefore emotionally grounded) are assertions depending on the relation among the sign, dynamic object and *dynamic* interpretant (S-DO-DI). If I am walking on a desert land and a stranger comes to me and says "water" with a weak and suffering voice, I am emotionally inclined to take it as a begging. If I am distractedly walking on the seashore and suddenly see someone pointing to the sea while shouting "water", I am inclined to interpret it as a warning of something dangerous such a huge wave coming towards me. All the habitual manners in which a community of users would employ the word "water" to reach pragmatic purposeful effects (esthetic, ethical and logical ones) involve the relation among sign, dynamic object and final interpretants (S-DO-FI). It is the communicative sum of all lessons to be learned from it.

Conclusion

Biosemiotics is an emergent, even effervescent interdisciplinary field of research that promises to bridge the gap opened in biology when it chose a too-materialistic approach by following the path taken by chemistry and physics. This line of research led to important discoveries about the functioning of metabolism and the role of specific molecules in organisms. But the description of living processes is not all that biology is about. It is also about understanding how life itself is possible and how biological processes can produce evolution, consciousness, meaning, representation and communication. Biosemiotics is the branch of biology interested in understanding life as semeiosis, where meaning and interpretation play the central role. Applying Peirce's theory of signs to biosemiotics will certainly help the development of biosemiotics, but it might equally help the development of Peirce's proposal for a logical theory of reality. His admonishment for a logically grounded metaphysics to deal with the vital questions of life must be taken into consideration by biosemioticians. There is even a sense in which semiotics and biosemiotics are synonymous, for semeiosis was considered by Peirce a genuine triadic relation of the same nature as that of living processes. If this would prove correct, advancing Peirce's theory of signs is also advancing biosemiotics. This was the motivation of this chapter, which has certainly not fulfilled all that is needed to reach a complete taxonomy of all possible classes of signs and their mutual relations, but nevertheless took the risk of proposing an interpretation of Peirce's mature and very complex theory of signs in terms of biosemiosis. While this chapter only opens a narrow trail by explaining how the solenoid of semeiosis can be applied to living processes, I think there is a large road to be completed in this same direction. The scientific question about the biological meaning of *life* and the metaphysical question about the logical *meaning* of life might have, at the end, the very same answer.

References

Bohm, D. (2000). *Wholeness and the implicate order*. London: Routledge.
Bohm, D., & Peat, F. D. (1987). *Science, order, and creativity*. New York: Bantam Books.
Brent, J. (1993) *Charles Sanders Peirce: A life*. Indiana University Press.
Collier, J. (1999). The dynamical basis of information and the origins of semiosis. In E. Taborsky (Ed.), *Semiosis evolution energy towards a reconceptualization of the sign* (Vol. 3, pp. 100–136). Aachen Shaker Verlag, Bochum Publications in Semiotics New Series.
Frieden, B. R. (2004/2000). *Science from Fisher information* (2nd ed.) Cambridge Univ. Press.
Frieden, R. B., & Romanini, V. (2008) Eluding the demon-how extreme physical information applies to semiosis and communication. *Cognitio-Estudos (PUC-SP), 5*, 52–63.
Maturana, H., & Varela, F. (1980). *Autopoiesis and cognition*. Dordrecht: Reidel Publishing Co.
Ogden, C. K., & Richards, I. A. (1923). *The meaning of meaning*. New York: Harcourt.
Posner, R. et al. (1997). *Semiotik/semiotics: A handbook on the sign-theoretical foundations of nature and culture* (Vol. 1). Berlin: Walter de Gruyter.
Romanini, V. (2006) Semiótica Minuta—Especulações sobre a Gramática dos Signos e da Comunicação a partir da Obra de C.S. Peirce. Doctoral dissertation, ECA/USP. Online: www.minutesemeiotic.org. [English and Portuguese]
Sebeok, T. A. (2001). *Global semiotics*. Bloomington: Indiana University Press.
Scerri, E. R. (1998).The evolution of the periodic system. *Scientific American,* September.
Turrisi, P. A. (1997). *Pragmatism as a principle and method of right thinking: The 1903 Harvard lectures on pragmatism*. Albany: State University of New York Press.
Uexküll, J. V. (1928). Theorische Biologie. Berlim: Springer.

Bibliography

Araujo, E. F. (2004, April 13–17). *A teoria da Umwelt de Jakob von Uexküll. Galáxia, Revista Transdisciplinar de Comunicação, Semiótica, Cultura. Educ São Paulo*. (in Portuguese).

Arnellos, A. et al. (in press). Anticipatory functions, digital-analog forms and biosemiotics: Integrating the tools to model information and normativity in autonomous biological agents. *Biosemiotics, 5*(3).

Ayim, M. (1974). Retroduction: The rational instinct. *Transactions of the Charles S. Peirce Society, 10*(1), 34–43.

Balashov, Y. (2002). Laws of physics and the universe. In Y. Balashov & V. Vizgin (Eds.), Einstein studies in Russia. Einstein studies (Vol. 10, pp. 107–148). Boston: Birkhäuser.

Balat, M. (1990). Type, Trace, Ton, le 'ton' peircien. *Semiosis,* 57–58. (in French).

Böhme, H. (1996). Über vormoderne Naturkonzepte. *Zeitschrift für Semiotik, 18*(1), 19–22. (in German).

Christiansen, P. V. (1997). Peirce and modern cosmology: Attractors and broken symmetry. *Transactions of the Charles S. Peirce Society, 33*(4), 941–957.

Colapietro, V. (2005). Conjectures concerning an uncertain faculty claimed for humans. *Semiotica, 153*(1/4), 413 ff.

Dewey, J. (1916/1985). *Democracy and education. The middle works of John Dewey* (Vol. 9). Carbondale: SIU Press. Cited as MW 9.

Ellis, G., et al. (2012). Introduction: Topdown causation: An integrating theme within and across the sciences? *Interface Focus, 2*(1), 1–3.

Fernández, E. (2010). Taking the relational turn: Biosemiotics and some new trends in biology. *Biosemiotics, 3*(2), 147–156.

Forster, P. (1997). The logical foundations of Peirce's indeterminism. In J. Brunning and & Forster (Eds.), *The rule of reason: The philosophy of Charles Sanders Peirce*. Toronto: University of Toronto Press.

Goldenfeld, N., & Woese, C. (2011). Life is physics: Evolution as a collective phenomenon far from equilibrium. *Annual Review of Condensed Matter Physics, 2,* 375–399.

Hacking, I. (1983). Nineteenth century cracks in the concept of determinism. *Journal of the History of Ideas, 44,* 455–475.

Hawking, S., & Leonard, M. (2010). *The grand design*. New York: Bantam Books.

Hoffmeyer, J. (1997). *Signs of meaning in the universe*. Bloomington: Indiana University Press.

Houser, N. (1991). A Peircean classification of models. In A. Myrdene & M. Floyd (Eds.), *On semiotic modeling* (pp. 431–439). Berlin: Mouton de Gruyter.

Hulswit, M. (2002). *From cause to causation: A Peircean perspective*. Dordrecht: Kluwer Academic Publishers.

James, W. (1890/1983). *Principles of psychology*. Cambridge: Harvard University Press.

James, W. (1892/1985). *Psychology: The briefer course*. Cambridge: Harvard University Press.

Juarrero, A. (2002). *Dynamics in action: Intentional behaviour as a complex system.* Cambridge: MIT Press.

V. Romanini, E. Fernández (eds.), *Peirce and Biosemiotics,* Biosemiotics 11,
DOI 10.1007/978-94-007-7732-3,

Kortlandt, F. (2003). The origin and nature of the linguistic parasite. In B. L. M. Bauer & G.-J. Pinault (Eds.), *Language in time and space: A festschrift for Werner Winter on the occasion of his 80th birthday* (pp. 241–244). Berlin: Mouton de Gruyter.

Lizska, J. (1996). *A general introduction to the semeiotic of Charles Sanders Peirce*. Indiana: Indiana University Press.

Lloyd, J. E. (1980). Male Photuris fireflies mimic sexual signals of their females' prey. *Science, 210,* 669–671.

Loewald, H. (1989). *Papers on psycho-analysis*. New Haven: Yale University Press.

Longo, G., & Montévil, M. (2011). From physics to biology by extending criticality and symmetry breakings. *Progress in Biophysics and Molecular Biology, 106*(2), 340–347.

Machado, I., Romanini, V. (2011). Semiotics of communication: From semiosis of nature to culture. *Biosemiotics, 5,* 47–60.

Magnani, L. (2001). *Abduction, reason, and science: Processes of discovery and explanation.* Kluwer.

Magnani, L. (2005). An abductive theory of scientific reasoning. *Semiotica, 153*(1/4), 261 ff.

Marty, R. (1982). C. S. Peirce's phaneroscopy and semiotics. *Semiotica, 41*(1/4), 169–181.

Mitchell, R. W. (1986). A framework for discussing deception. In R. W. Mitchell & N. S. Thompson (Eds.), *Deception: Perspectives on human and nonhuman deceit* (pp. 3–40). Albany: SUNY Press.

Moore, E., & Robin, R. (1964). *Studies in the philosophy of Charles Sanders Peirce (Second Series).* Cambridge: Harvard University Press.

Murphey, M. G. (1961). *The development of Peirce's philosophy*. Cambridge: Harvard University Press. Reprinted 1993, Indianapolis: Hackett.

Müller, R. (1994). On the principles of construction and the order of Peirce's trichotomies of signs. *Transactions of Charles S. Peirce Society, 30*(1), 135–153.

Nöth, W. (2004). Semiogenesis in the evolution from nature to culture. In M. Alač & P. Violi (Eds.), *In the beginning: Origins of semiosis* (pp. 69–82). Turnhout: Brepols.

Nöth, W. (2008). Natural signs from a synechistic perspective. In K.-M. Hingst & M. Liatsi (Eds.), *Pragmata. Festschrift für Klaus Oehler zum 80. Geburtstag* (pp. 130–140). Tübingen: Narr.

Nöth, W. (2010). The criterion of habit in Peirce's Definitions of the Symbol. *Transactions of the Charles S. Peirce Society, 46*(1), 82–93.

Nöth, W. (2012b). Signs from the life of organisms, species, languages, and the media. In T. Maran, K. Lindström, R. Magnus, & M. Tønnessen (Eds.), *Semiotics in the wild: Essays in honour of Kalevi Kull on the occasion of his 60th birthday* (pp. 123–130). Tartu: Tartu University Press.

Pennisi, A, & Falzone, A. (2010). *Il Prezzo del linguaggio*. Bologna: Saggi.

Romanini, V. (2010). A new aspect of sign and its implications for the theory of communication. *Semeiosis—Semiótica e Transdisciplinaridade em Revista, 1,* 1–12.

Romanini, V. (2012). Co-relations. In *A more developed sign—Interpreting the work of Jesper Hoffmeyer*. University of Tartu: Tartu Semiotics Library.

Santaella, L. (2005). Abduction: The logic of guessing. *Semiotica, 153*(1/4), 175 ff.

Santayana, G. (1942). *The realms of being*. NY: Charles Scribner's Sons.

Savan, D. (1988). *An introduction to C.S.Peirce's full system of semiotic.* Monograph Series of the Toronto Semiotic Circle, Vol. 1: Victoria College.

Savan, D. (1977). Questions concerning certain classifications claimed for signs. *Semiotica, 19*(3/4), 179–195.

Sebeok, T. A. (1976). Notes on lying and prevarication. In T. A. Sebeok (Eds.), *Contributions to the doctrine of signs* (pp. 143–147). Lanham: University Press of America.

Sebeok, T. A. (2001). *Global semiotics*. Bloomington: Indiana University Press.

Serres, M. (1982). *The parasite*. (trans: L. R. Schehr). Baltimore: Johns Hopkins University Press.

Smith, J. E. (1981). Philosophical interpretation and the religious dimension of Experience. *Logos, 2*(1981), 5–20.

Smith, J. E. (1995). *Experience and god*. NY: Fordham University Press.

Stjernfelt, F. (2007). *Diagrammatology: An Investigation on the borderlines of phenomenology, ontology and semiotics*. New York: Springer.

Tursman, R. (1987). *Peirce's theory of scientific discovery: A system of logic. Conceived as Semiotic.* Bloomington: Indiana University Press.

Index

A
Abduction (retroduction), 5, 6, 36, 45, 107, 155, 118, 128, 151–168, 185, 189, 196, 222, 236
  abductive reasoning, 5, 153–164
  hypothesis, 154
Actuality, 113, 114
Agapism (agapasm, agapastic), 30
Agency, 5, 6, 31, 84, 129, 134, 138, 146, 224
  of language, 84
  purpose and autonomous, 175, 176
Anthropomorphism, 77, 146, 222
Aquinas, Thomas, 136, 137
Argument, 179, 180, 231
Aristotle, 13, 14, 16, 37, 56, 99, 100, 103, 133, 142, 147, 217
Assertion, 7, 9, 51, 95, 130, 133, 138, 155, 172, 229, 238
Autopoiesis (autopoietic), 176, 178, 229

B
Backgrounding doctrine, 41, 45
Barbarus, Hermolaus, 55, 56
Bateson, Gregory, 167
Behavior, 54, 89–91, 117, 156, 160, 165, 166, 168, 174, 186, 201–203, 210, 222, 232, 234, 235
Belief, 37, 88, 116, 129, 218
Berkeley, 36
Biosemeiosis, 7, 224, 225
Biosemiotic, 192–196, 217, 218, 234–236
Biosign, 4, 54, 55
  intermediate, 71–78
Bohm, David, 223, 232, 236–238
Breadth, 13, 159, 165
Buchler, Justus, 143
Burks, Arthur, 11, 30, 203

C
Cantor, Georg, 11
Category, 7, 8, 18, 45, 103, 110, 121, 132, 134, 141, 142, 149, 160, 187, 221
  cenopythagorean, 8
  intimations of, 142–145
  symmetry of, 40, 44, 47
Causal
  action, 103
  chain, 120
  conditioning, 114, 123
  determination, 88, 91, 92
  efficacy, 14
  explanation, 99, 118, 120
  operation, 88
  process, 105, 122
  propositions, 106
  relata, 106, 107, 125
  relations, 137, 146, 189, 202
  rules, 40
  sequence of events, 108
Causality, 3, 22, 25, 40, 98, 106–108, 113, 122, 125, 141
  Peirce's critique of, 99–102
Causation, 106–108
  characteristics of, 112, 113
  complexity, 88–90
  efficient, 5, 14, 22, 25, 27, 80, 99, 100, 103–106, 112–115, 119, 120, 122, 124, 125, 179
  final, 5, 14, 15, 22, 25, 26, 29, 31, 96, 98, 104–109, 112–115, 117–119, 121–123, 125, 175, 187, 217, 222, 227
  formal structure of, 111, 112
  form-aspect of, signs conveying, 117–120
  habits and, 88–91

V. Romanini, E. Fernández (eds.), *Peirce and Biosemiotics,* Biosemiotics 11,
DOI 10.1007/978-94-007-7732-3, © Springer Science+Business Media Dordrecht 2014

Peirce's conception of, 104–106, 123, 124
semeiotic approach to, 109, 112–123
Cause, 99, 100
efficient, 14, 99, 100, 103, 104–106, 112–115, 119, 120, 124, 125
final, 14, 15, 22, 96, 104–109, 112–115, 117–119, 121–123, 125, 175, 187, 217, 222
formal, 14, 98, 117, 121, 141
material, 14, 28
Chance, 3, 5, 7, 16, 26, 37, 38, 46, 98, 154, 167
absolute, 2, 23, 31, 82, 97
irreducible objective or novelty, 101, 104, 105
pure, 2, 17, 22
Chaos
of feelings, 18
primal, 22
primeval, 15, 25–28
primordial, 16, 18, 22, 25
of reactions, 9, 10, 15, 16
of symbols, 18
Classes of signs (types of signs), 201–207
Codes
genetic, 55
organic, 4, 92
Cognition, 46, 48, 54, 55, 110, 131, 164, 225
Coherence, 90, 108, 109, 113
Collateral experience (or collateral observation), 66, 72, 76, 107, 155, 158, 160, 164, 215, 225, 227, 228, 232–234, 237
Commens (commind), 233
Common ground, 225, 237
Communication
animal, 166, 199, 200, 209
complexity, 209
human, 6, 55, 61, 64, 65, 76, 166
human-to-animal, 54, 71, 73
period, 231–233
successful, 69–71
use of symbols in, 91
Complexity, 3, 7, 8, 34, 39, 45, 54, 81, 88, 89, 96, 101, 108, 109, 113, 166, 167, 195, 209, 216, 220, 221, 234
Comprehension, 97, 138
Conditional propositions, 92, 116, 124, 162, 201, 219, 220
Conduct, 5, 15, 41, 43, 44, 46–48, 131, 161, 164–166, 168, 173, 176, 186, 222, 225
Consciousness, 15, 29, 30, 96, 139
affective, 131, 132
animal, 132
polar, 132
synthetical, 127, 132
Consequences, 40–43, 138, 143, 152, 203, 218, 235
Continuum, 6, 26, 27, 29, 45, 97, 210
Correlate, 2, 191
Cosmology, 3, 11, 15, 17, 18, 21, 23, 25, 30, 31, 81, 83–85, 87, 172
Creatio ex nihilo, 81, 82

D
Darwin, C., 37, 38, 134, 135, 166
Davies, Paul, 19
Deacon, Terrence, 23, 192, 193
Decartes (cartesianism), 42
Deduction, 45, 155, 235
Definite, 51, 52, 56, 58, 64, 70, 115, 117, 119
Depth
informed, 159, 160
logical, 13
Determinism, 31, 37, 40, 80, 97
absolute, 102
ontological, 42
scientific, 20
Dewey, John, 128, 129, 135, 140, 141, 145, 148
Diagram, 107, 128, 174, 224, 235
Diagrammatic reasoning, 157, 235, 236
Dicisign (Dicent), 7, 59, 60, 66–69, 73, 75, 76, 161, 162, 166, 180, 200, 208, 210, 229, 231
Dilworth, David, 31, 43
Dyadic relation, 1, 104, 110, 166, 183, 189, 190, 204
Dynamic interpretant, 224, 230, 231, 233–235, 238
Dynamic object, 69, 70, 73, 98, 118, 123, 124, 163, 216, 226–235, 237, 238

E
Eddington, Arthur, 20
Efficient causation (efficient cause), 5, 14, 22, 25, 27, 80, 99, 100, 103–106, 112–115, 119, 120, 122, 124, 125, 179
Effort, 13
Eisele, Carolyn, 11, 147
Emergence, 4, 20, 21, 23, 26, 34, 80, 82, 83, 86, 89–91, 119, 130, 143, 154, 168, 202, 207
Emerson, Ralph Waldo, 128, 129, 134, 136, 138, 139
Emmeche, Claus, 171

Emotion, 96, 131, 185, 224, 225, 230, 234, 237
Entelechy, 10, 13, 187, 195, 222, 230
Equilibrium (thermodynamic), 88
Esthetics (aesthetics), 47, 148, 157, 160
Ethics, 47, 148, 157, 161
Euclide, 12
Events, 106–109
  anatomy of, 113–115
  facts versus, 106–108
Evolution, 87, 88
Existent, 41, 46, 47, 63, 70, 144, 154, 161, 175, 201, 204, 206, 211
Experience
  collateral, 66, 72, 76, 107, 155, 158, 160, 164, 215, 225, 227, 228, 232–234, 237
Extension, 92, 159, 180, 184
  informed, 160
  temporal, 108, 112

F
Facts, 9, 14, 15, 21, 36, 95, 97, 106–108, 113, 119, 122, 125, 134, 151
Fallibilism, 38, 47, 149, 229
Feeling, 18, 25–30, 42, 97, 98, 102, 114, 127, 131, 158, 160, 186, 220, 224, 232–235, 237, 238
Feynman, Richard, 20
Fichte, 36
Final causation (final cause), 5, 14, 15, 22, 25, 26, 29, 31, 96, 98, 104–109, 112–115, 117–119, 121–123, 125, 175, 187, 217, 222, 227
Final interpretant, 84, 98, 118, 124, 189, 190, 195, 215, 228–235, 238
Firstness (priman, can-be), 18, 26, 27, 30, 42, 45, 46, 81, 110, 113, 141, 144, 145, 160, 161, 185, 221, 230, 231
Fisch, Max, 1, 3, 34, 68, 77
Fisher, Ronald, 89, 228
Force, 97
Form, 117–120, 229
Formal cause, 14, 98, 117, 121, 141
Fuzzy boundaries, 53, 63

G
Galileo, 154
Game-Theoretical Semantics (GTS), 59–63
Gauge symmetry, 89
Generality, 4, 28, 39, 41, 52–54, 64, 65, 81, 97, 173, 179, 188, 220, 221, 229, 232
  antithetical analogue of, 61
  distributive, 71
  indeterminacy as, 55–58
  objective, 60
Gibson, J.J., 141
Graph, 174, 203
Grounding
  backgrounding doctrine, 41, 45
  period, 229, 230
Growth, 4, 25, 28, 31, 97, 101, 133, 159, 177–179, 225
Guessing, 34, 36
Guth, Alan, 87

H
Habit
  causation and, 88–91
  and concrete generalization, 82, 83
  and cosmological speculation, 80–82
  and laws, 173–175
  and semiosis, 91, 92
  and symbols, 83–85, 173–175
Hawking, Stephen, 19, 20, 26
Hintikka, Jaakko, 58
Hoffmeyer, Jesper, 92, 137, 147, 171, 192, 193, 195
Hubble, Edwin, 85
Hypostatic abstraction, 52

I
Icon, 13, 14, 28, 30, 60, 133, 161, 174, 178, 185–191, 194, 195, 200–205, 207–211, 216, 222, 228, 231, 234
Idea, 68–71, 97, 101, 229
Immediate interpretant, 9, 69, 118, 124, 190, 215, 223, 224, 228, 230, 232, 234, 235, 237
Immediate object, 4, 38, 54, 72, 75, 83, 118, 124, 163, 186, 215, 228, 229, 230, 232, 237
  and indeterminacy, 69–71
Immune system, 7, 225
Indefinite, 52, 61, 64, 67, 69, 73
Indefiniteness, 61, 69, 73
Indeterminacy, 10, 18, 25, 30
  as generality, 55–58
  immediate objects and, 69–71
  mature account of, 58, 59
  semiotic, 4, 51–78
Indeterminate, 4, 9, 10, 15, 16, 18, 23, 26, 38, 41, 43, 51, 53–58, 60–62, 64, 65, 66, 131, 166, 220, 223
  biosigns, 55, 71–78
Index, 13, 28, 36, 37, 59, 60, 146, 161, 185, 188–191, 195, 201, 204–206, 209, 211

Individual, 2, 7, 10, 55, 64, 127–149, 168
concrete, 56–58, 65, 104
definite, 70
propositions, 12
replica, 10, 28
strict, 56, 57
Induction, 37, 154, 155, 231, 235
Inference, 153, 155, 156, 168, 201, 209, 222, 224, 225, 237
Information, 6, 7, 59, 89, 153, 158–167
Instinct, 5, 6, 34, 129, 151–168, 224, 225, 232, 234
Intelligibility, 127, 130–134, 141, 142, 144, 145, 147, 149
Interference, 90, 96
Interpretant
dynamic, 224, 230, 231, 233–235, 238
final, 84, 98, 118, 124, 189, 190, 195, 215, 228–235, 238
immediate, 9, 69, 118, 124, 190, 215, 223, 224, 228, 230, 232, 234, 235, 237
Interpreter, 3, 4, 13, 51, 52, 58–62, 64–67, 69–74, 76, 77, 163, 166, 173–176, 201, 206, 227

J
James, William, 16, 20, 44, 60, 67, 128, 134, 135, 137, 138, 140, 142, 143, 145, 189, 216, 227

K
Kant, Immanuel, 40, 52, 147, 218–220, 222, 225
Knowledge, 35, 36, 38, 41, 66, 134, 143, 151, 156, 157, 159, 164, 223, 228, 232
Kull, Kalevi, 8, 54, 171, 176

L
Ladd Franklin, Christine, 17
Law, 14, 19, 105, 114
Learning, 7, 11, 30, 46, 47, 77, 91, 132, 156, 202, 206, 235
Legisign, 173, 174, 188, 199, 204–206, 237
Logic, 1, 13–15, 36, 52, 68, 83, 97, 111, 123, 159, 161, 165, 186, 216, 229
Logos, 15

M
Material cause, 14
Material implication, 114, 116, 121
Materialism, 42
Mathematics, 11, 12, 15, 20, 137
Matter (as effete mind), 43
Maturana, Humberto, 176, 229
Meaning, 13, 41, 194, 225–227
Mechanical (as physical), 31
Mediation, 41, 46–48, 92, 141, 161, 186, 203, 208
Merrell, Floyd, 166, 167, 205
Metabolism, 225, 232, 238
Metalaw, 21, 22
Metaphor, 6, 16, 23, 27, 35, 38, 128, 140, 172, 229
Metaphysics, 3, 9, 10, 15, 17, 33, 40, 45, 98, 111, 157, 218–223, 238
Mimicry, 7, 186, 199–211
Mind, 4, 6, 42–44
Murphey, Murray, 16
Murphy, Michael, 19

N
Necessity, 30, 45, 96, 101, 102, 113, 114, 123, 129, 132
Nichols, Herbert, 14
Nominalism, 16, 219, 227

O
Objective idealism, 3, 6, 18, 31, 33–48, 97, 98, 191, 196, 223, 226, 2311
Ontology, 38, 40, 45–47, 103, 107, 121, 153

P
Panpsychism, 29, 30
Particular, 52, 56, 75, 80, 97
Percept, 163
Perception, 5, 13, 96, 115, 135, 153, 155, 163, 189, 207, 218, 224, 227, 228, 230, 232–236
Perceptual judgment, 153–158, 163, 230
Phaneron, 230, 232, 235
Phaneroscopy, 98
Phenomenology, 40, 45, 47, 139, 157, 160, 203, 216
Phenomenon, 1, 2, 8, 37, 40, 97, 132, 142, 154, 155, 157, 160, 161, 202, 217, 218, 229
Plato, 36, 37, 217
Polonius, 139
Popper, Sir Karl, 38
Possibility, 3, 8, 20, 23, 29, 38, 40, 41, 43, 45, 46, 48, 68, 70, 72, 113, 114, 129, 133, 147, 154, 179, 184, 205, 222, 225
Pragmaticism, 58, 163, 216
Pragmatism (pragmatic), 7, 16, 27, 40–45
Precision, 34, 52
Presentation, 230
Prigogine, Ilya, 23

Processes
events and, 108, 109
mental, 101–103
natural, 101
Proposition, 2, 5, 7, 13, 14, 35, 44, 51, 52, 59–62, 67, 70, 72, 95, 99, 101, 106, 107, 208, 209
causal, 106
conditional, 92, 116, 124, 162, 201, 219, 220
ordinary, 59, 199, 205
quasi-, 59
Psychology, 98
Psychonomy, 98
Purpose, 175, 176, 215

Q
Qualisign, 175, 185, 188, 204, 230

R
Ransdell, Joseph, 11, 67, 120, 147, 176
Real, 68, 72
Realism, 3, 16, 29, 33, 36–40, 217, 226
Reason, 156
Reasonableness, 97, 98
Reasoning
abductive, 153–164, 222
deductive, 107
inductive, 222
scientific, 101
Relation
causal, 137, 146, 189, 202
dyadic, 1, 104, 110, 166, 183, 189, 190, 204
triadic, 1–3, 7, 8, 96, 104, 105, 110–112, 123, 125, 162, 164–166, 183, 186–191, 196, 204, 218, 224, 226, 229, 234, 238
Replica, 10, 14, 28, 124, 173, 176, 177, 180
Representamen, 186, 187, 191, 230, 232
Representation, 10, 22, 23, 26, 27, 35, 37–39, 67, 69, 113, 118, 119, 161, 177, 185, 193, 218, 222, 228, 231, 233
Reynolds, Andrew, 16–18, 23
Rhema, 13
Rhetoric, 114, 233
Rorty, Richard, 37, 38, 136
Russell, Bertrand, 12

S
Santaella, Lucia, 25, 26, 85, 175
Schelling, 33, 39, 43, 172
Schema, 96, 113, 215, 222
Schiller, 16, 33
Sebeok, Thomas, 23, 92, 179, 202, 217, 226
Secondness (twoness, secundanity), 22, 27, 30, 41, 43, 44, 110, 113, 114, 131, 141, 144, 145, 160, 177, 180, 183, 185, 221, 230, 231
Selection, 7, 21, 200, 209
Self-control, 129, 179, 180
Self-organization, 168, 196
Sellars, Wilfrid, 128
Semiogenesis, 234
Semiosis (semeiosis)
biosemeiosis, 7, 224, 225
causation structure and, 111, 112
formal characteristics of, 109–111
habits and, 91, 92
life as, 171–173
as living process, 215–239
solenoid of, 228, 229
Semiosy, 186
Semiotics (Semeiotic)
biosemiotics, 192–196, 217, 218, 234–236
causation, 95–125
indeterminacy, 51–78
Peirce's theory of causation to biosemiotics, 123, 124
Sensation, 29, 156, 186, 189, 219, 220, 224, 232, 237
Short, Thomas, 16, 18, 19, 21, 22, 23, 25–27, 29, 31, 47, 55, 56, 67, 74, 75, 85, 130, 174, 187, 203
Sign
biosign, 4, 54, 55, 71–78
classes of (types of), 201–207
dicisign (dicent), 7, 59, 60, 66–69, 73, 75, 76, 161, 162, 166, 180, 200, 208, 210, 229, 231
legisign, 173, 174, 188, 199, 204–206, 237
qualisign, 175, 185, 188, 204, 230
singular, 64–66, 70, 173
sinsign, 173, 175, 188, 204, 205
Signification, 6, 10, 13, 180, 210, 230
Singular, 52, 55, 56, 58
signs, 64–66, 70, 173
Sinsign, 173, 175, 188, 204, 205
Smolin, Lee, 20–23, 82
Solenoid of semeiosis, 228, 229
Space, 12, 17, 87, 89, 90, 128, 202, 209, 218
Speculative grammar, 160, 161
Spencer, Herbert, 82, 167
Subject, 51, 52
Syllabus, 11, 13, 200
Symbol, 83–85
Symmetry, 85–88
breaking of, 85–88
of categories, 40, 44, 47
gauge, 89

Synechism, 3, 30, 33, 46, 97, 98, 172, 217, 226
Syntax, 3, 95, 96, 107, 161
System
  immune, 7, 225
  semeiotic, 51–78, 192–196, 217, 218, 234–236

T
Taxonomy, 7, 218, 234, 239
Teleology, 5, 6, 18, 98, 104, 108, 109, 113, 121–125, 137, 175, 184, 227, 237
Thirdness, 18, 41, 42, 45, 46, 110–114, 116, 121, 127, 131, 132, 141, 144, 145, 160, 177, 180, 185, 221, 223, 230, 231
Time, 89, 90, 121, 122, 133, 202, 209, 218
Triadic relation, 1–3, 7, 8, 96, 104, 105, 110–112, 123, 125, 162, 164–166, 183, 186–191, 196, 204, 218, 224, 226, 229, 234, 238
Trichotomies, 11, 13, 200, 203–205
Truth, 64, 102, 134, 135, 155, 156, 159, 216, 220, 221, 226, 229, 233
  of conditional proposition, 162, 219
  revelation of, 179, 180
Tychism, 2, 3, 97, 98

U
Uberty, 143, 156, 157
Uexküll, Jacob von, 105, 168, 218, 219, 224
Universals, 7, 21, 29, 39, 41, 51, 52, 56, 61, 62, 88, 89, 91, 96, 190, 219
Utterer, 4, 14, 51, 58, 59, 61–77, 146, 230

V
Vague, 4, 9, 10, 14, 16, 19, 23, 25, 26, 29, 51–54, 58, 64–66, 69–73, 75, 76, 86, 143, 186, 187, 220, 223, 227, 232, 233
  propositions, 12
  signs, game-theoretical semantics for, 59–63
Vagueness, 4, 9, 10, 34, 52, 53, 54, 58, 61, 63–65, 70, 73, 187, 220, 233
Varela, Francisco, 176, 229
Virtual, 23, 87, 134, 155, 163

W
Weissman, David, 130
Welby, Victoria Lady, 142, 162, 163, 215, 216
Wittgenstein, Ludwig, 135, 148
Would-be, 74, 82, 84, 120, 223

Printed by Printforce, the Netherlands